100 Great Problems
of Elementary Mathematics

THEIR HISTORY AND SOLUTION

100 Great Problems
of Elementary Mathematics

THEIR HISTORY AND SOLUTION

BY HEINRICH DÖRRIE

TRANSLATED BY DAVID ANTIN

NEW YORK

DOVER PUBLICATIONS, INC.

Published in Canada by General Publishing Company, Ltd., 30 Lesmill Road, Don Mills, Toronto, Ontario.

Published in the United Kingdom by Constable and Company, Ltd.; 10 Orange Street, London WC 2.

This Dover edition, first published in 1965, is a new translation of the unabridged text of the fifth edition of the work published by the Physica-Verlag, Würzburg, Germany, in 1958 under the title *Triumph der Mathematik: Hundert berühmte Probleme aus zwei Jahrtausenden mathematischer Kultur.*

This authorized translation is published by special arrangement with the German-language publishers, Physica-Verlag, Würzburg.

Standard Book Number: 486-61348-8

Library of Congress Catalog Card Number:65-14030

Manufactured in the United States of America
Dover Publications, Inc.
180 Varick Street
New York, N.Y. 10014

Preface

A book collecting the celebrated problems of elementary mathematics that would commemorate their origin and, above all, present their solutions briefly, clearly, and comprehensibly has long seemed a necessary and attractive task to the author.

The restriction to problems of elementary mathematics was considered advisable in view of those readers who have neither the time nor the opportunity to acquaint themselves in any detail with higher mathematics. Nevertheless, in spite of this limitation a colorful and compelling picture has emerged, one that gives an idea of the amazing variety of mathematical methods and one that will—I hope—enchant many who are interested in mathematics and who take pleasure in characteristic mathematical thought processes. In the present work there are to be found many pearls of mathematical art, problems the solutions of which represent, in the achievements of a Gauss, an Euler, Steiner, and others, incredible triumphs of the mathematical mind.

Because the difficult economic situation at the present time barred the publication of a larger work, a limit had to be set to the scope and number of the problems treated. Thus, I decided on a round number of one hundred problems. Moreover, since many of the problems and solutions require considerable space despite the greatest concision, this had to be compensated for by the inclusion of a number of mathematical miniatures. Possibly, however, it may be just these little problems, which are, in their way, true jewels of mathematical miniature work, that will find the readiest readers and win new admirers for the queen of the sciences.

As we have indicated already, a knowledge of higher analysis is not assumed. Consequently, the Taylor expansion could not be used for the treatment of the important infinite series. I hope nonetheless that the derivations we have given, particularly the striking derivation of the sine and cosine series, will please and will not be found unattractive even by mathematically sophisticated readers.

On the other hand, in some of the problems, e.g., the Euler tetrahedron problem and the problem of skew lines, the author believed it necessary not to dispense with the simplest concepts of vector analysis. The characteristic advantages of brevity and elegance of the vector method are so obvious, and the time and effort required for mastering it so slight, that the vectorial methods presented here will undoubtedly spur many readers on to look into this attractive area.

For the rest, only the theorems of elementary mathematics are assumed to be known, so that the reading of the book will not entail significant difficulties. In this connection the inclusion of the little problems may in fact increase the acceptability of the book, in that it will perhaps lead the mathematically weaker readers, after completion of the simpler problems, to risk the more difficult ones as well.

So then, let the book go out and do its part to awaken and spread the interest and pleasure in mathematical thought.

Wiesbaden,   HEINRICH DÖRRIE
Fall, 1932

Preface to the Second Edition

The second edition of the book contains few changes. An insufficiency in the proof of the Fermat-Gauss Impossibility Theorem has been eliminated, Problem 94 has been placed in historical perspective and the Problem of the Length of the Polar Night, which in relation to the other problems was of less significance, has been replaced by a problem of a higher level: "André's Derivation of the Secant and Tangent Series."

Wiesbaden,   HEINRICH DÖRRIE
Spring, 1940

Contents

Arithmetical Problems

1   Archimedes' *Problema Bovinum*

The sun god had a herd of cattle consisting of bulls and cows, one part of which was white, a second black, a third spotted, and a fourth brown.

Among the bulls, the number of white ones was one half plus one third the number of the black greater than the brown; the number of the black, one quarter plus one fifth the number of the spotted greater than the brown; the number of the spotted, one sixth and one seventh the number of the white greater than the brown.

Among the cows, the number of white ones was one third plus one quarter of the total black cattle; the number of the black, one quarter plus one fifth the total of the spotted cattle; the number of the spotted, one fifth plus one sixth the total of the brown cattle; the number of the brown, one sixth plus one seventh the total of the white cattle.

What was the composition of the herd?

SOLUTION.   If we use the letters X, Y, Z, T to designate the respective number of the white, black, spotted, and brown bulls and x, y, z, t to designate the white, black, spotted, and brown cows, we obtain the following *seven* equations for these *eight* unknowns:

$$(1) \quad X - T = \tfrac{5}{6}Y, \qquad (4) \quad x = \tfrac{7}{12}(Y + y),$$
$$(2) \quad Y - T = \tfrac{9}{20}Z, \qquad (5) \quad y = \tfrac{9}{20}(Z + z),$$
$$(3) \quad Z - T = \tfrac{13}{42}X, \qquad (6) \quad z = \tfrac{11}{30}(T + t),$$
$$\qquad\qquad\qquad\qquad (7) \quad t = \tfrac{13}{42}(X + x).$$

From equations (1), (2), (3) we obtain $6X - 5Y = 6T$, $20Y - 9Z = 20T$, $42Z - 13X = 42T$, and taking these three equations as equations for the three unknowns X, Y, and Z, we find

$$X = \tfrac{742}{297}T, \qquad Y = \tfrac{178}{99}T, \qquad Z = \tfrac{1580}{891}T.$$

Since 891 and 1580 possess no common factors, T must be some whole multiple—let us say G—of 891. Consequently,

$$(I) \qquad X = 2226G, \quad Y = 1602G, \quad Z = 1580G, \quad T = 891G.$$

If these values are substituted into equations (4), (5), (6), (7), the following equations are obtained:

$$12x - 7y = 11214G, \qquad 20y - 9z = 14220G,$$
$$30z - 11t = 9801G, \qquad 42t - 13x = 28938G.$$

These equations are solved for the four unknowns x, y, z, t and we obtain

(II) $\qquad \begin{cases} cx = 7206360G, & cy = 4893246G, \\ cz = 3515820G, & ct = 5439213G, \end{cases}$

in which c is the prime number 4657. Since none of the coefficients of G on the right can be divided by c, then G must be an integral multiple of c:

$$G = cg.$$

If this value of G is introduced into (I) and (II), we finally obtain the following relationships:

(I′) $\qquad \begin{cases} X = 10366482g, & Y = 7460514g, \\ Z = 7358060g, & T = 4149387g, \end{cases}$

(II′) $\qquad \begin{cases} x = 7206360g, & y = 4893246g, \\ z = 3515820g, & t = 5439213g, \end{cases}$

where g may be any positive integer.

The problem therefore has an infinite number of solutions. If g is assigned the value 1, we obtain the following:

Solution in the Smallest Numbers

white bulls	10,366,482	white cows	7,206,360
black bulls	7,460,514	black cows	4,893,246
spotted bulls	7,358,060	spotted cows	3,515,820
brown bulls	4,149,387	brown cows	5,439,213

HISTORICAL. As the above solution shows, the problem of the cattle cannot properly be considered a very difficult problem, at least in terms of present concepts. Since, however, in ancient times a difficult problem was frequently referred to specifically as a *problema bovinum* or else as a *problema Archimedis*, one may assume that the form of the problem dealt with above does not represent the complete and original form of Archimedes' problem, especially when one considers the rest of Archimedes' brilliant achievements, as well as the fact that Archimedes dedicated the cattle problem to the Alexandrian astronomer Eratosthenes.

A "more complete" formulation of the problem is contained in a manuscript (in Greek) discovered by Gotthold Ephraim Lessing in the Wolfenbüttel library in 1773. Here the problem is posed in the following poetic form, made up of twenty-two distichs, or pairs of verses:

Number the sun god's cattle, my friend, with perfect precision.
  Reckon them up with great care, if any wisdom you'd claim:
How many cattle were there that once did graze in the meadows
  On the Sicilian isle, sorted by herds into four,
Each of these four herds differently colored: the first herd was milk-white,
  Whereas the second gleamed in a deep ebony black.
Brown was the third group, the fourth was spotted; in every division
  Bulls of respective hues greatly outnumbered the cows.
Now, these were the proportions among the cattle: the white ones
  Equaled the number of brown, adding to that the third part
Plus one half of the ebony cattle all taken together.
  Further, the group of the black equaled one fourth of the flecked
Plus one fifth of them, taken along with the total of brown ones.
  Finally, you must assume, friend, that the total with spots
Equaled a sixth plus a seventh part of the herd of white cattle,
  Adding to that the entire herd of the brown-colored kine.
Yet quite different proportions held for the female contingent:
  Cows with white-colored hair equaled in number one third
Plus one fourth of the black-hued cattle, the males and the females.
  Further, the cows colored black totaled in number one fourth
Plus one fifth of the whole spotted herd, in this computation
  Counting in each spotted cow, each spotted bull in the group.
Likewise, the spotted cows comprised the fifth and the sixth part
  Out of the total of brown cattle that went out to graze.
Lastly, the cows colored brown made up a sixth and a seventh
  Out of the white-coated herd, female and male ones alike.
If, my friend, you can tell me exactly what was the number
  Gathered together there then, also the accurate count
Color by color of every well-nourished male and each female,
  Then with right you'll be called skillful in keeping accounts.

But you will not be reckoned a wise man yet; if you would be,
  Come and answer me this, using new data I give:
When the entire aggregation of white bulls and that of the black bulls
  Joined together, they all made a formation that was
Equally broad and deep; the far-flung Sicilian meadows
  Now were thoroughly filled, covered by great crowds of bulls.
But when the brown and the spotted bulls were assembled together,
  Then was a triangle formed; one bull stood at the tip;
None of the brown-colored bulls was missing, none of the spotted,
  Nor was there one to be found different in color from these.
If this, too, you discover and grasp it well in your thinking,
  If, my friend, you supply every herd's make-up and count,
Then with justice proclaim yourself victor and march about proudly,
  For your fame will glow bright all through the world of the wise.

Lessing, however, disputed the authorship of Archimedes. So also did Nesselmann (*Algebra der Griechen*, 1842), the French writer Vincent (*Nouvelles Annales de Mathématiques*, vol. XV, 1856), the Englishman Rouse Ball (*A Short Account of the History of Mathematics*), and others.

The distinguished Danish authority on Archimedes J. L. Heiberg (*Quaestiones Archimedeae*), the French mathematician P. Tannery (*Sciences exactes dans l'antiquité*), as well as Krummbiegel and Amthor (*Schlömilchs Zeitschrift für Mathematik und Physik*, vol. XXV, 1880), on the other hand, are of the opinion that this complete form of the problem is to be attributed to Archimedes.

The two conditions set forth in the last seven distichs require that $X + Y$ be a square number U^2 and $Z + T$ a triangular number* $\frac{1}{2}V(V + 1)$, as a result of which we obtain the following relations:

(8) $X + Y = U^2$ and   (9) $2Z + 2T = V^2 + V.$

If we substitute in (8) and (9) the values X, Y, Z, T in accordance with (I), these equations are transformed into

$$3828G = U^2 \quad \text{and} \quad 4942G = V^2 + V.$$

If we replace 3828, 4942, and G, respectively, with $4a$ (a being equal to $3 \cdot 11 \cdot 29 = 957$), b, and cg, we obtain

(8') $U^2 = 4acg,$ (9') $V^2 + V = bcg.$

---

* A triangular number is a number n such that it is possible to construct with n points a lattice of congruent equilateral triangles whose vertexes are the points. The first triangle numbers are $1 = \frac{1}{2} \cdot 1 \cdot 2$, $3 = 1 + 2 = \frac{1}{2} \cdot 2 \cdot 3$, $6 = 1 + 2 + 3 = \frac{1}{2} \cdot 3 \cdot 4$, $10 = 1 + 2 + 3 + 4 = \frac{1}{2} \cdot 4 \cdot 5$, etc.

U is consequently an integral multiple of 2, a, and c:

$$U = 2acu,$$

so that

$$U^2 = 4a^2c^2u^2 = 4acg$$

and

(8″) $$g = acu^2.$$

If this value for g is introduced into (9′) we obtain

$$V^2 + V = abc^2u^2$$

or

$$(2V + 1)^2 = 4abc^2u^2 + 1.$$

If the unknown is designated as $2V + 1v$ and the product $4abc^2 = 4 \cdot 3 \cdot 11 \cdot 29 \cdot 2 \cdot 7 \cdot 353 \cdot 4657^2$ is abbreviated as d, the last equation is transformed into

$$v^2 - du^2 = 1.$$

This is a so-called Fermat equation, which can be solved in the manner described in Problem 19. The solution is, however, extremely difficult because d has the inconveniently large value

$$d = 410286423278424$$

and even the smallest solution for u and v of this Fermat equation leads to astronomical figures.

Even if u is assigned the smallest conceivable value 1, in solving for g the value of ac is 4456749 and the combined number of white and black bulls is over 79 billion. However, since the island of Sicily has an area of only 25500 km^2 = 0.0255 billion m^2, i.e., less than $\frac{1}{30}$ billion m^2, it would be quite impossible to place that many bulls on the island, which contradicts the assertion of the seventeenth and eighteenth distichs.

2 The Weight Problem of Bachet de Méziriac

A merchant had a forty-pound measuring weight that broke into four pieces as the result of a fall. When the pieces were subsequently weighed, it was found that the weight of each piece was a whole number of pounds and that the four pieces could be used to weigh every integral weight between 1 and 40 pounds.

What were the weights of the pieces?

This problem stems from the French mathematician Claude Gaspard Bachet de Méziriac (1581–1638), who solved it in his famous book *Problèmes plaisants et délectables qui se font par les nombres*, published in 1624.

We can distinguish the two scales of the balance as the weight scale and the load scale. On the former we will place only pieces of the measuring weight; whereas on the load scale we will place the load and any additional measuring weights. If we are to make do with as few measuring weights as possible it will be necessary to place measuring weights on the load scale as well. For example, in order to weigh one pound with a two-pound and a three-pound piece, we place the two-pound piece on the load scale and the three-pound piece on the weight scale.

If we single out several from among any number of weights lying on the scales, e.g., two pieces weighing 5 and 10 lbs each on one scale, and three pieces weighing 1, 3, and 4 lbs each on the other, we say that these pieces give the first scale a preponderance of 7 lbs.

We will consider only integral loads and measuring weights, i.e., loads and weights weighing a whole number of pounds.

If we have a series of measuring weights $A, B, C, \ldots$, which when properly distributed upon the scales enable us to weigh all the integral loads from 1 through n lbs, and if P is a new measuring weight of such nature that its weight p exceeds the total weight n of the old measuring weights by 1 more than that total weight:

$$p - n = n + 1 \quad \text{or} \quad p = 2n + 1,$$

it is then possible to weigh all integral loads from 1 through $p + n = 3n + 1$ by addition of the weight P to the measuring weights $A, B, C, \ldots$.

In fact, the old pieces are sufficient to weigh all loads from 1 to n lbs. In order to weigh a load of $(p + x)$ and/or $(p - x)$ lbs, where x is one of the numbers from 1 to n, we place the measuring weight P on the weight scale and place weights $A, B, C, \ldots$ on the scales in such a manner that these pieces give either the weight scale or the load scale a preponderance of x lbs.

This being established, the solution of the problem is easy.

In order to carry out the maximum possible number of weighings with *two* measuring weights, A and B, A must weigh 1 lb and B 3 lbs. These two pieces enable us to weigh loads of 1, 2, 3, 4 lbs.

If we then choose a third piece C such that its weight

$$c = 2 \cdot 4 + 1 = 9 \text{ lbs,}$$

it then becomes possible to use the three pieces A, B, C to weigh all integral loads from 1 to $c + 4 = 9 + 4 = 13$.

Finally, if we choose a fourth piece D such that its weight

$$d = 2 \cdot 13 + 1 = 27 \text{ lbs,}$$

the four weights A, B, C, D then enable us to weigh all loads from 1 to $27 + 13 = 40$ lbs.

CONCLUSION. *The four pieces weigh* 1, 3, 9, 27 *lbs.*

NOTE. Bachet's weight problem was generalized by the English mathematician MacMahon. In Volume 21 of the *Quarterly Journal of Mathematics* (1886) MacMahon determined all the conceivable sets of integral weights with which all loads of 1 to n lbs can be weighed.

3   Newton's Problem of the Fields and Cows

In Newton's *Arithmetica universalis* (1707) the following interesting problem is posed:

> a *cows graze* b *fields bare in* c *days,*
> a' *cows graze* b' *fields bare in* c' *days,*
> a" *cows graze* b" *fields bare in* c" *days;*

what relation exists between the nine magnitudes a *to* c"*?*

It is assumed that all the fields provide the same amount of grass, that the daily growth of the fields remains constant, and that all the cows eat the same amount each day.

SOLUTION. Let the initial amount of grass contained by each field be M, the daily growth of each field m, and the daily grass consumption of each cow Q.

On the evening of the first day the amount of grass remaining in each field is

$$bM + bm - aQ,$$

on the evening of the second day

$$bM + 2bm - 2aQ,$$

on the evening of the third day

$$bM + 3bm - 3aQ,$$

etc., so that on the evening of the cth day

$$bM + cbm - caQ.$$

And this value must be equal to zero, since the fields are grazed bare in c days. This gives rise to the equation

(1) $$bM + cbm = caQ.$$

In like manner the following equations are obtained:

(2) $$b'M + c'b'm = c'a'Q$$

and

(3) $$b''M + c''b''m = c''a''Q.$$

If (1) and (2) are taken as linear equations for the unknowns M and m, we obtain

$$M = \frac{cc'(ab' - ba')}{bb'(c' - c)} Q, \qquad m = \frac{bc'a' - b'ca}{bb'(c' - c)} Q.$$

If these values are introduced into equation (3) and the resulting equation is multiplied by $[bb'(c' - c)]/Q$, we obtain the desired relation:

$$b''cc'(ab' - ba') + c''b''(bc'a' - b'ca) = c''a''bb'(c' - c).$$

The solution is more easily seen when expressed in the form of determinants. If q represents the reciprocal of Q, equations (1), (2), (3) assume the form

$$bM + cbm + caq = 0,$$

$$b'M + c'b'm + c'a'q = 0,$$

$$b''M + c''b''m + c''a''q = 0.$$

According to one of the basic theorems of determinant theory, the determinant of a system of n (3 in this case) linear homogeneous equations possessing n unknowns that do not all vanish (M, m, q in this case) must be equal to zero. Consequently, the desired relation has the form

$$\begin{vmatrix} b & bc & ca \\ b' & b'c' & c'a' \\ b'' & b''c'' & c''a'' \end{vmatrix} = 0.$$

4 Berwick's Problem of the Seven Sevens

In the following division example, in which the divisor goes into the dividend without a remainder:

```
**7*******:****7* = **7**
******
*****7*
*******
  *7****
  *7****
  *******
  ****7**
    ******
    ******
```

the numbers that occupied the places marked with the asterisks (*) were accidentally erased. What are the missing numbers?

This remarkable problem comes from the English mathematician E. H. Berwick, who published it in 1906 in the periodical *The School World*.

SOLUTION. We will assign a separate letter to each of the missing numerals. The example then has the following appearance:

$$AB\ 7\ CDEL\,QWz: \quad \alpha\beta\gamma\delta7\varepsilon = \kappa\lambda7\mu\nu$$

$$a\ b\ \Delta\ c\ d\ e$$

$F\,GH\ IK\,7\,L$	Third line
$f\ g\ h\ i\ k\ \Xi\ l$	Fourth line
$M\,7NO\,PQ$	Fifth line
$m\ 7\ n\ o\ p\ q$	$\leftarrow 7\cdot\mathfrak{b}$
$R\ STU\Sigma VW$	Seventh line
$r\ s\ t\ u\ 7\ v\ w$	
$X\,Y\,Z\,x\ y\ z$	Ninth line
$X\,Y\,Z\,x\ y\ z$	

I. The *first* numeral (α) of the divisor $\mathfrak{b}$ must be 1, since $7\mathfrak{b}$, as the sixth line of the example shows, possesses six numerals, whereas if α equaled 2, $7\mathfrak{b}$ would possess seven numerals.

Since the remainders in the third and seventh lines possess *six* numerals, F must equal 1 and R must equal 1, as a result of which f and r must also equal 1 (according to the outline).

Since ḅ cannot exceed 199979, the maximum value of μ is 9, so that the product in the eighth line cannot exceed 1799811, and $s < 8$. And since S can only be 9 or 0, and since there is no remainder in the ninth line under s, only the second case is possible. Consequently, $S = 0$ and (since $R = 1$) s is also equal to 0. It also follows from $R = 1$ and $S = 0$ that $M = m + 1$, thus $m \leqq 8$, and the product 7ḅ of the sixth line cannot be higher than 87*nopq*.

II. Consequently, the only possible values for the *second* divisor numeral β are 0, 1, and 2. (7·130000 is already higher than 900000.) $\beta = 0$ is eliminated because even when multiplied by nine 109979 does not give a seven-figure number, which, for example, is required by the eighth line.

Let us then consider the case of $\beta = 1$. This requires γ to be equal to only 0 or 1. (If $\gamma \geqq 2$, on determination of the second figure of line 6 one would have to add to $7\beta = 7 \cdot 1 = 7$ the amount $\geqq 1$ coming from the product $7 \cdot \gamma$, whereas the second figure must be 7.)

$\gamma = 0$, however, is impossible as a result of the seven figures of line 8, since not even 9·110979 yields a seven-figure product.

In the event that $\gamma = 1$ the following conditions must be observed, as a glance at line 8 will show: δ, ε, and μ must be so chosen that $\mu \cdot 111\delta7\varepsilon$ results in a seven-place number, the third last figure of which is 7. The only hope of this is offered by the multiplier $\mu = 9$ (since even 8·111979 has only six places). Now the third last figure of 9·111δ7ε, as is easily seen by experiment, can be a seven only if $\delta = 0$ or $\delta = 9$. In the first case line 8 will not possess seven places even when 111079 is multiplied by 9, and in the second case line 6 is 7·11197* = 783***, which is impossible. Thus, the case of $\gamma = 1$ is also excluded. The possibility of β equaling 1 must, therefore, be discarded.

The only appropriate value for the second figure of the divisor is therefore $\beta = 2$. From this it follows that $m = 8$ and $M = 9$.

III. The *third* figure γ of the divisor can only be 4 or 5, since 7·126000 is greater and 7·124000 is smaller than the sixth line. Moreover, since 9·124000 is greater and 7·126000 is smaller than the eighth line (10*tu*7*vw*), μ must be equal to 8.

Since 8·124979 = 999832 < 1000000 the assumption that $\gamma = 4$ fails to satisfy the requirements of line 8, and γ therefore has to be equal to 5.

IV. Since the third last figure of 8·125δ7ε must be 7, we find by testing that δ is equal to either 4 or 9. $\delta = 9$ is eliminated because even $7 \cdot 125970 = 881790$ comes out greater than the sixth line, so that only $\delta = 4$ is suitable. Thus, ε can be considered one of numbers 0 to 4. However, whichever one of these is chosen, we find for the third figure of the sixth line $n = 8$ from $7 \cdot 12547\varepsilon = 878***$. Similarly, for the eighth line we obtain $8 \cdot 12547\varepsilon = 10037**$, and consequently $t = 0$ and $u = 3$.

Since $\lambda\mathfrak{b} = \lambda \cdot 12547\varepsilon$ results in a seven-place fourth line and only $8\mathfrak{b}$ and $9\mathfrak{b}$ have seven places, λ is either 8 or 9.

V. From $t = 0$ and $X \geqq 1$ (together with $R = r = 1$, $S = s = 0$) it follows that $T \geqq 1$, and from $n = 8$, $N \leqq 9$, it follows that $T \leqq 1$, so that $T = 1$. N is therefore equal to 9 and $X = 1$. Since $X = 1$ and $2 \cdot \mathfrak{b} > 200000$ (line 9), it follows that $\nu = 1$ and also that $Y = 2$, $Z = 5$, $x = 4$, $y = 7$, and $z = \varepsilon$. With the results obtained at this point the problem has the following appearance:

$$A \; B \; 7 \; C \; D \; E \; L \; Q \; W \; \varepsilon : 12547\varepsilon = \kappa\lambda 781$$
$$a \; b \; \Delta \; c \; d \; e$$
$$\overline{1 \; G \; H \; I \; K \; 7 \; L}$$
$$1 \; g \; h \; i \; k \; \Xi \; l$$
$$\overline{\quad 9 \; 7 \; 9 \; O \; P \; Q}$$
$$8 \; 7 \; 8 \; o \; p \; q$$
$$\overline{\quad 1 \; 0 \; 1 \; U \; \Sigma \; V \; W}$$
$$1 \; 0 \; 0 \; 3 \; 7 \; v \; w$$
$$\overline{\qquad\quad 1 \; 2 \; 5 \; 4 \; 7 \; \varepsilon}$$
$$1 \; 2 \; 5 \; 4 \; 7 \; \varepsilon$$

VI. In this case ε is one of the five numbers

$$0, \; 1, \; 2, \; 3, \; 4.$$

These five cases correspond to the number series

$$vw = \quad 60, \quad 68, \quad 76, \quad 84, \quad 92,$$

$$opq = 290, \quad 297, \quad 304, \quad 311, \quad 318$$

and, depending upon whether λ is equal to 8 or 9,

$$\Xi l = 60, \; 68, \; 76, \; 84, \; 92$$

or

$$\Xi l = 30, \; 39, \; 48, \; 57, \; 66.$$

This presents ten different possibilities. If we test each of them by going upward in three successive additions beginning from lines 9 and 8 to line 7, then from lines 7 and 6 to line 5, and finally from lines 5 and 4 to line 3, we find that only when $\varepsilon = 3$ and $\lambda = 8$ do we obtain the requisite 7 for the next to last figure of line 3. In this case $vw = 84$, $U\Sigma VW = 6331$, $opq = 311$, $OPQ = 944$, $ghik\Xi l = 003784$, and $GHIK7L = 101778$. This gives the problem the following appearance:

$$A\ B\ 7\ C\ D\ E\ 8\ 4\ 1\ 3 : 125473 = \kappa 8781$$

$$a\ b\ \Delta\ c\ d\ e$$

```
1 1 0 1 7 7 8
1 0 0 3 7 8 4
-------------
    9 7 9 9 4 4
    8 7 8 3 1 1
-------------
  1 0 1 6 3 3 1
  1 0 0 3 7 8 4
-------------
      1 2 5 4 7 3
      1 2 5 4 7 3
-------------
```

VII. Finally, since of all the multiples of ♭ only $5♭ = 627365$ added to the division remainder 110177 of the third line gives a number containing a 7 in the third place, we get $\kappa = 5$ and at the same time $ab\Delta cde = 627365$ and $AB7CDE = 737542$, which gives us all of the figures missing from the problem.

5 Kirkman's Schoolgirl Problem

In a boarding school there are fifteen schoolgirls who always take their daily walks in rows of threes. How can it be arranged so that each schoolgirl walks in the same row with every other schoolgirl exactly once a week?

This extraordinary problem was posed in the *Lady's and Gentleman's Diary* for 1850, by the English mathematician T. P. Kirkman.

Of the great number of solutions that have been found we will reproduce two. One is by the English minister Andrew Frost ("General Solution and Extension of the Problem of the 15 School-girls," *Quarterly Journal of Pure and Applied Mathematics*, vol. XI, 1871); the other is that of B. Pierce ("Cyclic Solutions of the School-girl Puzzle," *The Astronomical Journal*, vol. VI, 1859–1861).

FROST'S SOLUTION. Mathematically expressed the problem consists of arranging the fifteen elements x, a_1, a_2, b_1, b_2, c_1, c_2, d_1, d_2, e_1, e_2, f_1, f_2, g_1, g_2 in seven columns of five triplets each in such a way that any two selected elements always occur in one and only one of the 35 triplets. As the initial triplets of the seven columns we shall select:

$$xa_1a_2|xb_1b_2|xc_1c_2|xd_1d_2|xe_1e_2|xf_1f_2|xg_1g_2.$$

Then we have only to distribute the 14 elements a_1, a_2, b_1, b_2, ..., g_1, g_2 correctly over the other four lines of our system.

Using the seven letters a, b, c, d, e, f, g, we form a group of triplets in which each pair of elements occurs exactly once, specifically the group:

abc, ade, afg, bdf, beg, cdg, cef.   (The triplets are in alphabetical order.)

From this group it is possible to take for each column exactly four triplets that contain all the letters except those contained in the first line of the column. If we then place the appropriate triplets in alphabetical order in each column, we obtain the following preliminary arrangement:

Sun.	Mon.	Tues.	Wed.	Thurs.	Fri.	Sat.
xa_1a_2	xb_1b_2	xc_1c_2	xd_1d_2	xe_1e_2	xf_1f_2	xg_1g_2
bdf	ade	ade	abc	abc	abc	abc
beg	afg	afg	afg	afg	ade	ade
cdg	cdg	bdf	beg	bdf	beg	bdf
cef	cef	beg	cef	cdg	cdg	cef

Now we have to index the triplets *bdf, beg, cdg, cef, ade, afg, abc,* i.e., to provide them with the index numbers 1 and 2. We index them in the order just mentioned, i.e., first all the triplets *bdf,* then all the triplets *beg,* etc., observing the following rules:

I. When a letter in one column has received its index number, the next time that letter occurs in the same column it receives the other index number.

II. If two letters of a triplet have already been assigned index numbers, these two index numbers must not be used in the same sequence for the same letters in other triplets.

III. If the index number of a letter is not determined by rules I. and II., the letter is assigned the index number 1.

The letters are indexed in three steps.

First step. The triplets *bdf*, *beg*, *cdg*, and all the letters aside from *a* that can be indexed in accordance with this numbering system and rules I., II., and III. are successively indexed.

Second step. The missing index numbers (in boldface in the diagram) of the triplets *ade* and *afg*, as well as the index numbers obtained in accordance with rule I. for the last two *a*'s in line 2 are assigned.

Third step. The still missing index numbers of the *a*'s in columns 4 and 5 (in the empty spaces of the printed diagram) are inserted; these are 2 in line 2 and 1 in line 3.

This method results in the following completed diagram, which represents the solution of the problem.

Sun.	Mon.	Tues.	Wed.	Thurs.	Fri.	Sat.
xa_1a_2	xb_1b_2	xc_1c_2	xd_1d_2	xe_1e_2	xf_1f_2	xg_1g_2
$b_1d_1f_1$	$a_1d_2e_2$	$a_1d_1e_1$	ab_2c_2	ab_1c_1	$a_1b_2c_1$	$a_1b_1c_2$
$b_2e_1g_1$	$a_2f_2g_2$	$a_2f_1g_1$	af_2g_1	af_1g_2	$a_2d_2e_1$	$a_2d_1e_2$
$c_1d_2g_2$	$c_1d_1g_1$	$b_1d_2f_2$	$b_1e_1g_2$	$b_2d_1f_2$	$b_1e_2g_1$	$b_2d_2f_1$
$c_2e_2f_2$	$c_2e_1f_1$	$b_2e_2g_2$	$c_1e_2f_1$	$c_2d_2g_1$	$c_2d_1g_2$	$c_1e_1f_2$

PIERCE'S SOLUTION (judged the best by Sylvester). Let one girl, whom we will indicate as *, walk in the middle of the same row on all days; we will divide the other girls into two groups of 7 and designate the first group by the Arabic numbers 1 to 7 or else by lower-case letters and the second group by the Roman numbers I to VII or else by capital letters. We will let an equation such as $R = s$ indicate that the Roman number indicated by the letter R possesses the same numerical value as the Arabic numeral corresponding to the letter s. Also, we will designate the days of the week Sunday, Monday,..., Saturday by the numerals 0, 1, 2,..., 6.

Let the Sunday arrangement have the following order:

$$
\begin{array}{ccc}
a & \alpha & A \\
b & \beta & B \\
c & \gamma & C \\
d & * & D \\
E & F & G
\end{array}
$$

From this, by adding $r = R$ to each numeral, we obtain the arrangement

$$
\begin{array}{ccc}
a + r & \alpha + r & A + R \\
b + r & \beta + r & B + R \\
c + r & \gamma + r & C + R \\
d + r & * & D + R \\
E + R & F + R & G + R
\end{array}
$$

for the rth weekday. Here every figure thus obtained that exceeds 7, such as perhaps $c + r$ or $D + R$, will represent the girl who receives a number ($c + r - 7$ or $D + R - 7$), that is 7 below the figure and is subsequently converted into that number.

The arrangements thus obtained yield the solution of the problem if the following three conditions are satisfied:

I. The three differences $\alpha - a, \beta - b, \gamma - c$ are 1, 2, and 3.

II. The seven differences $A - a, A - \alpha, B - b, B - \beta, C - c,$ $C - \gamma, D - d$ form a complete residue system of incongruent numbers to the modulus 7 (cf. No. 19).

III. The three differences $F - E, G - F, G - E$ are 1, 2, 3.

PROOF. We take as a premise that the following congruences (cf. No. 19) are all related to the modulus 7.

1. Each girl x of the first group will come together exactly once with every other girl y of this group. The difference $x - y$ is then (according to I.) congruent to only one of the 6 differences $a - \alpha,$ $b - \beta, c - \gamma, \alpha - a, \beta - b, \gamma - c$. Let us assume $x - y \equiv \beta - b$ or $x - \beta \equiv y - b$. Thus, if r represents the number of the day of the week that is congruent to $x - \beta$ (or $y - b$), then

$$x \equiv \beta + r \quad \text{and} \quad y \equiv b + r,$$

so that the girls x and y walk in the same row on weekday r.

2. Each girl x of the first group comes together exactly once with each girl X of the second group.

The difference $X - x$ (according to II.) can be congruent to only one of the seven differences $A - a, A - \alpha, B - b, B - \beta, C - c,$ $C - \gamma, D - d$. Let us assume $X - x \equiv C - \gamma$ or $X - C \equiv x - \gamma$. If $s = S$ is the weekday number that is congruent to $X - C$ (or $x - \gamma$), then we have

$$X \equiv C + S \quad \text{and} \quad x \equiv \gamma + s,$$

so that the girls X and x walk in the same row on weekday s.

3. Each girl X of the second group comes together exactly once with every other girl Y of this group.

The difference $X - Y$ is (according to III.) congruent to only one of the differences $F - E$, $G - F$, $G - E$, $E - F$, $F - G$, $E - G$. Let us assume that $X - Y \equiv G - F$ or $X - G \equiv Y - F$. Then if R represents the weekday number that is congruent to $X - G$ (or $Y - F$), we obtain

$$X \equiv G + R \quad \text{and} \quad Y \equiv F + R,$$

so that the girls X and Y walk in the same row on weekday R.

Thus, we need only satisfy conditions I., II., and III. to obtain the Sunday arrangement.

We choose $a = 1$, $\alpha = 2$, $b = 3$, consequently $\beta = 5$, and then $c = 4$, so that $\gamma = 7$ and $d = 6$. We then select $A = 1$, and thus $B = VI$, $C = II$, and $D = III$, so that the differences mentioned in condition II. are the numbers 0, -1, 3, 1, -2, -5, which are incongruent to the modulus 7. The numbers IV, V, and VII then remain for the letters E, F, G.

The Sunday arrangement is therefore

1	2	I
3	5	VI
4	7	II
6	*	III
IV	V	VII

The weekday rows, in order, are arranged in the following manner:

2	3	II	3	4	III	4	5	IV
4	6	VII	5	7	I	6	1	II
5	1	III	6	2	IV	7	3	V
7	*	IV	1	*	V	2	*	VI
V	VI	I,	VI	VII	II,	VII	I	III,

5	6	V	6	7	VI	7	1	VII
7	2	III	1	3	IV	2	4	V
1	4	VI	2	5	VII	3	6	I
3	*	VII	4	*	I	5	*	II
I	II	IV,	II	III	V,	III	IV	VI.

6 The Bernoulli-Euler Problem of the Misaddressed Letters

To determine the number of permutations of n *elements in which no element occupies its natural place.*

This problem was first considered by Niclaus Bernoulli (1687–1759), the nephew of the two great mathematicians Jacob and Johann Bernoulli. Later Euler became interested in the problem, which he called a *quaestio curiosa ex doctrina combinationis* (a curious problem of combination theory), and he solved it independently of Bernoulli.

The problem can be stated in a somewhat more concrete form as the *problem of the misaddressed letters:*

Someone writes n *letters and writes the corresponding addresses on* n *envelopes. How many different ways are there of placing all the letters in the wrong envelopes?*

This problem is particularly interesting because of its ingenious solution.

Let the letters be known as *a*, *b*, *c*, . . ., the corresponding envelopes as *A*, *B*, *C*, ... Let the number of misplacements, which we are seeking, be designated as $\bar{n}$.

Let us first consider all the cases in which *a* finds its way into *B* and *b* into *A* as one group, and all the cases in which *a* gets into *B* and *b* does *not* get into *A* as a second group.

The first group obviously includes $\overline{n-2}$ cases.

The number of cases falling into the second group can be determined if instead of *b, c, d, e, . . .* and *A, C, D, E, . . .* we write, say, *b', c', d', e', . . .* and *B', C', D', E', ...* Accordingly, the number is $\overline{n-1}$.

The number of all the cases in which *a* ends up in *B* is then $\overline{n-1} + \overline{n-2}$. Since each operation of placing "*a* in *C*," "*a* in *D*," . . . yields an equal number of cases, the total number $\bar{n}$ of all the possible cases is

$$\bar{n} = (n-1)[\overline{n-1} + \overline{n-2}].$$

We write this recurrence formula

$$\bar{n} - n \cdot \overline{n-1} = \iota[\overline{n-1} - (n-1) \cdot \overline{n-2}],$$

in which ι represents -1 and apply it to the letter numbers 3, 4, 5, ... up to n. Thus, we obtain

$$\bar{3} - 3\cdot\bar{2} = \iota[\bar{2} - 2\cdot\bar{1}],$$
$$\bar{4} - 4\cdot\bar{3} = \iota[\bar{3} - 3\cdot\bar{2}],$$
$$\vdots$$
$$\bar{n} - n\cdot\overline{n-1} = \iota[\overline{n-1} - (n-1)\cdot\overline{n-2}].$$

By multiplying these $(n-2)$ equations we obtain

$$\bar{n} - n\cdot\overline{n-1} = \iota^{n-2}[\bar{2} - 2\cdot\bar{1}],$$

or, since $\bar{1} = 0$, $\bar{2} = 1$, and $\iota^{n-2} = \iota^n$,

$$\bar{n} - n\cdot\overline{n-1} = \iota^n.$$

We then divide this equation by $n!$, which gives

$$\frac{\bar{n}}{n!} - \frac{\overline{n-1}}{(n-1)!} = \frac{\iota^n}{n!}.$$

If we replace n in this formula by the series $2, 3, 4, \ldots, n$, we obtain

$$\frac{\bar{2}}{2!} - \frac{\bar{1}}{1!} = \frac{\iota^2}{2!},$$
$$\frac{\bar{3}}{3!} - \frac{\bar{2}}{2!} = \frac{\iota^3}{3!},$$
$$\vdots$$
$$\frac{\bar{n}}{n!} - \frac{\overline{n-1}}{(n-1)!} = \frac{\iota^n}{n!}.$$

Addition of these $(n-1)$ equations results (since $\bar{1} = 0$) in

$$\frac{\bar{n}}{n!} = \frac{\iota^2}{2!} + \frac{\iota^3}{3!} + \cdots + \frac{\iota^n}{n!}.$$

From this we are finally able to obtain the desired number $\bar{n}$:

$$\bar{n} = n!\left(\frac{1}{2!} - \frac{1}{3!} + \frac{1}{4!} - + \cdots + \frac{\iota^n}{n!}\right).$$

If $\mathfrak{z}$ represents a symbol such that the application of the binomial theorem (cf. No. 9) to $(\mathfrak{z} - 1)^n$ allows $\nu!$ to be written for each power $\mathfrak{z}^\nu$ of the binomial expansion, the number can be expressed in the simpler form

$$\bar{n} = (\mathfrak{z} - 1)^n.$$

For a value such as $n = 4$, for example, we obtain $\bar{4} = (\mathfrak{z} - 1)^4 =$ $\mathfrak{z}^4 - 4\mathfrak{z}^3 + 6\mathfrak{z}^2 - 4\mathfrak{z} + 1 = 4! - 4\cdot3! + 6\cdot2! - 4\cdot1! + 1 = 9$, which is easily checked by testing.

Similarly, the number of permutations that can be formed from n elements in which no element is in its natural place is $(\mathfrak{z} - 1)^n$.

For the four elements 1, 2, 3, 4, for example, there are the nine permutations 2143, 2341, 2413, 3142, 3412, 3421, 4123, 4312, 4321.

Note. The result obtained also contains the solution of the determinant problem:

In how many constituents of an n-degree determinant do no principal diagonal elements occur?

This is immediately seen if the rth element of the sth column is called c_r^s. The elements of the principal diagonal are then

$$c_1^1, c_2^2, c_3^3, \ldots, c_n^n.$$

The determinant therefore contains $(\mathfrak{z} - 1)^n$ constituents outside the principal diagonal elements.

7   Euler's Problem of Polygon Division

In how many ways can a (plane convex) polygon of n sides be divided into triangles by diagonals?

Leonhard Euler posed this problem in 1751 to the mathematician Christian Goldbach. For the number to be found, E_n, the number of possible divisions, Euler developed the formula:

$$(1) \qquad E_n = \frac{2\cdot6\cdot10\ldots(4n - 10)}{(n - 1)!}.$$

This problem is of the greatest interest because it involves many difficulties in spite of its innocuous appearance, as many a surprised reader will discover if he attempts to derive the Euler formula without outside assistance. Euler himself said, "The process of induction I employed was quite laborious."

In the simplest cases $n = 3, 4, 5, 6$ the various divisions

$$E_3 = 1, \qquad E_4 = 2, \qquad E_5 = 5, \qquad E_6 = 14$$

are easily obtained from the graphic representations. But this method soon becomes impossible as the number of angles is increased.

In 1758 Segner, to whom Euler had communicated the first seven division numbers 1, 2, 5, 14, 42, 132, 429, established a recurrence formula for E_n (*Novi Commentarii Academiae Petropolitanae pro annis 1758 et 1759*, vol. VII) which we will begin by deriving.

Let the angles of any convex polygon of n angles be $1, 2, 3, \ldots, n$. For every possible division E_n of the polygon of n angles we may take the side $n1$ as the base line of a triangle the apex of which is situated at one of the angles $2, 3, 4, \ldots, n - 1$ in accordance with the division selected. If the apex is, for example, situated at angle r, on one side of the triangle $n1r$ there is a polygon of r angles and on the other a polygon of s angles, $r + s$ being equal to $n + 1$ (since the apex r belongs to both the polygon of r angles and the polygon of s angles).

Since the polygon of r angles (or r-gon) permits E_r divisions and the s-gon permits E_s divisions, and since each division of the r-gon can be connected with every division of the s-gon toward a division of the given n-gon, the mere choice of the apex r results in $E_r \cdot E_s$ different divisions of the given n-gon.

Since, then, r can possess successively every value of the series $2, 3, \ldots, n - 1$ and s can accordingly possess successively every value of the series $n - 1, n - 2, \ldots, 3, 2$, it follows that

$$(2) \qquad E_n = E_2 E_{n-1} + E_3 E_{n-2} + \cdots + E_{n-1} E_2,$$

where the factor E_2, which is merely added for better appearance, has the value 1.

Formula (2) is Segner's recurrence formula. It confirms the previously given values for E_3 to E_6 as well as giving

$$E_7 = E_2 E_6 + E_3 E_5 + E_4 E_4 + E_5 E_3 + E_6 E_2 = 42,$$

$$E_8 = E_2 E_7 + E_3 E_6 + E_4 E_5 + E_5 E_4 + E_6 E_3 + E_7 E_2 = 132,$$

etc.

As the index number is increased Segner's formula, in contrast with Euler's, grows more and more unwieldy, as Goldbach has already indicated.

We can obtain the Euler formula (1) most simply if we consider Euler's division problem or Segner's recurrence formula in the light of an idea of Rodrigues (*Journal de Mathématiques*, 3 [1838]) and connect it with a problem treated by the French mathematician Catalan in the year 1838 in the *Journal de Mathématiques*.

CATALAN'S PROBLEM has the form:

How many different ways can a product of n different factors be calculated by pairs?

We say that a product is calculated by pairs when it is always only two factors that are multiplied together and when the product arising from such a "paired" multiplication is used as one factor in the continuation of the calculation. Calculation by pairs of the product $3 \cdot 4 \cdot 5 \cdot 7$, for example, is carried out in the following manner: $3 \cdot 5 = 15$, $4 \cdot 15 = 60$, $7 \cdot 60 = 420$. For the four-membered product *abcd* an alphabetical arrangement of the factors gives the following five paired multiplications:

$$[(a \cdot b) \cdot c] \cdot d, \quad [a \cdot (b \cdot c)] \cdot d, \quad (a \cdot b) \cdot (c \cdot d), \quad a[(b \cdot c) \cdot d], \quad a \cdot [b \cdot (c \cdot d)].$$

A product in which the paired multiplications that are to be carried out are marked by brackets or the like will be referred to in abbreviated form as "paired."

$\{[(a \cdot b) \cdot c] \cdot [(d \cdot e) \cdot (f \cdot g)]\} \cdot \{(h \cdot i) \cdot k\}$ is therefore a paired product of the ten factors *a* to *k*. It is immediately seen that a paired product of *n* factors contains $(n - 1)$ multiplication signs and correspondingly involves $(n - 1)$ paired multiplications (for every two factors).

Catalan's problem requires the answers to two questions:

1. *How many paired products of n different prescribed factors are there?*

2. *How many paired products can be formed from n factors if the sequence of the factors (e.g., an alphabetical sequence) is prescribed?*

The first number we will designate as R_n and the second as C_n.

The simplest method of obtaining R_n (according to Rodrigues) is by means of a recurrence formula. We will imagine the R_n *n*-membered paired products to be formed of the *n* given factors $f_1, f_2, \ldots, f_n$; we will add to this an $(n + 1)$th factor $f_{n+1} = f$ and form from the available R_n *n*-membered products all the R_{n+1} $(n + 1)$-membered products of the factors $f_1, f_2, \ldots, f_{n+1}$.

Now each of the R_n *n*-membered products P includes $(n - 1)$ paired multiplications of the form $A \cdot B$. If we use *f* once as the multiplier in front of *A*, once as the multiplicand after *A*, once as the multiplier before *B* and once as the multiplicand after *B*, we thereby obtain from $A \cdot B$ four new paired products $(f \cdot A) \cdot (B)$, $(A \cdot f) \cdot (B)$, $(A) \cdot (f \cdot B)$, and $(A) \cdot (B \cdot f)$.

Since these four arrangements of the factor *f* can be effected for each of the $n - 1$ paired subproducts of *P*, we obtain from *P*

$4(n - 1)$ $(n + 1)$-membered paired products. Moreover, we also obtain from P the *two* $(n + 1)$-membered paired products $f \cdot P$ and $P \cdot f$. The described arrangement of the factors f thus yields from only one (P) of the R_n n-membered products $(4n - 2)$ $(n + 1)$-membered products. From all R_n n-membered paired products we therefore obtain $R_n \cdot (4n - 2)$ $(n + 1)$-membered paired products. The sought-for recurrence formula accordingly reads

$$(3) \qquad\qquad R_{n+1} = (4n - 2)R_n.$$

To obtain an independent representation of R_n we begin with $R_2 = 2$ (two factors a and b yield only two products: $a \cdot b$ and $b \cdot a$) and we infer from (3) $R_3 = 6R_2 = 2 \cdot 6$, $R_4 = 10R_3 = 2 \cdot 6 \cdot 10$, $R_5 = 14R_4 = 2 \cdot 6 \cdot 10 \cdot 14$, etc., and finally

$$(4) \qquad\qquad R_n = 2 \cdot 6 \cdot 10 \cdot 14 \ldots (4n - 6).$$

The second question can also be answered by returning to a recurrence formula.

Let the n factors f_v in the prescribed order be $\varphi_1, \varphi_2, \ldots, \varphi_n$. We will take from the C_n paired n-membered products belonging to this series those having the form

$$(\;) \cdot (\;),$$

where the parenthesis on the left includes the r members $\varphi_1, \varphi_2, \ldots, \varphi_n$ and the one on the right the $s = n - r$ members $\varphi_{r+1}, \varphi_{r+2}, \ldots, \varphi_{r+s} = \varphi_n$. Since the left parenthesis, in accordance with its r members, can possess C_r different forms and the right correspondingly can possess C_s different forms, while each form belonging to the left parenthesis can combine with each form included in the right parenthesis, the above main form yields $C_r \cdot C_s$ different n-membered paired products.

Since, moreover, r can have every value from 1 to $n - 1$, it follows that

$$(5) \qquad C_n = C_1 C_{n-1} + C_2 C_{n-2} + \cdots + C_{n-1} C_1.$$

By using this recurrence formula and beginning from $C_1 = 1$ and $C_2 = 1$, we obtain the following sequence

$$C_3 = C_1 C_2 + C_2 C_1 = 2,$$
$$C_4 = C_1 C_3 + C_2 C_2 + C_3 C_1 = 5,$$
$$C_5 = C_1 C_4 + C_2 C_3 + C_3 C_2 + C_4 C_1 = 14,$$

etc.

To obtain an independent representation of C_n we can imagine that there are $n!$ different sequences (permutations) of the factors $f_1, f_2, \ldots, f_n$, that each of these sequences possesses C_n paired n-membered products and that all the sequences together possess R_n such products. Then $R_n = C_n \cdot n!$ or

$$(6) \qquad C_n = \frac{R_n}{n!} = \frac{2 \cdot 6 \cdot 10 \ldots (4n - 6)}{n!}.$$

Formulas (4) and (6) solve Catalan's problem
Now for Euler's formula!
From the indicated values

$$E_2 = 1, \qquad E_3 = 1, \qquad E_4 = 2, \qquad E_5 = 5,$$

$$C_1 = 1, \qquad C_2 = 1, \qquad C_3 = 2, \qquad C_4 = 5$$

and formulas (2) and (5) it immediately follows that in general

$$(7) \qquad E_n = C_{n-1}.$$

[The proof is by induction. We assume that (7) is true for all indices through n, so that $E_2 = C_1, E_3 = C_2, \ldots, E_n = C_{n-1}$.
According to (2) and (5)

$$E_{n+1} = E_2 E_n + E_3 E_{n-1} + \cdots + E_n E_2,$$

$$C_n = C_1 C_{n-1} + C_2 C_{n-2} + \cdots + C_{n-1} C_1.$$

Since the right sides of the two last equations correspond member for member, it also follows that

$$E_{n+1} = C_n;$$

i.e., formula (7) is valid for *every* index.]
(6) and (7) give us Euler's formula immediately:

$$(8) \qquad E_n = \frac{2 \cdot 6 \cdot 10 \ldots (4n - 10)}{(n - 1)!}.$$

In conclusion we would like to give a slight simplification of Euler's formula. It is

$$E_n = \frac{2^{n-2} \cdot 1 \cdot 3 \cdot 5 \ldots (2n - 5)}{(n - 1)!} = \frac{2^{n-2}(2n - 3)!}{(n - 1)! \, 2^{n-2} \cdot (n - 2)!(2n - 3)},$$

and consequently

$$E_n = k_f/k,$$

where $f = n - 2$ is the number of triangles into which the n-gon can always be divided and $k = 2n - 3$ is the number of sides bounding these triangles.

Recently (*Zeitschrift für math. und naturw. Unterricht*, 1941, vol. 4) H. Urban derived Euler's formula in the following manner.

He first calculated E_5, E_6, E_7 by means of the Segner recurrence formula and "inferred" the following:

$$E_2 = 1, \quad E_3 = 1, \quad E_4 = 2, \quad E_5 = 5, \quad E_6 = 14, \quad E_7 = 42,$$

$$\frac{E_3}{E_2} = \frac{2}{2}, \quad \frac{E_4}{E_3} = \frac{6}{3}, \quad \frac{E_5}{E_4} = \frac{10}{4}, \quad \frac{E_6}{E_5} = \frac{14}{5}, \quad \frac{E_7}{E_6} = \frac{18}{6},$$

on the strength of which he surmised that E_n would have to be

(I) $$E_n = \frac{4n - 10}{n - 1} E_{n-1}.$$

(Unfortunately, he does not say whether it was *Euler's recurrence formula* or some other idea that led him to his "inference.")

This recurrence formula is certainly correct for the first values of the index n. To prove its general validity the conclusion for n is applied to $n + 1$: it is assumed that the recurrence formula (I) is true for all index numbers from 1 to $n - 1$ and it is demonstrated that it is therefore also true for n.

The proof is carried out by means of the expression

(II) $$S = 1 \cdot E_2 \cdot E_{n-1} + 2 \cdot E_3 \cdot E_{n-2} + 3 \cdot E_4 \cdot E_{n-3} + \cdots$$
$$+ (n - 2) \cdot E_{n-1} \cdot E_2$$

or, written in the reverse order,

(III) $$S = (n - 2) \cdot E_{n-1} \cdot E_2 + (n - 3) \cdot E_{n-2} \cdot E_3$$
$$+ (n - 4) \cdot E_{n-3} \cdot E_4 + \cdots + 1 \cdot E_2 \cdot E_{n-1}.$$

Columnar addition of these two equations gives

$$2S = (n - 1)[E_2 E_{n-1} + E_3 E_{n-2} + \cdots + E_{n-1} E_2]$$

or, since in accordance with Segner's recurrence formula the value of the expression within the brackets is equal to E_n,

(IV) $$2S = (n - 1)E_n.$$

Now the left-hand factor E_r in each product $E_r \cdot E_s$ of (II) and (III) (except the case in which $r = 2$) is replaced in accordance with the recurrence formula (I) by $\lambda_{r-1} E_{r-1}/(r - 1)$ with $\lambda_v = 4_v - 6$. This gives us

(II') $S = E_2 E_{n-1} + \lambda_2 E_2 E_{n-2} \quad + \lambda_3 E_3 E_{n-3} + \cdots$
$$+ \lambda_{n-2} E_{n-2} E_2,$$

(III') $S = \qquad\qquad \lambda_{n-2} E_{n-2} E_2 + \lambda_{n-3} E_{n-3} E_3 + \cdots$
$$+ \lambda_2 E_2 E_{n-2} + E_2 E_{n-1}$$

and by columnar addition of these two lines, since $\lambda_v + \lambda_{n-v} = 4n - 12$, we obtain

$$2S = E_{n-1} + (4n - 12) [E_2 E_{n-2} + E_3 E_{n-3} + \cdots + E_{n-2} E_2] + E_{n-1}$$

or, since the expression within brackets is equal to E_{n-1},

(V) $$2S = (4n - 10) E_{n-1}.$$

Equations (IV) and (V) give us

$$E_n = \frac{4n - 10}{n - 1} E_{n-1},$$

so that Euler's recurrence formula (I) is thereby shown to be valid for the index number n, also, and thus generally valid.

8   Lucas' Problem of the Married Couples

How many ways can n married couples be seated about a round table in such a manner that there is always one man between two women and none of the men is ever next to his own wife?

This problem appeared (probably for the first time) in 1891 in the *Théorie des Nombres* of the French mathematician Edouard Lucas (1842–1891), author of the famous work *Récréations mathématiques*. The English mathematician Rouse Ball has said of this problem, "The solution is far from easy."

The problem has been solved by the Frenchmen M. Laisant and M. C. Moreau and by the Englishman H. M. Taylor. A solution based upon modern viewpoints is to be found in MacMahon's *Combinatory Analysis*. The approach adopted here is essentially that of Taylor (*The Messenger of Mathematics*, 32, 1903).

We will number the series of circularly arranged chairs from 1 through $2n$. The wives will then all have to be seated on the even- or odd-numbered chairs. In each of these two cases there are $n!$ different possible seating arrangements, so that there are $2 \cdot n!$ different possible seating arrangements for the women alone.

We will assume that the women have been seated in one of these arrangements and we will maintain this seating arrangement throughout the following. The *nucleus of the problem* then consists of determining the number of possible ways of seating the men between the women.

Let us designate the women in the assumed seating sequence as $F_1, F_2, \ldots, F_n$, their respective husbands $M_1, M_2, \ldots, M_n$, the couples $(F_1, M_1), (F_2, M_2), \ldots,$ as $1, 2, \ldots$ and arrangements in which there are n married couples as n-pair arrangements. Let us designate the husbands about whom we have no further information as $X_1, X_2, \ldots$.

Let

$$F_1 X_1 F_2 X_2 \ldots F_n X_n F_{n+1} X_{n+1}$$

be an $(n + 1)$-pair arrangement in which none of the husbands sits beside his own wife. (It must be remembered that the arrangement is circular, so that X_{n+1} is seated between F_{n+1} and F_1.) If we take F_{n+1} and $M_{n+1} = X_\nu$ out of the arrangement and replace X_ν with $X_{n+1} = M_\mu$, we obtain the n-pair arrangement

$$F_1 X_1 F_2 X_2 \ldots F_\nu M_\mu F_{\nu+1} \ldots F_n X_n.$$

This arrangement can occur in three ways:

1. No man sits next to his wife

(thus $\qquad M_\mu \neq M_\nu, \ M_\mu \neq M_{\nu+1}, \ X_n \neq M_1$).

2. One man sits next to his own wife (namely when

$$M_\mu = M_\nu \quad \text{or} \quad M_\mu = M_{\nu+1} \quad \text{or else} \quad X_n = M_1).$$

3. Two men sit next to their own wives (when $M_\mu = M_\nu$ or $M_\mu = M_{\nu+1}$ and at the same time $X_n = M_1$, that is, when in our arrangement the order $M_1 F_1$ occurs).

Thus, we must consider other seating arrangements in addition to the one prescribed in the problem.

In the following we will distinguish between three types of arrangements: arrangements A, B, and C. An A-arrangement will be one in which no man sits next to his wife. A *B*-arrangement will be one in which a certain man sits on a certain side of his wife. Finally, a *C*-arrangement will be one in which a certain man sits on a certain side of his wife and another man—which one, is not prescribed—sits alongside his wife—but the side is likewise not prescribed.

We will designate the number of *n*-pair *A*-, *B*-, *C*-arrangements as A_n, B_n, C_n, respectively.

First we will try to determine the relationships among the six magnitudes A_n, B_n, C_n, A_{n+1}, B_{n+1}, C_{n+1}; we will begin with the simplest of these relationships.

Let us consider B_{n+1} *B*-arrangements

$$F_1 X_1 F_2 X_2 \ldots F_n X_n F_{n+1} M_{n+1}$$

of the pairs $1, 2, \ldots, (n + 1)$, in which M_{n+1} sits next to F_{n+1} on her right. We will divide the arrangements into two groups in accordance with whether $X_n = M_1$ or $X_n \neq M_1$. We then remove the pair $F_{n+1} M_{n+1}$ from all of them. The first group then gives us all B_n *n*-pair *B*-arrangements, and the second all A_n *n*-pair *A*-arrangements, so that

(1) $$B_{n+1} = B_n + A_n.$$

We can obtain a second relationship by considering the C_{n+1} $(n + 1)$-pair *C*-arrangements

$$M_1 F_1 X_1 F_2 X_2 \ldots F_n X_n F_{n+1},$$

in which one of the men $X_1, X_2, \ldots, X_n$ sits next to his own wife. We also divide these arrangements into two groups in accordance with whether or not X_1 is or is not equal to M_{n+1}.

The second group then contains $(2n - 1)$ subgroups. In the first, M_2 is seated on the left of F_2, in the second on her right; in the third, M_3 sits on the left of F_3, in the fourth on her right, etc.; in the $(2n - 1)$th, M_{n+1} is seated on the left of F_{n+1}.

If we leave the pair $M_1 F_1$ out of all of the C_{n+1} *C*-arrangements, we obtain from the first group all C_n *C*-arrangements of the pairs $2, 3, 4, \ldots, (n + 1)$ in which M_{n+1} is seated on the right of F_{n+1}, and from each subgroup of the second group we obtain B_n *B*-arrangements of the pairs $2, 3, \ldots, (n + 1)$, so that

(2) $$C_{n+1} = C_n + (2n - 1)B_n.$$

As we found above, if we remove the pair F_{n+1}, M_{n+1} from an $(n + 1)$-pair A-arrangement $F_1X_1F_2X_2 \ldots F_{n+1}X_{n+1}$ and replace the M_{n+1} that has been removed with X_{n+1}, the arrangement is transformed into an n-pair A-, B-, or C-arrangement.

Conversely, we obtain an A-arrangement of the $(n + 1)$ pairs $1, 2, \ldots, (n + 1)$ when we insert $F_{n+1}M_{n+1}$ before F_1 of an A-, B-, or C-arrangement of the n pairs $1, 2, \ldots, n$ and then exchange the places of M_{n+1} and some other man (in such a manner that none of the men is seated next to his own wife after the exchange of places). It is also clear that this method gives us all the A-arrangements of the $(n + 1)$ pairs $1, 2, \ldots, (n + 1)$.

In order to find A_{n+1} it is therefore only necessary to determine the number of ways in which this insertion and the subsequent exchange can be accomplished for all possible A-, B-, and C-arrangements of the n pairs 1 through n.

We accomplish the described formation of the $(n + 1)$-pair A-arrangements in three steps.

I. *Formation from* A-*arrangements.*

After the insertion:

$$F_1X_1F_2X_2 \ldots F_nX_nF_{n+1}M_{n+1}$$

we can exchange the places of M_{n+1} and any other man except X_n and M_1, so that from each of the A_n n-pair A-arrangements we obtain $(n - 2)$ $(n + 1)$-pair A-arrangements. Consequently, we obtain a total of

$$(n - 2)A_n \quad (n + 1)\text{-pair } A\text{-arrangements.}$$

II. *Formation from* B-*arrangements.*

The n-pair B-arrangements exhibit the following $2n$ forms:

1.	$\ldots F_1M_1 \ldots$
2.	$\ldots F_1M_2F_2 \ldots$
3.	$\ldots F_1X_1F_2M_2 \ldots,$
	$\vdots$
$(2n - 2).$	$\ldots M_nF_nX_nF_1 \ldots,$
$(2n - 1).$	$\ldots F_nM_nF_1 \ldots,$
$2n.$	$\ldots F_nM_1F_1 \ldots.$

And there are B_n of each of these forms.

Our process of formation is not applicable to the first and the $(2n - 1)$th of these forms (since the inserted M_{n+1} would have to be exchanged with M_1 or M_n, as a result of which, however, M_1 would end up on the left side of F_1, or M_{n+1} would be on the left side of F_{n+1}).

In the second, third, ..., $(2n - 2)$th form, the exchange of the inserted M_{n+1} with M_2, M_2, M_3, M_3, ..., M_{n-1}, M_{n-1}, M_n transforms the n-pair B-arrangements into $(n + 1)$-pair A-arrangements, as a result of which a total of

$$(2n - 3)B_n \quad (n + 1)\text{-pair } A\text{-arrangements}$$

are obtained.

In the $(2n)$th form, the inserted M_{n+1} can be exchanged with any of the men M_2, M_3, ..., M_n, as a result of which a total of

$$(n - 1)B_n \quad (n + 1)\text{-pair } A\text{-arrangements}$$

are obtained.

III. *Formation from C-arrangements.*

Our method transforms any one of the C_n n-pair C-arrangements:

$$M_1F_1X_2F_2X_3F_3 \ldots X_nF_n$$

into an $(n + 1)$-pair A-arrangement if we switch the places of M_{n+1} and the man M_ν seated next to his wife (ν being one of the values $2, 3, 4, \ldots, n$). In this manner we obtain from every n-pair C-arrangement an $(n + 1)$-pair A-arrangement, which corresponds to a total of

$$C_n \quad (n + 1)\text{-pair } A\text{-arrangements.}$$

Thus, the methods of formation described under I., II., and III. give us all of the $(n + 1)$-pair A-arrangements, or a total of

$$[(n - 2)A_n + (3n - 4)B_n + C_n],$$

arrangements, so that

$$(3) \qquad A_{n+1} = (n - 2)A_n + (3n - 4)B_n + C_n.$$

In order to obtain formulas in which only the *same* capital letters occur, we infer from (1)

$$A_n = B_{n+1} - B_n \quad \text{and} \quad A_{n+1} = B_{n+2} - B_{n+1}$$

and introduce these values into (3). This gives

$$B_{n+2} = (n - 1)B_{n+1} + (2n - 2)B_n + C_n.$$

If we then replace n by $n + 1$, it follows that

$$B_{n+3} = nB_{n+2} + 2nB_{n+1} + C_{n+1}.$$

If we subtract the next to the last equation from the last one and take (2) into consideration, we get

$$B_{n+3} = (n + 1)[B_{n+2} + B_{n+1}] + B_n$$

or, if we replace $n + 1$ here by n,

(4) $$B_{n+2} = n(B_{n+1} + B_n) + B_{n-1}.$$

This simple *recurrence formula* for the B's enables us to calculate from *three* successive B's the B that follows immediately.

It is also possible to derive a recurrence formula in which only *three* successive B's are connected, i.e., a formula having the form

(5) $$e_nB_{n+1} + f_nB_n + g_nB_{n-1} = c,$$

in which the coefficients e_n, f_n, g_n represent known functions of n and c is a constant.

In order to find it we replace n in (5) with $(n + 1)$ and obtain

$$e_{n+1}B_{n+2} + f_{n+1}B_{n+1} + g_{n+1}B_n = c.$$

Subtraction of this equation from (5) gives

$$-e_{n+1}B_{n+2} + (e_n - f_{n+1})B_{n+1} + (f_n - g_{n+1})B_n + g_nB_{n-1} = 0.$$

In order to find the equations of condition for the coefficients e, f, g which are still unknown, we compare the formula obtained with equation (4) after equation (4) has been multiplied by g_n:

$$-g_nB_{n+2} + ng_nB_{n+1} + ng_nB_n + g_nB_{n-1} = 0.$$

Thus, we are able to obtain e, f, g and satisfy the three conditions

(I) $e_{n+1} = g_n$,   (II) $e_n - f_{n+1} = ng_n$,   (III) $f_n - g_{n+1} = ng_n$,

giving us the sought-for recurrence formula.

From (III) it follows that

$$f_n = g_{n+1} + ng_n \quad \text{or} \quad f_{n+1} = g_{n+2} + (n + 1)g_{n+1},$$

and from (II) and (I)

$$f_{n+1} = e_n - ng_n = g_{n-1} - ng_n.$$

By equating the two values obtained for f_{n+1} we get

$$(n + 1)g_{n+1} + ng_n = g_{n-1} - g_{n+2}.$$

It is easily seen that

$$g_n = n\iota^n \qquad\qquad (\iota = -1)$$

is a solution of this equation.   This, according to (I), yields

$$e_n = g_{n-1} = -(n-1)\iota^n$$

and, according to (III),

$$f_n = g_{n+1} + ng_n = \iota^n(n^2 - n - 1).$$

Equation (5) is thereby transformed into

$$(n-1)B_{n+1} - (n^2 - n - 1)B_n - nB_{n-1} = -c\iota^n.$$

In order to determine the constant c, we set n equal to 4, we observe that $B_3 = 0$, $B_4 = 1$, and $B_5 = 3$, and we obtain $c = 2$.

The sought-for recurrence formula consequently reads

(6) $(n-1)B_{n+1} = (n^2 - n - 1)B_n + nB_{n-1} - 2\iota^n.$

In order to obtain a recurrence formula for the A's as well, we express A_{n-1}, A_n, and A_{n+1}, in accordance with (1) and (6), by B_n and B_{n+1}.   Thus we obtain

$$A_{n-1} = \frac{1-n}{n}B_{n+1} + \frac{n^2-1}{n}B_n - \frac{2\iota^n}{n},$$

$$A_n = B_{n+1} - B_n,$$

$$A_{n+1} = \frac{n^2-1}{n}B_{n+1} + \frac{n+1}{n}B_n + \frac{2\iota^n}{n},$$

and from this by elimination of B_n and B_{n+1} we obtain

(7) $(n-1)A_{n+1} = (n^2 - 1)A_n + (n+1)A_{n-1} + 4\iota^n.$

This is *Laisant's recurrence formula*.   It makes possible the calculation of each A from the two immediately preceding A's.

Thus, from $A_3 = 1$, $A_4 = 2$, and (7), it follows that $A_5 = 13$, which is still easy to check directly.   Moreover, the whole series $A_6 = 80$, $A_7 = 579$, $A_8 = 4738$, $A_9 = 43387$, $A_{10} = 439792$, $A_{11} = 4890741$, $A_{12} = 59216642$, etc. can then be derived from (7).   The difficult point in the calculation of A can therefore be considered as eliminated.

The problem is solved.

The number of possible seating arrangements of n *married couples is* $2A_n \cdot n!$, *in which* A_n *can be calculated from Laisant's recurrence formula.*

9 Omar Khayyam's Binomial Expansion

To obtain the n*th power of the binomial* a + b *in powers of* a *and* b *when* n *is any positive whole number.*

SOLUTION. In order to determine the binomial expansion we write

$$(a + b)^n = (a + b)(a + b) \ldots (a + b),$$

where the right side consists of a product of n identical parentheses $(a + b)$. As is known, the multiplication of parentheses consists of choosing one term from each parenthesis and obtaining the product of the terms chosen, and continuing this process until all the possible choices are exhausted. Finally, the resulting products are added together.

A product of this sort has the following appearance:

$$P = a^{\alpha_1} b^{\beta_1} a^{\alpha_2} b^{\beta_2} a^{\alpha_3} b^{\beta_3} \ldots,$$

in which the factor a is taken from the first α_1 parentheses, the factor b from the next β_1 parentheses, the factor a from the next α_2 parentheses, etc.   In this case $\alpha_1 + \beta_1 + \alpha_2 + \beta_2 + \cdots$ equals the number of parentheses present, i.e., n.

If we set $\alpha_1 + \alpha_2 + \alpha_3 + \cdots$ equal to α and $\beta_1 + \beta_2 + \cdots$ equal to β, the expression can be written in the simpler form

$$P = a^\alpha b^\beta \quad \text{with} \quad \alpha + \beta = n.$$

Now the product P can generally be obtained in many other ways than the one described, for example, by taking a from the first α parentheses and b from the last β parentheses, or by taking b from the first β parentheses, and a from the last α parentheses, etc. If we assume that the product P occurs exactly C times in the method described, C being understood to represent an initially unknown whole number, then

$$G = Ca^\alpha b^\beta$$

represents one term of the binomial expansion.   The other terms have the same form, except that the exponents α and β and the coefficients C are different.   However, $\alpha + \beta$ always equals n.

The core of the problem is to determine the so-called *binomial coefficient* C, i.e., to answer the question: How many times does the product $P = a^\alpha b^\beta$ appear in the binomial expansion?

To answer this question we first write the factors a and b of the product one after another in the order in which we initially selected them from the parentheses:

$$\underbrace{aa \ldots a}_{} \cdot \underbrace{bb \ldots b}_{} \cdot \underbrace{aa \ldots a}_{} \cdot \ldots.$$

$$\text{totaling} \quad \text{totaling} \quad \text{totaling}$$
$$\alpha_1 \qquad \beta_1 \qquad \alpha_2$$

This is a permutation of n elements in which α identical elements a and β identical elements b occur. There are as many possible permutations of these elements as there are terms P resulting from the multiplication of the n parentheses $(a + b)$.

But the number of permutations of n elements among which there appear α identical elements of one kind and β identical elements of the other is $n!/\alpha!/\beta!$. This is how often the product $a^\alpha b^\beta$ appears in the binomial expansion. Consequently,

$$C = \frac{n!}{\alpha!\beta!}.$$

An apparent exception to this formula is presented by the terms a^n and b^n of the expansion, each of which occurs only once. To eliminate this exception let us agree to let the symbol 0! represent unity; we are then able to write the coefficients of a^n and b^n as $n!/n!0!$ and $n!/0!n!$, respectively, in agreement with the formula.

The individual possibilities of forming the product P can also be represented geometrically. We can, for example, represent the first possibility considered above in the following way: We mark off a horizontal distance of α_1 successive segments a, and from the end of this distance extend a vertical distance of β_1 successive segments b, from the end of this vertical line a third horizontal distance of α_2 successive segments a, etc. In a similar manner we represent the other possibilities of forming the product P; however, we begin all C zigzag traces, which represent the C possibilities, from the same point. Thus, for example, if we are concerned with finding the number ν of all the products of the form $a^{11}b^7$ in the binomial expansion of $(a + b)^{18}$, we draw a rectangular network of $11 \cdot 7$ rectangular compartments possessing a horizontal side a and a vertical side b and lying in seven 11-compartment rows one below the other. The possibility $a^4b^3a^7b^4$ (a from the first four parentheses, b from the following three, a from the next seven parentheses and b from the last four) is then represented by the unbroken heavy line, and the possibility $b^2a^6b^3a^2b^2a^3$ by the line of dashes. The sought-for number ν is therefore equal to the number of all the possible direct paths leading from the corner E of the network to the opposite corner F.

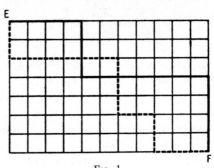

FIG. 1.

The formula previously found for C thus also provides us with the solution to the interesting problem:

A city has m *streets that run from east to west and* n *that run from north to south; how many ways (without detours) are there of getting from the north-west corner of the city to the southeast corner?*

Since there are $(n - 1)$ west–east partial paths a and $(m - 1)$ north–south partial paths b, the number of all the possible paths is

$$\frac{(m + n - 2)!}{(m - 1)!(n - 1)!}.$$

Now back to the binomial theorem!

Determination of the binomial coefficient C gives us immediately the sought-for *binomial expansion*:

$$(a + b)^n = \sum Ca^\alpha b^\beta \quad \text{with} \quad C = \frac{n!}{\alpha!\beta!}.$$

Here α and β pass through all the possible integral non-negative values that satisfy the condition $\alpha + \beta = n$.

The expansion of $(a + b)^5$, for example, gives

$$(a + b)^5 = a^5 + \frac{5!}{4!1!} a^4b + \frac{5!}{3!2!} a^3b^2 + \frac{5!}{2!3!} a^2b^3 + \frac{5!}{1!4!} ab^4 + b^5$$

or

$$(a + b)^5 = a^5 + 5a^4b + 10a^3b^2 + 10a^2b^3 + 5ab^4 + b^5.$$

Instead of $n!/\alpha!\beta!$ one usually writes

$$\frac{n(n-1)(n-2)\ldots(n-\alpha+1)}{1\cdot2\cdot3\ldots\alpha}$$

and also abbreviates this coefficient n_α (read as n sub α). The expansion then takes on a somewhat simpler appearance:

$$(a+b)^n = a^n + n_1 a^{n-1}b + n_2 a^{n-2}b^2 + \cdots + b^n.$$

The coefficient n_ν is known as the binomial coefficient to the base n with index ν.

The binomial theorem was probably discovered by the Persian astronomer Omar Khayyam, who lived during the eleventh century. At least he prided himself on having discovered the expansion "for all (integral positive) exponents n, which no one had been able to accomplish before him."

NOTE. The derivation given above is easily extended to give the nth power expansion of a polynomial $a + b + c + \cdots$. The polynomial theorem for a polynomial consisting of three terms, for example, is

$$(a+b+c)^n = \sum \frac{n!}{\alpha!\beta!\gamma!} a^\alpha b^\beta c^\gamma,$$

where the sum $\sum$ includes all possible terms for which the integral non-negative exponents α, β, γ satisfy the condition $\alpha + \beta + \gamma = n$.

10   Cauchy's Mean Theorem

The geometric mean of several positive numbers is smaller than the arithmetic mean of these numbers.

Augustin Louis Cauchy (1789–1857) was one of the greatest French mathematicians. The theorem concerning the arithmetic and geometric means occurs in his *Cours d'Analyse* (pp. 458–9), which appeared in 1821.

The proof of the theorem that will be presented here is based upon the solution of the *fundamental problem: When does the product of* n *positive numbers of constant sum attain its maximum value?*

We will call the n numbers $a, b, c, \ldots$, their constant sum K, and their product P. Experimentation with various numbers suggests that the product P reaches its maximal value when the numbers $a, b, c, \ldots$ all possess the same value $M = K/n$.

To determine the accuracy of this hypothesis, we use the

AUXILIARY THEOREM: *Of two pairs of numbers of equal sum the pair possessing the greater product is the one whose numbers exhibit the smaller difference.*

[If X and Y represent one pair and x and y the other, and $X + Y = x + y$, the auxiliary theorem follows from the equations

$$4XY = (X + Y)^2 - (X - Y)^2, \qquad 4xy = (x + y)^2 - (x - y)^2,$$

in which the minuends of the right sides are equal and the greater right side is the one in which the subtrahends are smaller.]

If the n numbers $a, b, c, \ldots$ are not all equal, then at least one, a, for example, must be greater than M, and at least one, let us say b, must be smaller than M. Let us form a new system of n numbers $a', b', c', \ldots$ in such a manner that (1) $a' = M$, (2) the pairs a, b and a', b' have the same sum, (3) the other numbers $c', d', e', \ldots$ correspond to $c, d, e, \ldots$. The new numbers then have the same sum K as the old ones, but a greater product $P'(= a'b'c' \ldots)$, since in accordance with the auxiliary theorem $a'b' > ab$.

If the numbers $a', b', c', \ldots$ are not all equal to M, then at least one, let us say b', is greater (smaller) and at least one, say c', is smaller (greater) than M. Let us form a new system of n numbers $a'', b'', c'', d'', \ldots$ in such a manner that (1) $a'' = a' = M$, (2) $b'' = M$, (3) the pairs b', c' and b'', c'' possess the same sum, (4) $d'', e'', \ldots$ correspond to $d', e', \ldots$. The numbers $a'', b'', c'', \ldots$ then have the same sum K as the numbers $a', b', c', \ldots$, but possess a greater product $P'' = a''b''c'' \ldots$, since in accordance with the auxiliary theorem $b''c'' > b'c'$.

We continue in this fashion and obtain a series of increasing products $P, P', P'', \ldots$ each successive member of which is greater than the immediately preceding one by at least one more multiple of the factor M. The last product obtained in this manner is the greatest of all and consists of n equal factors M. Consequently,

$$P < M^n,$$

which gives us the theorem:

The product of n *positive numbers whose sum is constant attains its maximal value when the numbers are equally great.*

If we extract the nth root of the last inequality and express P and M in terms of the magnitudes $a, b, c, \ldots$, we obtain *Cauchy's formula*:

$$\sqrt[n]{abc \ldots} < \frac{a + b + c + \cdots}{n}.$$

This is expressed verbally as follows:

THE THEOREM OF THE ARITHMETIC AND GEOMETRIC MEAN: *The geometric mean of several numbers is always smaller than the arithmetic mean of the numbers*, except when the numbers are equal, in which case the two means are also equal.

NOTE 1. Cauchy's theorem leads directly to the converse of the above extreme theorem:

The sum of n positive numbers whose product is constant attains its minimal value when the numbers are equal.

PROOF. Let us call the n numbers $x, y, z, \ldots$, their given product k, their variable sum s, and let us designate by m the nth root of k.

According to Cauchy,

$$\frac{x + y + z + \cdots}{n} \geqq \sqrt[n]{xyz\ldots};$$

consequently

$$s \geqq nm,$$

where the equality sign applies only in the event that $x = y = z$. Q.E.D.

The two preceding extreme theorems form the basis for a simple solution of many problems concerning maximum and minimum (cf. Nos. 54, 92, 96, 98).

NOTE 2. Cauchy's theorem also furnishes us directly with the important *exponential inequality* for the exponential function x^c.

If a is any positive number not equal to 1, n a whole number > 0, m a whole number $> n$, then the geometric mean of the m numbers of which n possess the value a and the $(m - n)$ others possess the value 1 is smaller than the arithmetic mean $(na + m - n)/m$ of these m numbers or

$$\sqrt[m]{a^n} < 1 + \frac{n}{m}(a - 1),$$

or, if we write ε in place of n/m,

$$(1) \qquad a^\varepsilon < 1 + \varepsilon(a - 1).$$

In this inequality ε is any rational, positive proper fraction. We will now show that this inequality is also true for any *irrational* proper fraction i.

First, it is clear that $a^J > 1 + J(a - 1)$ cannot be true for any irrational proper fraction J. If that were the case it would be possible to find a rational proper fraction $R < J$ so close to J that a^R would differ from a^J, and $1 + R(a - 1)$ from $1 + J(a - 1)$, by less than—

let us say—$\frac{1}{4}$ of the difference $a^J = [1 + J(a - 1)]$. In that event a^R would still be $> 1 + R(a - 1)$, which is, however, impossible according to (1).

Now let z be so small that $i + z$ and $i - z$ are both positive proper fractions. Then we have

$$a^i < a^i \cdot \frac{a^z + a^{-z}}{2}$$

(since the arithmetic mean of the numbers a^z and a^{-z} is greater than 1 according to Cauchy) or

$$a^i < \frac{a^{i+z} + a^{i-z}}{2}.$$

According to the above relation, however,

$$a^{i+z} \leqq 1 + (i + z)(a - 1), \quad a^{i-z} \leqq 1 + (i - z)(a - 1),$$

therefore

$$\frac{a^{i+z} + a^{i-z}}{2} \leqq 1 + i(a - 1);$$

thus, it is certain that

$$a^i < 1 + i(a - 1).$$

Inequality (1) is therefore true for *any* proper fraction ε.

If we replace ε in (1) by $1/\mu$, $1 + \varepsilon(a - 1)$ by b, i.e., a by $1 + \mu(b - 1)$, (1) is transformed into

(2) $b^\mu > 1 + \mu(b - 1)$,

where μ is any positive improper fraction, b any positive number.

CONCLUSION. *The exponential inequality*. If x is any positive magnitude and c any positive exponent, the exponential inequality is:

$$x^c \lessgtr 1 + c(x - 1),$$

in which proper fractional exponents require the use of the upper sign and improper fractional exponents require the use of the lower sign.

11   Bernoulli's Power Sum Problem

Determine the sum

$$S = 1^p + 2^p + 3^p + \cdots + n^p$$

of the p powers of the first n natural numbers for integral positive exponents p.

The problem, posed in this general form, was first solved in the *Ars Conjectandi* (Probability Computation), which appeared in 1713. It was the work of the Swiss mathematician Jacob Bernoulli (1654–1705).

The following elegant solution is based upon the binomial theorem.

By resorting to the device of considering the magnitudes $\mathfrak{S}^1$, $\mathfrak{S}^2$, $\mathfrak{S}^3, \ldots, \mathfrak{S}^\nu$ resulting from the binomial expansion of $(x + \mathfrak{S})^\nu$ as unknowns subject to ν certain conditions rather than as powers of $\mathfrak{S}$, we obtain an amazingly short derivation of S.

According to the binomial theorem, if P is understood to represent the number $p + 1$,

$$(\nu + \mathfrak{S})^P = \nu^P + P\nu^p\mathfrak{S}^1 + P_2\nu^{p-1}\mathfrak{S}^2 + \cdots$$

and

$$(\nu + \mathfrak{S} - 1)^P = \nu^P + P\nu^p(\mathfrak{S} - 1)^1 + P_2\nu^{p-1}(\mathfrak{S} - 1)^2 + \cdots.$$

Subtraction of these two equations gives us

$$(I) \quad \begin{cases} (\nu + \mathfrak{S})^P - (\nu - 1 + \mathfrak{S})^P = P\nu^p + P_2\nu^{p-1}[\mathfrak{S}^2 - (\mathfrak{S} - 1)^2] \\ \qquad\qquad + P_3\nu^{p-3}[\mathfrak{S}^3 - (\mathfrak{S} - 1)^3] + \cdots. \end{cases}$$

We now define the unknowns $\mathfrak{S}^1$, $\mathfrak{S}^2$, $\mathfrak{S}^3, \ldots$ by the equations

$$(1) \ (\mathfrak{S} - 1)^2 = \mathfrak{S}^2, \quad (2) \ (\mathfrak{S} - 1)^3 = \mathfrak{S}^3, \quad (3) \ (\mathfrak{S} - 1)^4 = \mathfrak{S}^4,$$

etc. This results in the simplification of (I) to

$$(Ia) \qquad\qquad P\nu^p = (\nu + \mathfrak{S})^P - (\nu - 1 + \mathfrak{S})^P.$$

This equation is formed for $\nu = 1, 2, 3, \ldots, n$, and we thereby obtain

$$P \cdot 1^p = (1 + \mathfrak{S})^P - \mathfrak{S}^P,$$
$$P \cdot 2^p = (2 + \mathfrak{S})^P - (1 + \mathfrak{S})^P,$$
$$\vdots$$
$$P \cdot n^p = (n + \mathfrak{S})^P - (n - 1 + \mathfrak{S})^P.$$

Addition of these n equations gives us

$$(II) \qquad\qquad PS = (n + \mathfrak{S})^P - \mathfrak{S}^P$$

or

$$(II) \quad 1^p + 2^p + \cdots + n^p = \frac{(n + \mathfrak{S})^P - \mathfrak{S}^P}{P} \quad \text{with} \quad P = p + 1.$$

This formula, in which the magnitudes $\mathfrak{S}^1$, $\mathfrak{S}^2$, $\mathfrak{S}^3, \ldots$ on the right side of the equation, obtained from expansion of the binomial $(n + \mathfrak{S})^P$, are defined by equations $(1), (2), (3), \ldots$, gives us the sought-for power sum.

In order to apply it to the cases $n = 1, 2, 3, 4$, we first determine the unknowns $\mathfrak{S}^1$, $\mathfrak{S}^2$, $\mathfrak{S}^3$, and $\mathfrak{S}^4$ in accordance with equations (1), (2), ...

From (1) it follows that $-2\mathfrak{S}^v + 1 = 0$, i.e., $\mathfrak{S}^1 = \frac{1}{2}$. Then, from (2), $-3\mathfrak{S}^2 + 3\mathfrak{S}^1 - 1 = 0$, i.e., $\mathfrak{S}^2 = \frac{1}{6}$. And from (3), $-4\mathfrak{S}^3 + 6\mathfrak{S}^2 - 4\mathfrak{S}^1 + 1 = 0$, i.e., $\mathfrak{S}^3 = 0$. Finally, from $(\mathfrak{S} - 1)^5 = \mathfrak{S}^5$ we obtain $\mathfrak{S}^4 = -\frac{1}{30}$. The numbers $\mathfrak{S}^1 = \frac{1}{2}$, $\mathfrak{S}^2 = \frac{1}{6}$, $\mathfrak{S}^3 = 0$, $\mathfrak{S}^4 = -\frac{1}{30}$, etc., are known as *Bernoulli numbers*.

Then from (II) we obtain

$$1 + 2 + 3 + \cdots + n = \frac{(n + \mathfrak{S})^2 - \mathfrak{S}^2}{2} = \frac{n^2 + 2n\mathfrak{S}^1}{2}$$

$$= n\frac{n + 1}{2},$$

$$1^2 + 2^2 + 3^2 + \cdots + n^2 = \frac{(n + \mathfrak{S})^3 - \mathfrak{S}^3}{3} = \frac{n^3 + 3n^2\mathfrak{S}^1 + 3n\mathfrak{S}^2}{3}$$

$$= \tfrac{1}{6}n(n + 1)(2n + 1),$$

$$1^3 + 2^3 + 3^3 + \cdots + n^3 = \frac{(n + \mathfrak{S})^4 - \mathfrak{S}^4}{4}$$

$$= \frac{n^4 + 4n^3\mathfrak{S}^1 + 6n^2\mathfrak{S}^2 + 4n\mathfrak{S}^3}{4}$$

$$= \left(n\frac{n + 1}{2}\right)^2,$$

$$1^4 + 2^4 + 3^4 + \cdots + n^4 = \frac{(n + \mathfrak{S})^5 - \mathfrak{S}^5}{5}$$

$$= \frac{n^5 + 5n^4\mathfrak{S}^1 + 10n^3\mathfrak{S}^2 + 10n^2\mathfrak{S}^3 + 5n\mathfrak{S}^4}{5}$$

$$= \frac{pst}{30},$$

with $p = n(n + 1)$, $s = 2n + 1$, $t = 3p - 1$.

If n in (II) increases without limit, S also increases without limit, but the quotient S/n^P possesses a finite value. In fact, in accordance with the binomial theorem, (II) is written

$$PS = n^P + P_1\mathfrak{S}^1 n^{P-1} + P_2\mathfrak{S}^2 n^{P-2} + \cdots,$$

so that

$$\frac{S}{n^P} = \frac{1}{P} + \frac{P_1\mathfrak{S}^1}{Pn} + \frac{P_2\mathfrak{S}^2}{Pn^2} + \cdots.$$

Now, if n increases infinitely all the fractions on the right-hand side with the exception of the first become infinitely small, and we obtain the limit equation of the power sum:

(III) $$\lim_{n \to \infty} \frac{1^p + 2^p + \cdots + n^p}{n^{p+1}} = \frac{1}{p+1}.$$

This important limit equation can also be derived from the exponential inequality (No. 10)

$$x^P > 1 + P(x - 1).$$

This derivation has the advantage over the one just given that it is true for *any positive exponent p*, not only for integral positive exponents!

If we first replace x in the exponential inequality with the improper fraction V/v, then by the proper fraction v/V, after elimination of the denominators we obtain

$$V^P > v^P + Pv^p(V - v) \quad \text{and} \quad v^P > V^P - PV^p(V - v)$$

or

$$Pv^p < \frac{V^P - v^P}{V - v} < PV^p.$$

Into this new inequality we introduce the series $1|0, 2|1, 3|2, \ldots,$ $n|n - 1$ for the pair of values $V|v$ and we obtain

$$P \cdot 0^p < 1^P - 0^P < P \cdot 1^p,$$
$$P \cdot 1^p < 2^P - 1^P < P \cdot 2^p,$$
$$\vdots$$
$$P \cdot (n - 1)^p < n^P - (n - 1)^P < P \cdot n^p.$$

Addition of these n inequalities results in

$$P(S - n^p) < n^P < PS$$

or

$$\frac{1}{P} < \frac{S}{n^P} < \frac{1}{P} + \frac{1}{n}.$$

Since both boundaries between which the quotient S/n^P is situated assume the value $1/P$ when $n = \infty$,

(III) $$\lim_{n \to \infty} \frac{1^p + 2^p + \cdots + n^p}{n^{p+1}} = \frac{1}{p+1},$$

where p represents *any* positive magnitude.

If the mean value of the function x^p is introduced, the limit equation of the power sum can be obtained in still another form.

The *mean value of a function over an interval* is commonly understood to mean the limiting value toward which the mean value of n values of the function uniformly distributed over the interval tends if n increases without limit. The mean value M of the function $f(x)$ over the interval 0 to x, if δ represents the nth part of x, is thus the limiting value of the quotient

$$\mu = \frac{f(\delta) + f(2\delta) + \cdots + f(n\delta)}{n}$$

for $n = \infty$. We write this mean value as $\overset{x}{\underset{0}{\mathfrak{M}}} f(x)$.

Thus, the mean value of the function x^p over the interval 0 through x is the limiting value of

$$\mu = \frac{\delta^p + (2\delta)^p + \cdots + (n\delta)^p}{n} = \delta^p \cdot \frac{1^p + 2^p + \cdots + n^p}{n},$$

i.e., since $\delta = x/n$, the limiting value of

$$\mu = x^p \cdot \frac{1^p + 2^p + \cdots + n^p}{n^{p+1}}.$$

Since the fractional factor of the right side according to (III) has the limiting value $1/(p + 1)$, it follows that the sought-for *mean value of the function x^p* is

(IIIa) $$\overset{x}{\underset{0}{\mathfrak{M}}} x^p = \frac{x^p}{p + 1};$$

this formula, however, is basically no different from (III).

Formula (III) or (IIIa) has found many applications in geometry and physics.

12   The Euler Number

Find the limiting values of the functions

$$\varphi(x) = \left(1 + \frac{1}{x}\right)^x \quad \text{and} \quad \Phi(x) = \left(1 + \frac{1}{x}\right)^{x+1}$$

for an infinitely increasing x.

The simplest solution of this very interesting problem is based upon the exponential inequality

$$x^\varepsilon < 1 + \varepsilon(x - 1)$$

(cf. No. 10), in which x is any positive magnitude and ε is any proper fraction between 0 and 1.

Let us introduce two arbitrary positive numbers a and b, the first of which is larger than the second and the second > 0, and introduce into the exponential inequality first

$$x = 1 + \frac{1}{b}, \qquad \varepsilon = \frac{b}{a},$$

and then

$$x = 1 - \frac{1}{b + 1}, \qquad \varepsilon = \frac{b + 1}{a + 1}.$$

In the first case we obtain $\left(1 + \frac{1}{b}\right)^{b/a} < 1 + \frac{1}{a}$ or

(1)
$$\left(1 + \frac{1}{b}\right)^{b} < \left(1 + \frac{1}{a}\right)^{a},$$

in the second $\left(1 - \frac{1}{b + 1}\right)^{b + 1/a + 1} < 1 - \frac{1}{a + 1}$ or

$$\left(\frac{b}{b + 1}\right)^{b + 1} < \left(\frac{a}{a + 1}\right)^{a + 1}$$

or, finally,

(2)
$$\left(1 + \frac{1}{b}\right)^{b + 1} > \left(1 + \frac{1}{a}\right)^{a + 1}.$$

The two inequalities obtained, (1) and (2), contain the remarkable theorem:

With an increasingly positive argument x *the function* $\varphi(x) = \left(1 + \frac{1}{x}\right)^{x}$ *increases while the function* $\Phi(x) = \left(1 + \frac{1}{x}\right)^{x + 1}$ *decreases.*

Thus, for $X > x$

$$\varphi(X) > \varphi(x), \quad \text{whereas} \quad \Phi(X) < \Phi(x).$$

Since, on the other hand, for the *same* values of the argument the function Φ exceeds the function φ

$$\left[\Phi(x) = \left(1 + \frac{1}{x}\right)\cdot\varphi(x)\right],$$

we obtain the inequalities

$$\varphi(x) < \varphi(X) < \Phi(X) \quad\text{and}\quad \varphi(X) < \Phi(X) < \Phi(x),$$

i.e., every value of the function Φ is greater than every value of the function φ. (Only positive values of the argument will be considered.)

Let us imagine two movable points p and P on the positive number axis which are situated at distances $\varphi(t)$ and $\Phi(t)$ from the zero point at time t and begin their movements in the instant $t = 1$. Point p, beginning from $\varphi(1) = 2$, then moves continuously toward the right, while point P, which begins at $\Phi(1) = 4$, moves continuously toward the left. However, since $\Phi(t)$ is always greater than $\varphi(t)$, i.e., P is *always* to the right of p, the points can never meet. Nevertheless, the distance between them is diminished

$$d = \Phi(t) - \varphi(t) = \varphi(t)/t,$$

since $\varphi(t) < 4$, and thus $d < 4/t$ without limit with increasing time, so that they finally are separated by an infinitely small distance.

The only way to explain this situation is to assume that on the number axis (between the numbers 2 and 4) there exists a fixed point that the moving points p and P approach infinitely closely from the left and from the right, respectively, without ever touching. The distance of this fixed point from the zero point is the so-called *Euler number e*. The proposal to designate this number, which also forms the base of the natural logarithmic system (No. 14), by the letter e stems from Euler (*Commentarii Academiae Petropolitanae ad annum 1739*, vol. IX).

The important inequality

$$\text{(I)}\qquad\qquad \left(1 + \frac{1}{x}\right)^{x} < e < \left(1 + \frac{1}{x}\right)^{x+1}$$

is true for Euler's number (x represents any positive number > 0).

If we choose $x = 1,000,000$, this inequality gives us the number e exactly to five decimal places. However, the use of the series for e (No. 13) is a better method of computation.

Then we obtain
$$e = 2.718281828459045\ldots.$$
The sought-for limiting values, however, are

$$\lim_{x \to \infty} \left(1 + \frac{1}{x}\right)^{x} = e \quad \text{and} \quad \lim_{x \to \infty} \left(1 + \frac{1}{x}\right)^{x+1} = e,$$

the first of which is an *upper* limit, while the second is a *lower* limit.

NOTE.   From the inequality (I) for the number e the inequality for the exponential function e^x follows directly.

1. In the inequality

$$\left(1 + \frac{1}{x}\right)^{x} < e$$

we replace x by $1/P$, where P is any positive number > 0; we assign to e the power P and obtain

(1) $$e^{P} > 1 + P.$$

2. In the inequality

$$e < \left(1 + \frac{1}{x}\right)^{x+1}$$

we replace $x + 1$ by $-1/n$, thus $1 + \frac{1}{x}$ by $\frac{1}{1 + n}$, n being a negative proper fraction $\neq 0$; we assign to e the power n and obtain

(2) $$e^{n} > 1 + n.$$

3. We consider that for every negative improper fraction N $(1 + N)$ is negative, and consequently we have

(3) $$e^{N} > 1 + N.$$

Combining the inequalities (1), (2), (3), we obtain the *inequality of the exponential function*:

$$e^{x} > 1 + x,$$

which is true for every finite real value of x and only becomes an *equation* when $x = 0$.

The inequality obtained leads directly to the so-called limit equation of the exponential function.

Let x be any finite real magnitude and n a positive number of such magnitude that $1 \pm \dfrac{x}{n}$ is positive. In accordance with the inequality of the exponential function,

$$e^{x/n} > 1 + \frac{x}{n} \quad \text{and} \quad e^{-x/n} > 1 - \frac{x}{n}.$$

We assign these inequalities the power n, in the case of the second, however, only after we have multiplied it by $1 + \dfrac{x}{n}$. This results in

$$e^x > \left(1 + \frac{x}{n}\right)^n \quad \text{and} \quad \left(1 + \frac{x}{n}\right)^n e^{-x} > \left(1 - \frac{x^2}{n^2}\right)^n.$$

Since the right-hand side of the second inequality, in accordance with the exponential inequality (No. 10), is greater than $1 - \dfrac{x^2}{n}$, then actually

$$\left(1 + \frac{x}{n}\right)^n e^{-x} > 1 - \frac{x^2}{n} \quad \text{or} \quad \left(1 + \frac{x}{n}\right)^n > \left(1 - \frac{x^2}{n}\right) e^x.$$

Combining the inequalities obtained, we get

$$\left(1 - \frac{x^2}{n}\right) e^x < \left(1 + \frac{x}{n}\right)^n < e^x.$$

If n is then allowed to increase infinitely, the left side of this inequality is transformed into e^x and we obtain the *limit equation of the exponential function:*

$$\lim_{n \to \infty} \left(1 + \frac{x}{n}\right)^n = e^x,$$

in which x represents any finite real number and n is an infinitely increasing magnitude.

13   Newton's Exponential Series

Transform the exponential function e^x *into a progression in terms of powers of* x.

This power progression, the so-called exponential series, which may in fact be the most important series in mathematics, was discovered by the great English mathematician and physicist Isaac Newton

(1642–1727). The famous treatise that contains the sine series, the cosine series, the arc sine series, the logarithmic series, and the binomial series as well as the exponential series was written in 1665 and bears the title *De analysi per aequationes numero terminorum infinitas.* Newton's derivation of the exponential series is, however, not rigorous and rather complicated.

The following derivation is based upon the mean values of the functions x^c (No. 11) and e^x.

We find the mean value of the function e^x with the help of the inequality of the exponential function

$$(1) \qquad\qquad e^u > 1 + u. \qquad\qquad \text{(No. 12)}$$

We will consider two arbitrary values v and $V = v + \varphi > v$ of the argument of the exponential function and first set $u = \varphi$ and then $u = -\varphi$ in (1). This gives us

$$e^\varphi > 1 + \varphi \quad \text{and} \quad e^{-\varphi} > 1 - \varphi, \quad \text{respectively.}$$

Multiplication with e^v and e^V, respectively, results in

$$e^V > e^v + \varphi e^v \quad \text{and} \quad e^v > e^V - \varphi e^V, \quad \text{respectively;}$$

combining, we obtain:

$$(2) \qquad\qquad e^v < \frac{e^V - e^v}{V - v} < e^V.$$

The mean value M of e^x over the interval 0 to x is the limiting value of the quotient

$$\mu = \frac{e^\delta + e^{2\delta} + e^{3\delta} + \cdots e^{n\delta}}{n} \qquad\qquad \left(\delta = \frac{x}{n}\right)$$

for an unlimitedly increasing n. In order to find μ, for a positive x we set down in (2) for the pair of values $v | V$ in succession

$$0|\delta, \ \delta|2\delta, \ 2\delta|3\delta, \ldots, \ (n-1)\delta|n\delta$$

and add the resulting n inequalities. This gives

$$n\mu + 1 - e^x < \frac{e^x - 1}{\delta} < n\mu$$

or, solved for μ,

$$\frac{e^x - 1}{x} < \mu < \frac{e^x - 1}{x} + \frac{e^x - 1}{n} \qquad\qquad (x > 0).$$

For a negative x we put down successively for $v|V$ in (2)

$$\delta|0,\ 2\delta|\delta,\ 3\delta|2\delta, \ldots, n\delta|(n-1)\delta.$$

Summation of the resulting n inequalities then leads to the same final inequality; only in this case the extremes are reversed, so that this time it reads

$$\frac{e^x - 1}{x} + \frac{e^x - 1}{n} < \mu < \frac{e^x - 1}{x} \qquad (x < 0).$$

If we then allow n to become infinite in the two inequalities obtained, we get for the lim μ the value

$$(3) \qquad\qquad \overset{x}{\underset{0}{\mathfrak{M}}} e^x = \frac{e^x - 1}{x},$$

whether x is positive or negative.

Now for the series expansion of e^x!

We begin with the inequality

$$e^x > 1 + x.$$

We assume initially that x is positive and obtain the mean values of both sides. This gives us

$$\frac{e^x - 1}{x} > 1 + \frac{x}{2} \quad \text{or} \quad e^x > 1 + x + \frac{x^2}{2!}.$$

Repeated mean formation gives rise to

$$\frac{e^x - 1}{x} > 1 + \frac{x}{2} + \frac{x^2}{3!} \quad \text{or} \quad e^x > 1 + x + \frac{x^2}{2!} + \frac{x^3}{3!}.$$

We continue in this manner and obtain

$$(4) \qquad e^x > 1 + x + \frac{x^2}{2!} + \frac{x^3}{3!} + \cdots + \frac{x^n}{n!}.$$

In order to obtain an upper limit for e^x also we begin with the inequality

$$e^{-x} > 1 - x,$$

multiply by e^x and obtain $1 > e^x - xe^x$ or

$$e^x < 1 + xe^x.$$

In the subsequent mean formations we employ the self-evident theorem: "The mean of the product of two (positive) functions u and v is smaller than the product of the mean value of u and the maximum value of v over the interval considered."

In the first step ($u = x$, $v = e^x$) we obtain

$$\frac{e^x - 1}{x} < 1 + \frac{x}{2} e^x \quad \text{or} \quad e^x < 1 + x + \frac{x^2}{2!} e^x,$$

in the second $\left(v = \frac{x^2}{2}, \; v = e^x \right)$

$$\frac{e^x - 1}{x} < 1 + \frac{x}{2} + \frac{x^2}{3!} e^x \quad \text{or} \quad e^x < 1 + x + \frac{x^2}{2!} + \frac{x^3}{3!} e^x,$$

etc., and finally

(5) $$e^x < 1 + x + \frac{x^2}{2!} + \frac{x^3}{3!} + \cdots + \frac{x^n}{n!} e^x.$$

If we then consider the case in which x is negative, the situation is somewhat simpler.

From $e^x > 1 + x$ it follows as above that

$$\frac{e^x - 1}{x} > 1 + \frac{x}{2};$$

however, since x is now negative,

$$e^x < 1 + x + \frac{x^2}{2!}.$$

The next mean formation yields

$$\frac{e^x - 1}{x} < 1 + \frac{x}{2} + \frac{x^2}{3!} \quad \text{or} \quad e^x > 1 + x + \frac{x^2}{2!} + \frac{x^3}{3!},$$

the next

$$e^x < 1 + x + \frac{x^2}{2!} + \frac{x^3}{3!} + \frac{x^4}{4!},$$

etc., and finally

(6) $$e^x > 1 + x + \frac{x^2}{2!} + \frac{x^3}{3!} + \cdots + \frac{x^{2v-1}}{(2v-1)!}$$

and

(7) $$e^x < 1 + x + \frac{x^2}{2!} + \frac{x^3}{3!} + \cdots + \frac{x^{2v}}{(2v)!}.$$

From inequalities (4), (5), (6), and (7) it follows that: When x is positive e^x lies between

$$1 + x + \frac{x^2}{2!} + \cdots + \frac{x^n}{n!} \quad \text{and} \quad 1 + x + \frac{x^2}{2!} + \cdots + \frac{x^n}{n!} e^x,$$

and when x is negative between

$$1 + x + \frac{x^2}{2!} + \cdots + \frac{x^n}{n!} \quad \text{and} \quad 1 + x + \frac{x^2}{2!} + \cdots + \frac{x^{n+1}}{(n+1)!}.$$

Then if we write

$$(8) \qquad\qquad e^x = 1 + x + \frac{x^2}{2!} + \cdots + \frac{x^n}{n!},$$

the error encountered for a positive value of x is less than

$$\frac{x^n}{n!}(e^x - 1),$$

and for a negative value of x less than $\left| \dfrac{x^{n+1}}{(n+1)!} \right|.$

But for a finite value of x and for an infinitely increasing n the fraction $x^n/n!$ approaches zero. [In accordance with No. 10 each of the products $2(n-1), 3(n-2), \ldots, (n-1) \cdot 2$ is greater than $1 \cdot n$. The product of these products is therefore greater than n^{n-2}, i.e., $(n-1)!^2 > n^{n-2}$ or $n!^2 > n^n$ or $n! > \sqrt{n^n}$. Thus, it follows that

$$\left| \frac{x^n}{n!} \right| < \left| \frac{x}{\sqrt{n}} \right|^n.$$

If n is assigned a value such that $\sqrt{n}$ is greater than $|2x|$, then

$$\left| \frac{x^n}{n!} \right| < \left| \frac{x}{\sqrt{n}} \right|^n < \left(\frac{1}{2}\right)^n \quad \text{and} \quad \lim_{n \to \infty} \frac{x^n}{n!} = 0.]$$

The error encountered with formula (8) thus disappears as x increases infinitely. Consequently:

The progression

$$(9) \qquad\qquad e^x = 1 + x + \frac{x^2}{2!} + \frac{x^3}{3!} + \cdots$$

is true for every finite x.

NOTE. The series obtained is particularly well suited for computation of the Euler number e. If, for example, we set x equal to 1,

$$e = 1 + \frac{1}{1!} + \frac{1}{2!} + \cdots + \frac{1}{10!} = 2.7182818012$$

and the encountered error is

$$F = \frac{1}{11!} + \frac{1}{12!} + \frac{1}{13!} + \cdots = \frac{1}{11!}\left(1 + \frac{1}{12} + \frac{1}{12 \cdot 13} + \cdots\right),$$

which is smaller than

$$\frac{1}{11!}\left(1 + \frac{1}{12} + \frac{1}{12^2} + \frac{1}{12^3} + \cdots\right)$$

or smaller than

$$\frac{1}{11!} \cdot \frac{12}{11} < 0.00000008.$$

The exact value is $e = 2.71828182845904523536\ldots$.

Formula (9), which applies to every finite real value of x, suggests the further extension of the concept of the exponential function to include the *complex* argument values z.

The exponential function e^z *for the complex argument* z *is defined by the formula*

(10) $$e^z = 1 + z + \frac{z^2}{2!} + \frac{z^3}{3!} + \cdots \quad \text{to infinity.}$$

It is easily seen that the infinite power series on the right-hand side of (10) has a definite finite value for *every* finite z, or, in other words, that the series *converges* for every finite z:

We set

$$1 + z + \frac{z^2}{2!} + \cdots + \frac{z^n}{n!} = E_n(z),$$

$$\frac{z^{n+1}}{(n+1)!} + \frac{z^{n+2}}{(n+2!)} + \cdots + \frac{z^{n+v}}{(n+v)!} = R_v(z),$$

so that

$$E_{n+v}(z) - E_n(z) = R_v(z).$$

If ζ represents the absolute magnitude of z, then the absolute magnitude of $R_v(z)$ must certainly be smaller than

$$\frac{\zeta^{n+1}}{(n+1)!} + \frac{\zeta^{n+2}}{(n+2!)} + \cdots \frac{\zeta^{n+v}}{(n+v)!},$$

and consequently considerably smaller than

$$\frac{\zeta^{n+1}}{(n+1)!} + \frac{\zeta^{n+2}}{(n+2)!} + \cdots \text{ to infinity} = e^\zeta - E_n(\zeta).$$

Since, in accordance with (8) or (9), $e^\zeta - E_n(\zeta)$ can be made as small as desired with the selection of a sufficiently high value for n, $R_v(z)$ can certainly be made as small as desired for such an n, no matter how great the value of v. However, this means that the series

$$1 + z + \frac{z^2}{2!} + \frac{z^3}{3!} + \cdots$$

converges. (It is in fact absolutely convergent, i.e., it still converges when z is converted into its absolute magnitude ζ.)

Moreover, let a and b be two arbitrary real or complex values, α and β their absolute magnitudes, and $\alpha + \beta = \gamma$. By multiplication of

$$E_n(a) = 1 + \frac{a}{1!} + \frac{a^2}{2!} + \cdots + \frac{a^n}{n!}$$

and

$$E_n(b) = 1 + \frac{b}{1!} + \frac{b^2}{2!} + \cdots + \frac{b^n}{n!}$$

we obtain $E_n(a)E_n(b) = 1 + C_1 + C_2 + \cdots + C_{2n}$, C_v representing the sum of all the members of the form $\dfrac{a^r b^s}{r!s!}$ in which the exponents r and s have the sum v. As long as v does not exceed the value of n, all $v + 1$ positive index pairs (r, s) occur in C_v with the sum v, whereas when $v > n$ only some of them do. Consequently, according to the binomial theorem (No. 9)

for $v \leqq n$ $$\qquad\qquad C_v = \frac{1}{v!}(a + b)^v,$$

for $v > n$ $$\qquad\qquad |C_v| < \frac{1}{v!}\gamma^v.$$

The sum of the first $(n + 1)$ terms of $E_n(a)E_n(b)$ is therefore equal to $E_n(a + b)$, and the sum of the absolute magnitudes of the following n terms is smaller than $R_n(\gamma)$, i.e., is certainly smaller than

$$\frac{\gamma^{n+1}}{(n + 1)!} + \frac{\gamma^{n+2}}{(n + 2)!} + \cdots + \text{to infinity} = e^\gamma - E_n(\gamma) = \delta,$$

so that we can set it equal to $\varepsilon\delta$, where $|\varepsilon| < 1$.

Accordingly, we obtain the equation

$$E_n(a) \cdot E_n(b) = E_n(a + b) + \varepsilon\delta.$$

If we then allow n to become infinite in this equation, δ becomes equal to zero, and the equation is converted into

$$(11) \qquad e^a \cdot e^b = e^{a+b}.$$

This fundamental formula justifies our previous suggestion of designating the series

$$1 + z + \frac{z^2}{2!} + \frac{z^3}{3!} + \cdots$$

as e^z.

Now let $z = x + iy$, where x and y are real. According to (11), $e^z = e^x \cdot e^{iy}$ or

$$e^z/e^x = e^{iy} = 1 + iy - \frac{y^2}{2!} - i\frac{y^3}{3!} + \frac{y^4}{4!} + i\frac{y^5}{5!} - \frac{y^6}{6!} + \cdots$$

$$= \left(1 - \frac{y^2}{2!} + \frac{y^4}{4!} - \frac{y^6}{6!} + - \cdots \right)$$

$$+ i\left[y - \frac{y^3}{3!} + \frac{y^5}{5!} - \frac{y^7}{7!} + - \cdots \right].$$

The brackets appearing here are, in accordance with No. 15, $\cos y$ and $\sin y$, and we obtain the *Euler formula*:

$$(12) \qquad e^{x+iy} = e^x(\cos y + i \sin y),$$

which when $x = 0$ takes the form

$$(12a) \qquad e^{iy} = \cos y + i \sin y.$$

If in (12a) $y = \pi$, we obtain the remarkable *Euler relation*

$$e^{i\pi} = -1$$

between the two significant numbers e *and* π.

If we then replace y by $-y$ in (12a), we obtain

$$(12b) \qquad e^{-iy} = \cos y - i \sin y$$

and subsequent addition and subtraction of (12a) and (12b) yields the equally remarkable pair of formulas

$$\cos y = \frac{e^{iy} + e^{-iy}}{2}, \qquad \sin y = \frac{e^{iy} - e^{-iy}}{2i}.$$

14   **Nicolaus Mercator's Logarithmic Series**

To calculate the logarithm of a given number without the use of the logarithmic table.

This fundamental problem, which forms the basis for the construction of the logarithmic tables, is solved simply and conveniently by logarithmic series. The simplest logarithmic series:

$$x - \tfrac{1}{2}x^2 + \tfrac{1}{3}x^3 - \tfrac{1}{4}x^4 + - \cdots,$$

which represents the natural log of $1 + x$, is found for the first time in the *Logarithmotechnia* (London, 1668) of the Holstein mathematician Nicolaus Mercator (1620–1687) (whose real name was Kaufmann). For the derivation of the logarithmic series we will make use of the mean value of the function $f(x) = \dfrac{1}{1 + x}$, which we will therefore determine first.

We will begin with the inequality (2) for the above number; we begin by converting this inequality into an inequality for the logarithmic function nat log x (nat log x, abbreviated as lx, is the logarithm of x when Euler's number e is taken as the base of the logarithmic system, i.e., the logarithm is the power of e required to obtain x).

Consequently, we replace v and V with lu and lU, where $U > u > 0$, and, correspondingly, e^v and e^V with u and U. This gives us

$$u < \frac{U - u}{lU - lu} < U$$

or

$$(1) \qquad \frac{1}{U} < \frac{lU - lu}{U - u} < \frac{1}{u} \qquad (U > u > 0).$$

The mean value of the function $f(x) = 1/(1 + x)$ is the limiting value of the fraction

$$\mu = \frac{f(\delta) + f(2\delta) + \cdots + f(n\delta)}{n}$$

for an infinitely increasing n and $\delta = x/n$.

To determine $\lim \mu$ for positive and negative values of x, respectively, we write $1 + \nu\delta \,|\, 1 + (\nu - 1)\delta$ in (1) for the pairs $U \,|\, u$ and $u \,|\, U$, respectively, and then form (1) for $\nu = 1, 2, 3, \ldots, n$. Addition of the resulting n inequalities gives in both cases:

$$\frac{l(1 + x)}{\delta} \text{ lies between } n\mu \text{ and } n\mu + \frac{x}{1 + x},$$

in other words,

$$\mu \text{ lies between } \frac{l(1 + x)}{x} \text{ and } \frac{l(1 + x)}{x} - \frac{x}{n(1 + x)},$$

Thus, if n becomes infinite, it follows that

$$(2) \qquad \underset{0}{\overset{x}{\mathfrak{M}}} \frac{1}{1 + x} = \frac{l(1 + x)}{x},$$

where $(1 + x)$ is naturally to be considered positive.

Now for the derivation of the series for $l(1 + x)$!

If we replace f on the *right-hand* side of this equation with $1 - xf$, we obtain

$$f = 1 - x + x^2 f.$$

If we again replace f on the right-hand side by $1 - xf$, we obtain

$$f = 1 - x + x^2 - x^3 f.$$

Similarly, from this we obtain

$$f = 1 - x + x^2 - x^3 + x^4 f,$$

etc., and in general:

$$f = 1 - x + x^2 - x^3 + x^4 - + \cdots - \varepsilon x^{n-1} + \varepsilon x^n f,$$

where ε is equal to $+1$ for even values of n and -1 for uneven values of n.

Obtaining the mean value from this formula, we have

$$(3) \qquad \frac{l(1 + x)}{x} = 1 - \frac{x}{2} + \frac{x^2}{3} - \frac{x^3}{4} + - \cdots - \varepsilon \frac{x^{n-1}}{n} + \varepsilon \mathfrak{M} x^n f.$$

If F represents the maximum value assumed by f over the interval 0 to x (thus $F = 1$ for positive values of x, $F = 1/(1 + x)$ for negative values of x), then in terms of the absolute value the mean value of $x^n f$ must be smaller than the F-value of the mean value $[x^n/(n + 1)]$ of x^n. Accordingly, we are able to write

$$\mathfrak{M} x^n f = \Theta F \frac{x^n}{n + 1},$$

where Θ is a definite positive proper fraction.

This converts (3) into

$$l(1 + x) = x - \frac{x^2}{2} + \frac{x^3}{3} - \frac{x^4}{4} + - \cdots - \varepsilon \frac{x^n}{n} + R$$

$$\text{with} \quad R = \varepsilon \Theta F \frac{x^{n+1}}{n+1}.$$

As n approaches infinity, if x is a *proper fraction* (also when $x = +1$) the "residue" R tends toward zero.

Consequently, the following progression is valid when x is a *proper fraction* and when $x = 1$:

(4) $$l(1 + x) = x - \frac{x^2}{2} + \frac{x^3}{3} - \frac{x^4}{4} + - \cdots.$$

The series on the right-hand side of the equation is Mercator's series.

Since it is only valid for proper fractional values of x, it is not suited for computing the logarithms of any number whatever. In order to obtain the series required for this, we substitute in (4) $-x$ for x and obtain

(5) $$l(1 - x) = -x - \frac{x^2}{2} - \frac{x^3}{3} - \frac{x^4}{4} - \cdots.$$

Subtracting (5) from (4) gives us

$$l \frac{1 + x}{1 - x} = 2 \left[x + \frac{x^3}{3} + \frac{x^5}{5} + \cdots \right].$$

For every positive or negative proper fractional value of x, $X = \dfrac{1 + x}{1 - x}$ is positive, while at the same time $x = \dfrac{X - 1}{X + 1}$, and the formula obtained is written

(6) $$lX = 2[x + \tfrac{1}{3}x^3 + \tfrac{1}{5}x^5 + \cdots] \quad \text{with} \quad x = \frac{X - 1}{X + 1}.$$

This new series converges for every positive X.

In this series we substitute for X the quotient Z/z of two arbitrary positive numbers (>0). This gives us

(7) $$\begin{cases} lZ - lz = 2[Q + \tfrac{1}{3}Q^3 + \tfrac{1}{5}Q^5 + \tfrac{1}{7}Q^7 + \cdots] \\ \text{with} \quad Q = \dfrac{Z - z}{Z + z}. \end{cases}$$

This series, in which Z and z may be any two positive numbers, is the logarithmic series from which the logarithmic tables can be computed.

In order, for example, to compute $l2$ we set z equal to 1 and Z to 2, which gives us

$$l2 = 2\left(\frac{1}{3} + \frac{1}{3 \cdot 3^3} + \frac{1}{5 \cdot 3^5} + \frac{1}{7 \cdot 3^7} + \cdots\right).$$

In order to compute $l5$ we set $z = 125 = 5^3$, and $Z = 128 = 2^7$, and this gives us

$$7l2 - 3l5 = 2(Q + \tfrac{1}{3}Q^3 + \tfrac{1}{5}Q^5 + \cdots) \text{ with } Q = \tfrac{3}{253}.$$

To compute $l3$ we assume that $z = 80 = 5 \cdot 2^4$, $Z = 81 = 3^4$, so that $lz = l5 + 4l2$, $lZ = 4l3$. This gives us

$$4l3 - l5 - 4l2 = 2(Q + \tfrac{1}{3}Q^3 + \tfrac{1}{5}Q^5 + \cdots) \quad \text{with} \quad Q = \tfrac{1}{161}.$$

To compute $l7$ we set z equal to $2400 = 2^5 \cdot 5^2 \cdot 3$, $Z = 2401 = 7^4$, and obtain

$$4l7 - 5l2 - 2l5 - l3 = 2(Q + \tfrac{1}{3}Q^3 + \tfrac{1}{5}Q^5 + \cdots)$$
$$\text{with} \quad Q = \tfrac{1}{4801}.$$

The series in the parentheses converge very rapidly, i.e., we require relatively few terms to obtain their sum fairly exactly.

NOTE. The common logarithms to the base 10 are computed from the natural logarithms. From

$$10^{\log x} = e^{lx} \quad (= x)$$

it follows in terms of the natural logarithms that

$$\log x \cdot l10 = lx$$

or

$$\log x = Mlx,$$

where

$$M = \frac{1}{l10} = 0.4342944819$$

is the so-called modulus by which the natural logarithm must be multiplied to give the common logarithm.

15   Newton's Sine and Cosine Series

Compute the circular functions sine and cosine of a given angle without the use of tables.

The simplest way of carrying out the required computation is with the use of the sine and cosine series.

The series for sin x and cos x first appeared in Newton's treatise *De analysi per aequationes numero terminorum infinitas* (1665–1666). (No. 13.) The sine series appears there as the converse of the arc sine series, which today is a very uncommon approach.

The derivation of the sine and cosine series presented here is based upon the mean values of the functions sin x and cos x over the interval 0 through x. (All of the angles mentioned in what follows are considered in *circular measure*.)

The mean value M of the function sin x over the interval 0 through x is the limiting value of the quotient

$$\mu = \frac{\sin \delta + \sin 2\delta + \cdots + \sin n\delta}{n}$$

for an infinitely increasing integral positive n, where δ represents the nth part of x.

But the numerator of the quotient* possesses the value

$$\sin m \cdot \frac{\sin n \dfrac{\delta}{2}}{\sin \dfrac{\delta}{2}},$$

where m is the arithmetic mean of the n argument values $\delta, 2\delta, \ldots, n\delta$, i.e.,

$$\frac{n+1}{2} \delta = \frac{x}{2} + \frac{\delta}{2}.$$

Consequently,

$$\mu = \frac{\sin m \sin \dfrac{x}{2}}{n \sin \dfrac{\delta}{2}}.$$

Since the denominator of the fraction on the right-hand side tends toward the limit $\frac{1}{2}x$ as n becomes infinitely great,* and the lim m is also equal to $\frac{1}{2}x$, we obtain

$$M = \lim_{n \to \infty} \mu = \frac{\sin \dfrac{x}{2} \sin \dfrac{x}{2}}{\dfrac{x}{2}}$$

---

* The reader who is unfamiliar with this fact will find the proof in note 2 at the end of this number, p. 63.

or

(1)
$$\overset{x}{\underset{0}{\mathfrak{M}}} \sin x = \frac{1 - \cos x}{x}.$$

By the same route, with the use of the formula

$$\cos \delta + \cos 2\delta + \cdots + \cos n\delta = \cos m \cdot \frac{\sin \dfrac{n\delta}{2}}{\dfrac{\delta}{2}},$$

we obtain

(2)
$$\overset{x}{\underset{0}{\mathfrak{M}}} \cos x = \frac{\sin x}{x}.$$

The series for $\sin x$ and $\cos x$ are now very easily found. Starting with the inequality

$$\cos x < 1,$$

we obtain the mean value for both sides and we have

$$\frac{\sin x}{x} < 1 \quad \text{or} \quad \sin x < x.$$

If we once again obtain the mean values (Formula [1] and No. 11) we obtain

$$\frac{1 - \cos x}{x} < \frac{1}{2} x \quad \text{or} \quad \cos x > 1 - \frac{x^2}{2}.$$

By again obtaining the mean value we get

$$\frac{\sin x}{x} > 1 - \frac{x^2}{3!} \quad \text{or} \quad \sin x > x - \frac{x^3}{3!},$$

etc. This results in:

$\cos x < 1$	$\sin x < x$
$\cos x > 1 - \dfrac{x^2}{2!}$	$\sin x > x - \dfrac{x^3}{3!}$
$\cos x < 1 - \dfrac{x^2}{2!} + \dfrac{x^4}{4!}$	$\sin x < x - \dfrac{x^3}{3!} + \dfrac{x^5}{5!}$
$\cos x > 1 - \dfrac{x^2}{2!} + \dfrac{x^4}{4!} - \dfrac{x^6}{6!}$	$\sin x > x - \dfrac{x^3}{3!} + \dfrac{x^5}{5!} - \dfrac{x^7}{7!},$

etc.

The integral rational functions on the right-hand side of these inequalities are the 1st, 2nd, 3rd, ..., νth *approximations* of the functions $\sin x$ and $\cos x$. They are called approximations because the degree of their deviation from the correct circular function grows progressively smaller as the index ν becomes higher and can be made as small as desired if ν is sufficiently great. Specifically, each of the two circular functions lies between two successive approximations of the true value. Thus, if we set them equal to one of these two approximations, the error incurred is smaller than the difference between the approximations, which has the form $x^\nu/\nu!$. The fraction $x^\nu/\nu!$, however, tends toward zero as ν becomes infinitely great (No. 13).

Accordingly, the following progressions

$$\sin x = x - \frac{x^3}{3!} + \frac{x^5}{5!} - \frac{x^7}{7!} + - \cdots,$$

$$\cos x = 1 - \frac{x^2}{2!} + \frac{x^4}{4!} - \frac{x^6}{6!} + - \cdots.$$

are valid for finite values of x.

If one of these series is interrupted at any point the error thereby incurred is smaller than the first disregarded term.

With these series it is possible to compute the sine and cosine of any given angle. They were used to draw up the sine and cosine tables found in logarithmic handbooks.

In order to illustrate the degree of approximation let us compute, for example, the $\sin 1^0 = \sin x$ (where $x = \pi/180$). We set

$$\sin 1^0 = \sin x = x - \frac{x^3}{6}.$$

The error thereby incurred is smaller than $x^5/120$, and this fraction is smaller than 0.000 000 000 02, so that, calculated exactly to 10 places, $\sin 1^0 = 0.0174524064$.

NOTE 1. *Summation of the series*

$$S = \sin \alpha + \sin (\alpha + \delta) + \sin (\alpha + 2\delta) + \cdots + \sin (\alpha + \overline{n - 1}\delta).$$

We multiply both sides by $2 \sin \delta/2$ and transform each of the products on the right in accordance with the formula

$$2 \sin \frac{\delta}{2} \sin (\alpha + \nu\delta) = \cos \left(\alpha + \frac{2\nu - 1}{2} \delta\right) - \cos \left(\alpha + \frac{2\nu + 1}{2}\right)\delta.$$

We are then left with

$$2S \sin \frac{\delta}{2} = \cos\left(\alpha - \frac{\delta}{2}\right) - \cos\left(\alpha + \frac{2n-1}{2}\delta\right).$$

Since the right side of this equation is

$$2 \sin\left(\alpha + \frac{n-1}{2}\right) \sin n \frac{2}{\delta},$$

we obtain

$$S = \sin m \cdot \frac{\sin n \dfrac{\delta}{2}}{\sin \dfrac{\delta}{2}},$$

where $m = \alpha + \dfrac{n-1}{2}\delta$ represents the mean value of all n angles α, $\alpha + \delta, \ldots, \alpha + \overline{n-1}\delta$.

In order to obtain the sum of the series

$$\sum = \cos \alpha + \cos(\alpha + \delta) + \cdots + \cos(\alpha + \overline{n-1}\delta)$$

we again multiply both sides by $2 \sin \dfrac{\delta}{2}$, but on the right-hand side we write

$$2 \sin \frac{\delta}{2} \cos(\alpha + \nu\delta) = \sin\left(\alpha + \frac{2\nu+1}{2}\delta\right) - \sin\left(\alpha + \frac{2\nu-1}{2}\delta\right).$$

We are then left with

$$2\sum \cdot \sin \frac{\delta}{2} = \sin\left(\alpha + \frac{2n-1}{2}\delta\right) - \sin\left(\alpha - \frac{\delta}{2}\right)$$

$$= 2 \cos\left(\alpha + \frac{n-1}{2}\delta\right) \sin n \frac{\delta}{2},$$

and we obtain

$$\sum = \cos m \cdot \frac{n \sin \dfrac{\delta}{2}}{\sin \dfrac{\delta}{2}}.$$

NOTE 2. *Proof* that $\lim\limits_{n \to \infty} n \sin \dfrac{w}{n} = w.$

$$\text{Sin } w = 2 \sin \frac{w}{2} \cos \frac{w}{2} = 2 \tan \frac{w}{2} \cos^2 \frac{w}{2} = 2 \tan \frac{w}{2} \cdot \left(1 - \sin^2 \frac{w}{2}\right).$$

However, since $\sin w < w$ and $\tan w > w$, it follows that

$$\sin w > 2 \cdot \frac{w}{2} \cdot \left(1 - \frac{w^2}{4}\right) \quad \text{or} \quad \sin w > w - \frac{1}{4} w^3.$$

Then $\sin \dfrac{w}{n}$ lies between $\dfrac{w}{n}$ and $\dfrac{w}{n} - \dfrac{1}{4} \dfrac{w^3}{n^3}$, i.e., $n \sin \dfrac{w}{n}$ lies between w

and $w - \dfrac{1}{4} \dfrac{w^3}{n^2}$. Thus,

$$\lim_{n \to \infty} n \sin \frac{w}{n} = w.$$

16   André's Derivation of the Secant and Tangent Series

Perhaps the most convenient and certainly the most attractive way of deriving the exponential series of the functions sec x and tan x is the *method of zigzag permutations* devised by the French mathematician André (*Comptes Rendus*, 1879, and *Journal de Mathématiques*, 1881).

A zigzag permutation—called by André an "alternating permutation"—of the n numbers $1, 2, 3, \ldots, n$ is an arrangement $c_1, c_2, \ldots, c_n$ of these numbers in which no element c_v possesses a magnitude such that it lies between its two neighbors c_{v-1} and c_{v+1}. If the points $P_1, P_2, \ldots, P_n$ are marked off on a system of coordinates such that their respective abscissas are $1, 2, \ldots, n$ and their respective ordinates $c_1, c_2, \ldots, c_n$, and each two successive points P_v and P_{v+1} are connected by a line segment, the zigzag line by which the permutation gets its name is obtained.

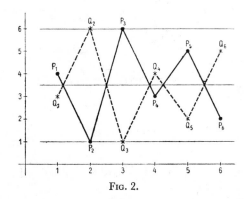

FIG. 2.

A zigzag line or zigzag permutation can begin either by rising or falling. We assert:

There are as many zigzag permutations (among n *elements) that begin by rising as by falling.*

PROOF. Let $P_1 P_2 \ldots P_n$ be the zigzag line corresponding to one zigzag permutation. Let us draw, through their highest and lowest point, parallels to the abscissa axis and a parallel midway between them. If we construct a mirror image of the zigzag line upon the middle parallel, the mirror image gives us a new zigzag line $Q_1 Q_2 \ldots Q_n$ or zigzag permutation, which begins either by falling or rising, depending upon whether the first zigzag line begins by rising or falling. Thus, for every zigzag permutation which begins by rising (or falling) we can obtain a corresponding zigzag permutation which begins by falling (or rising). Consequently, there is an equal number of each type.

Naturally there are just as many zigzag permutations that end by rising as by falling.

Let us, therefore, designate the number of zigzag permutations of n elements as $2A_n$, so that A_n represents the number of zigzag permutations of n elements that begin (or end) by rising (or falling).

The number A_n can be determined by a periodic formula. Let us consider all the $2A_n$ zigzag permutations of the n elements $1, 2, \ldots, n$ as written down and let us single out one of them, in which the highest element n occupies the $(r + 1)$th place (counting from the left). To the left of n there are then the r elements $\alpha_1, \alpha_2, \ldots, \alpha_r$, while to the right of n there are the s numbers $\beta_1, \beta_2, \ldots, \beta_s$, with $r + s = m = n - 1$. The permutation $\alpha_1 \alpha_2 \ldots \alpha_r$ ends by falling, since α_r is followed by n, which is higher; the permutation $\beta_1 \beta_2 \ldots \beta_s$ begins by rising, since β_1 follows n, which is higher.

Now let there be formed from the r elements $\alpha_1, \alpha_2, \ldots, \alpha_r$ a total of A_r zigzag permutations with falling ends and, similarly, from the s elements $\beta_1, \beta_2, \ldots, \beta_s$ a total of A_s zigzag permutations with rising beginnings. Consequently, there are $A_r \cdot A_s$ zigzag permutations of n elements in which n occupies the $(r + 1)$th position and in which to the left of n there are r elements $\alpha_1, \alpha_2, \ldots, \alpha_r$. However, since there are many other combinations of m elements to the rth class aside from the considered combination $\alpha_1, \alpha_2, \ldots, \alpha_r$—as is commonly known, there are a total of $C_m^r = m_r = m!/r!s!$—there are consequently a total of

$$p_r = m_r A_r A_s \qquad (r + s = m)$$

zigzag permutations of n elements in which the highest element (n) occupies the $(r + 1)$th place. It is also easily seen that this formula is also valid for the indices $r = 0, 1, 2$ if one sets $A_0 = A_1 = A_2 = 1$.

In order to obtain all the possible zigzag permutations we must obtain the expression p_r for all the values from $r = 0$ through $r = m = n - 1$ and add the resulting products. This gives us

$$2A_n = \sum_0^m p_r = \sum_r^{0,m} m_r A_r A_s.$$

In order to simplify this formula somewhat further, we write $m!/r!s!$ instead of m_r and set

(1)
$$\frac{A_v}{v!} = a_v.$$

It is then transformed into

$$2na_n = a_0 a_{n-1} + a_1 a_{n-2} + \cdots + a_{n-2} a_1 + a_{n-1} a_0,$$

or, utilizing the symbol for the sum, into

(2)
$$2na_n = \sum a_r a_s,$$

where r and s pass through all the possible integral numbers ≥ 0, for which $r + s = n - 1$.

Using the periodic formula (2) it is possible to compute, beginning with a_2, each number of the series $a_0, a_1, a_2, a_3, a_4, \ldots$ from the numbers preceding it.

From a_n, when it is multiplied by $n!$, *it is possible to obtain half the number of zigzag permutations of* n *elements.*

We can draw up a table for the simplest cases:

$n =$	0	1	2	3	4	5	6	7	8
$a_n =$	1	1	$\frac{1}{2}$	$\frac{1}{3}$	$\frac{5}{24}$	$\frac{2}{15}$	$\frac{61}{720}$	$\frac{17}{315}$	$\frac{277}{8064}$
$A_n =$	1	1	1	2	5	16	61	272	1385

We are able to confirm, for example, that the four elements 1, 2, 3, 4 yield $2 \cdot A_4 = 10$ zigzag permutations

$$1324, \quad 2143, \quad 3142, \quad 4132,$$
$$1423, \quad 2314, \quad 3241, \quad 4231,$$
$$2413, \quad 3412.$$

It is but a short step from the zigzag permutations to the series for sec x and tan x.

First we establish that starting with the index 3 all a_v are proper fractions $< \frac{1}{2}$. Since the number of zigzag permutations of n elements for $n > 2$ is smaller than the number of all the permutations of n elements, then $2A_n$ must be $< n!$, and consequently,

$$a_n < \tfrac{1}{2}.$$

Therefore, the infinite series

$$y = a_0 + a_1 x + a_2 x^2 + a_3 x^3 + \cdots$$

converges absolutely and is uniform over every interval $-h$ through $+h$ where $h < 1$. It therefore represents over this interval a continuous function with differentiable terms. The derivative of y is

$$y' = a_1 + 2a_2 x + 3a_3 x^2 + \cdots.$$

Since, moreover, the series for y converges absolutely, we can square it and thereby obtain

$$y^2 = \sum_{n}^{1,\infty} b_n x^{n-1},$$

where $b_1 = 1$ and for all $n \geq 2$

$$b_n = a_0 a_{n-1} + a_1 a_{n-2} + a_2 a_{n-3} + \cdots + a_{n-1} a_0.$$

In accordance with (2), therefore, whenever $n \geq 2$,

$$b_n = 2n a_n,$$

and then

$$y^2 = 1 + 2 \cdot 2a_2 x + 2 \cdot 3a_3 x^2 + 2 \cdot 4a_4 x^3 + \cdots.$$

If we then add one to both sides we obtain

$$1 + y^2 = 2[a_1 + 2a_2 x + 3a_3 x^2 + 4a_4 x^3 + \cdots]$$

or

$$1 + y^2 = 2y'.$$

We write this equation

$$\frac{y'}{1 + y^2} - \frac{1}{2} = 0$$

and reflect that the left side is the derivative of the function

$$Y = \arctan y - \tfrac{1}{2}x,$$

but that the derivative of a function (Y) can be zero only if this function is a constant. Thus we have

$$Y = \text{arc tan } y - \tfrac{1}{2}x = \text{const.}$$

In order to determine the constant, we set x equal to zero and obtain for this value of the argument x

$$y = 1, \quad \text{arc tan } y = \frac{\pi}{4}, \quad \text{and} \quad Y = \frac{\pi}{4}.$$

The constant therefore has the value $\pi/4$, and our equation is transformed into

$$\text{arc tan } y = \frac{\pi}{4} + \frac{x}{2}.$$

From this it follows that

$$y = \tan\left(\frac{\pi}{4} + \frac{x}{2}\right),$$

and we have the progression

$$(3) \qquad \tan\left(\frac{\pi}{4} + \frac{x}{2}\right) = a_0 + a_1 x + a_2 x^2 + a_3 x^3 + \cdots$$

which is true in any case for every proper fractional positive or negative value of x.

We replace x in (3) by $-x$ and obtain

$$(4) \qquad \tan\left(\frac{\pi}{4} - \frac{x}{2}\right) = a_0 - a_1 x + a_2 x^2 - a_3 x^3 + - \cdots.$$

As is easily seen, however, the two trigonometric formulas

$$2 \sec x = \tan\left(\frac{\pi}{4} + \frac{x}{2}\right) + \tan\left(\frac{\pi}{4} - \frac{x}{2}\right)$$

and

$$2 \tan x = \tan\left(\frac{\pi}{4} + \frac{x}{2}\right) - \tan\left(\frac{\pi}{4} - \frac{x}{2}\right)$$

are true.

If we introduce on the right-hand side here the series indicated in (3) and (4) we obtain the progressions for $\sec x$ and $\tan x$ which we were seeking:

$$\sec x = a_0 \ + a_2 x^2 + a_4 x^4 + a_6 x^6 + \cdots,$$

$$\tan x = a_1 x + a_3 x^3 + a_5 x^5 + a_7 x^7 + \cdots$$

or, if we return to half the number of zigzag permutations, A_n,

$$\sec x = A_0 \ + A_2 \frac{x^2}{2!} + A_4 \frac{x^4}{4!} + A_6 \frac{x^6}{6!} + \cdots,$$

$$\tan x = A_1 x + A_3 \frac{x^3}{3!} + A_5 \frac{x^5}{5!} + A_7 \frac{x^7}{7!} + \cdots.$$

These two progressions are true in all cases for every proper fractional value of x.

However, since $\sec x$ and $\tan x$ as functions of the complex argument x are analytic functions of x and the individual position closest to zero is $x = \pi/2$, the convergence circle has the radius $\pi/2$.

The two exponential series for sec x *and tan* x *consequently converge for every* x *the absolute value of which lies below $\pi/2$.*

17   Gregory's Arc Tangent Series

Determine the angles of a triangle from the sides without the use of tables.

If a, b, c are the given sides of the triangle, α, β, γ the angles (given in arc measure), the following relations, as is well known, are obtained:

$$\tan \frac{\alpha}{2} = \frac{\rho}{u}, \qquad \tan \frac{\beta}{2} = \frac{\rho}{v}, \qquad \tan \frac{\gamma}{2} = \frac{\rho}{w},$$

where $\rho^2 = uvw/s$, $u = s - a$, $v = s - b$, $w = s - c$, $2s = a + b + c$. Thus, $\alpha/2$, $\beta/2$, $\gamma/2$ are the arcs whose tangents are ρ/u, ρ/v, ρ/w. We write

$$\frac{\alpha}{2} = \text{arc tan } \frac{\rho}{u}, \qquad \frac{\beta}{2} = \text{arc tan } \frac{\rho}{v}, \qquad \frac{\gamma}{2} = \text{arc tan } \frac{\rho}{w}.$$

Arc tan x is understood to represent the arc whose tangent is x. The function arc tan x is called a *cyclometric function.*

We can consider our problem solved if we can succeed in *calculating the cyclometric function arc tan* x *for any given* x. This can be calculated by means of the exponential series for the arc tangent function obtained in 1671 by the English mathematician James Gregory (1638–1675).

To derive the arc tangent series we make use of the mean value of the function $f(x) = \dfrac{1}{1 + x^2}$, which we must consequently compute beforehand.

On a tangent of a unit circle $\Re$ we mark off from the point of tangency A the two segments $Ap = v$ and $AP = V$ in such a manner that $Pp = \varphi = V - v$; we connect p and P with the center of the circle O and designate the distances Op and OP as r and R, their intersections with $\Re$ as q and Q, and the arcs Aq, AQ, qQ in that order as w, W, ω. This gives us the equations $w = $ arc tan v, $W = $ arc tan V, $\omega = $ arc tan $V - $ arc tan v.

We would like to divide the area $(\tfrac{1}{2}\varphi)$ of the triangle OPp into two sections and for this purpose we draw the two arcs ph and PH concentric to qQ so that they meet OP and the extension of Op at h and H. The area of the triangle is then greater than the area $(\tfrac{1}{2}r^2\omega)$ of the sector Oph but smaller than the area $(\tfrac{1}{2}R^2\omega)$ of the sector OPH, so that

$$r^2\omega < \varphi < R^2\omega.$$

It follows from this that

$$\frac{1}{R^2} < \frac{\omega}{\varphi} < \frac{1}{r^2}$$

or, if instead of φ, ω, r^2, and R^2 we write in the same order $V - v$, arc tan $V - $ arc tan v, $1 + v^2$, $1 + V^2$ (Pythagoras),

(1) $$\frac{1}{1 + V^2} < \frac{\text{arc tan } V - \text{arc tan } v}{V - v} < \frac{1}{1 + v^2}.$$

In order to determine the mean value of the function $F(x) \dfrac{1}{1 + x^2}$ over the interval 0 through x, i.e., the limiting value of

$$\mu = \frac{F(\delta) + F(2\delta) + \cdots + F(n\delta)}{n}$$

(where $\delta = x/n$), in (1) we substitute successively $0|\delta$, $\delta|2\delta$, $2\delta|3\delta$, ..., $(n - 1)\delta|n\delta$ for the value pair $v|V$, add the resulting inequalities, and obtain

$$n\mu < \frac{\text{arc tan } x}{\delta} < n\mu + 1 - \frac{1}{1 + x^2}$$

or

$$\frac{\text{arc tan } x}{x} - \frac{x^2}{n(1 + x^2)} < \mu < \frac{\text{arc tan } x}{x}.$$

As the limit $n = \infty$ is approached this inequality is transformed into

(2) $$\mathop{\mathfrak{M}}_{0}^{x} \frac{1}{1 + x^2} = \frac{\text{arc tan } x}{x}.$$

Now for the derivation of the arc tangent series!

It is

$$\frac{1}{1+x^2} = 1 - \frac{x^2}{1+x^2}$$

or

$$F = 1 - x^2 F,$$

if for the sake of brevity we write F for $F(x)$. If we replace the F on the *right-hand* side of this equation with $1 - x^2 F$, we obtain

$$F = 1 - x^2 + x^4 F.$$

If here we once again write $1 - x^2 F$ for F on the right-hand side, we obtain

$$F = 1 - x^2 + x^4 - x^6 F.$$

In a similar manner, from this we obtain

$$F = 1 - x^2 + x^4 - x^6 + x^8 F,$$

$$F = 1 - x^2 + x^4 - x^6 + x^8 - x^{10} F,$$

etc. Consequently, we obtain the inequality

$$1 - x^2 + x^4 - x^6 + - \cdots - x^{4n-2} < F$$
$$< 1 - x^2 + x^4 - x^6 + - \cdots + x^{4n}.$$

Obtaining the mean value here gives us

$$1 - \frac{x^2}{3} + \frac{x^4}{5} - \frac{x^6}{7} + - \cdots - \frac{x^{4n-2}}{4n-1}$$
$$< \frac{\arctan x}{x} < 1 - \frac{x^2}{3} + \frac{x^4}{5} - \frac{x^6}{7} + - \cdots + \frac{x^{4n}}{4n+1}$$

or

$$(3) \quad x - \frac{x^3}{3} + \frac{x^5}{5} - \frac{x^7}{7} + - \cdots - \frac{x^{4n-1}}{4n-1} < \arctan x$$
$$< x - \frac{x^3}{3} + \frac{x^5}{5} - \frac{x^7}{7} + - \cdots + \frac{x^{4n+1}}{4n+1}.$$

If we then set

$$\arctan x = x - \frac{x^3}{3} + \frac{x^5}{5} - \frac{x^7}{7} + - \cdots - \frac{x^{4n-1}}{4n-1}$$

or rather

$$\arctan x = x - \frac{x^3}{3} + \frac{x^5}{5} - \frac{x^7}{7} + - \cdots - \frac{x^{4n-1}}{4n-1} + \frac{x^{4n+1}}{4n+1},$$

the error thereby incurred is smaller than the difference $x^{4n+1}/(4n+1)$ of the boundaries of (3). Since, however, this difference tends toward zero when n becomes infinitely great and x is a *proper fraction* (also when $x = 1$), we obtain the progression

$$(4) \qquad \text{arc tan } x = x - \frac{x^3}{3} + \frac{x^5}{5} - \frac{x^7}{7} + - \cdots \qquad \text{(for } x \leq 1\text{).}$$

This is *Gregory's formula*. *If the progression is interrupted at any point the error incurred is smaller than the first disregarded term.*

The series cannot be used when x is an improper fraction, because it no longer converges. In order to calculate arc tan x in this case we introduce $y = 1/x$, the reciprocal value of x, and make use of the formula

$$(5) \qquad \text{arc tan } x + \text{arc tan } y = \frac{\pi}{2}.$$

[If arc tan $x = \alpha$, i.e., $x = \tan \alpha$, then from

$$\tan \left(\frac{\pi}{2} - \alpha \right) = \frac{1}{\tan \alpha} = y$$

we obtain by inversion

$$\frac{\pi}{2} - \alpha = \text{arc tan } y \quad \text{or} \quad \frac{\pi}{2} = \text{arc tan } x + \text{arc tan } y.]$$

We then obtain arc tan y in accordance with Gregory's formula and arc tan x in accordance with (5).

But even if x is a proper fraction the arc tangent series is not advisable when x is very close to 1. In this case we introduce $z = \dfrac{1-x}{1+x}$, the half reciprocal value of x, and make use of the formula

$$(6) \qquad \text{arc tan } x + \text{arc tan } z = \frac{\pi}{4}.$$

[If arc tan $x = \alpha$, i.e., $x = \tan \alpha$, then from

$$\tan \left(\frac{\pi}{4} - \alpha \right) = \frac{1 - \tan \alpha}{1 + \tan \alpha}$$

we obtain by inversion

$$\frac{\pi}{4} - \alpha = \text{arc tan } \frac{1-x}{1+x} \quad \text{or} \quad \frac{\pi}{4} = \text{arc tan } x + \text{arc tan } z.]$$

Thus we obtain arc tan z with Gregory's formula and then arc tan x with (6).

Note. If in (4) we set $x = 1$, we obtain the so-called *Leibniz series*:

$$\frac{\pi}{4} = 1 - \frac{1}{3} + \frac{1}{5} - \frac{1}{7} + - \cdots,$$

which was discovered by Leibniz independently of Gregory in 1674.

It is not advisable, however, to use this series to *calculate* π. The series discovered by the English mathematician John Machin († 1751), which was published by him in 1706, is much better suited for this purpose. Machin made use of the auxiliary angle λ whose tangent is $\frac{1}{5}$. From $\tan \lambda = \frac{1}{5}$ it follows that $\tan 2\lambda = 2 \tan \lambda / (1 - \tan^2 \lambda) = \frac{5}{12}$, and from this, similarly, that

$$\tan 4\lambda = 2 \tan 2\lambda/(1 - \tan^2 2\lambda) = \frac{120}{119}$$

Inversion gives us $4\lambda = \text{arc tan } \frac{120}{119}$ or

$$\text{arc tan } \frac{120}{119} = 4 \text{ arc tan } \frac{1}{5}.$$

The left side of this equation, according to (5), has the value $\frac{\pi}{2} - \text{arc tan } \frac{119}{120}$; arc tan $\frac{119}{120}$, however, according to (6), has the value $\frac{\pi}{4} - \text{arc tan } \frac{1}{239}$, so that the left side is $\frac{\pi}{4} + \text{arc tan } \frac{1}{239}$. Consequently,

$$\frac{\pi}{4} = 4 \text{ arc tan } \frac{1}{5} - \text{arc tan } \frac{1}{239},$$

or written out completely:

$$\frac{\pi}{4} = 4\left(\frac{1}{5} - \frac{1}{3 \cdot 5^3} + \frac{1}{5 \cdot 5^5} - + \cdots\right) - \left(\frac{1}{239} - \frac{1}{3 \cdot 239^3} + \frac{1}{5 \cdot 239^5} - + \cdots\right).$$

Using this series, Machin calculated π to 100 decimal places.

18   Buffon's Needle Problem

On a table at d *intervals parallels are drawn. A needle of length* l *smaller than* d *is thrown at random on the table. What is the probability that the needle will touch one of the parallels?*

This remarkable problem stems from Georges Louis Leclerc, Comte de Buffon (1707–1788), who was the first man to clothe probability problems in geometric form.

The probability of an event is commonly understood to mean the ratio of the number of cases favoring an event to the total number of possible cases.

Let the probability we are seeking be W.

Let the needle have the terminal points A and B. Let us imagine the parallels extended horizontally. Let us single out two such adjacent parallels I and II (below I) and from any point P on line I let us drop a perpendicular PQ $(= d)$ to line II.

Let us begin by considering the *special positions* $\mathfrak{L}$ of the needle which are characterized by the following three conditions: (1) the terminal point A lies on the segment PQ; (2) the needle lies to the right of QP; (3) AP forms an acute angle: the inclination of the needle toward QP.

Let the probability that the needle touches parallel I in any of the special positions be w.

First we will show that

$$W = w.$$

If we consider all of the positions $\mathfrak{L}'$ in which the needle touches with its terminal point A either end of the segment PQ but is otherwise arbitrarily situated (i.e., touching either I or II or neither) this quadruples (as compared to the number of positions $\mathfrak{L}$) both the number of all the possible cases and the number of all the favorable cases.

The probability of touching one of the two parallels I and II in all of the positions $\mathfrak{L}'$ is, therefore, likewise w.

If to the cases $\mathfrak{L}'$ we add those positions in which the terminal point B instead of terminal point A comes to rest on the segment PQ, we obtain a total of $\mathfrak{L}''$ positions, which doubles the number of possible cases as well as the number of favorable cases.

Consequently, the probability of touching one of the parallels I and II in the positions $\mathfrak{L}''$ is also w.

Now if instead of taking one perpendicular PQ we take a very great number—ν—of very closely situated equidistant successive perpendiculars between I and II and consider all the positions of the needle in which one end of the needle comes to rest upon one of these ν perpendiculars, we thereby multiply by ν (with respect to $\mathfrak{L}''$) the number of all the possible as well as that of all the favorable cases.

Consequently, the probability of touching one of the parallels I and II by a needle position in which one needle end lies between I and II is again *w*.

The addition of still a third parallel III representing a mirror image of I on II (or of II on I), as well as the addition of the needle positions in which one end of the needle lies between III and II (or between III and I), again give us a probability of *w*.

In short, we have shown that

$$W = w.$$

Consequently, our problem has been limited to the task of determining the probability *w* of the needle touching line I *in a special position.*

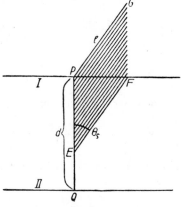

Fig. 3.

To obtain a better view of the infinitely great number of special positions, let us divide the above segment *PQ* into a very great number —$N > 1000^{1000}$—of equal parts and let us consider all of the cases in which the needle end *A* cuts one of the dividing points. For each dividing point there are an infinitely great number of possibilities corresponding to the infinitely great number of possible needle angles. For convenience in considering these possibilities also, let us consider only the *M* angles

$$\theta_0 = 0, \ \theta_1 = \varepsilon, \ \theta_2 = 2\varepsilon, \ \theta_3 = 3\varepsilon, \ldots, \ \theta_{M-1} = (M-1)\varepsilon,$$

where *M* likewise represents a very great number (e.g., $M > 2^{273}$) and ε is the *M*th part of $\pi/2$.

In this manner our consideration involves N points and M angles, thus, a total of NM needle positions.

However, only a certain fraction—just w—of these positions are favorable.   In order to determine this fraction we begin by obtaining the total number of only those favorable positions in which the angle of inclination of the needle has the selected value θ_s, as illustrated in Figure 3.   These positions form a parallelogram $EFGP$ with the sides $EF = l$ and $EP = l \cos \theta_s$.   Since there are

$$N \cdot \frac{EP}{PQ} = N \cdot \frac{l}{d} \cos \theta_s$$

dividing points on the segment EP, our overall total comprises

$$N \cdot \frac{l}{d} \cos \theta_s$$

favorable positions (with the common needle angle θ_s).   The number n of *all* the favorable positions altogether is consequently

$$n = N \cdot \frac{l}{d} (\cos \theta_0 + \cos \theta_1 + \cos \theta_2 + \cdots + \cos \theta_{M-1}).$$

The probability that we are seeking is, therefore,

$$w = \frac{n}{NM} = \frac{l}{d} \cdot \frac{\cos \theta_0 + \cos \theta_1 + \cos \theta_2 + \cdots + \cos \theta_{M-1}}{M}.$$

There remains then only the task of determining the value of the fraction

$$m = \frac{\cos \theta_0 + \cos \theta_1 + \cos \theta_2 + \cdots + \cos \theta_{M-1}}{M}.$$

The fraction m is no different from the mean value of the cosine function over the interval 0 through $\pi/2$.

Those who are familiar with the elements of integral calculus will immediately be able to write this mean value; it is

$$m = \int_0^{\pi/2} \cos x \, dx \bigg/ \frac{\pi}{2} = \frac{2}{\pi}.$$

Those readers who are not familiar with this type of calculation can obtain m just as easily in the following adroit manner.

Draw a quadrant of a circle with a radius of 1, designating the horizontal arm as OH and the vertical as OK.   If this is rotated about the radius OK it forms a hemisphere the area of whose surface is commonly known to be 2π.

The area of this surface can be expressed in a different form.

For this purpose let us move the above angles of inclination $\theta_0, \theta_1, \theta_2, \ldots, \theta_{M-1}$ so that the angles are formed at O with OH. The resulting free arms divide the quadrant into M very small arcs with the common length ε. Let us select from among them the one lying between the free arms of the angles θ_s and θ_{s+1}. On being rotated it forms a very small spherical zone, which when flattened out to a strip possesses the length $2\pi \cos \theta_s$ and the height ε, so that the area is then $2\pi\varepsilon \cos \theta_s$.

Since the sum of all the spherical zones obtained in this manner gives the hemisphere, we obtain the equation

$$2\pi\varepsilon(\cos \theta_0 + \cos \theta_1 + \cos \theta_2 + \cdots + \cos \theta_{M-1}) = 2\pi$$

or, since $M\varepsilon = \pi/2$,

$$\frac{\cos \theta_0 + \cos \theta_1 + \cos \theta_2 + \cdots + \cos \theta_{M-1}}{M} = \frac{2}{\pi}.$$

Thus, we have obtained the mean value that we were seeking.

The mean value of the cosine function (naturally that of the sine function also) over the interval 0 through $\pi/2$ is $2/\pi$.

[This also follows from formulas (1) and (2) of No. 15.]

At the same time we obtain

$$w = \frac{l}{d} m = \frac{l}{d} \cdot \frac{2}{\pi}$$

or

$$W = \frac{2}{\pi} \cdot \frac{l}{d}.$$

This formula gives us the probability we were seeking.

NOTE. Wolf in Zurich (1850) arrived at the original idea of using the obtained formula to calculate the number π. Experimentally, by a great number (5000) of throws with a needle 36 mm long and a distance of 45 mm between the parallels, he found the probability W to be (approximately) 0.5064, and obtained

$$\pi = \frac{2l}{dW} = 3.1596.$$

The Englishmen Smith (1855) and Fox (1864) repeated the experiment and found with 3200 and 1100 throws, respectively, values of 3.1553 and 3.1419 for π.

19   The Fermat-Euler Prime Number Theorem

Every prime number of the form 4n + 1 *can be represented in only one manner as the sum of two squares.*

This famous theorem was discovered about 1660 by Pierre de Fermat (1601–1665), the greatest French mathematician of the seventeenth century. It was not published, however, until 1670, when it appeared, unfortunately without proof, in the notes to the works of Diophantus, edited by Fermat's son. It is not certain whether or not Fermat had obtained the proof.

The first proof of the theorem was presented almost 100 years later by Leonhard Euler in his treatise "Demonstratio theorematis Fermatiani, omnem numerum primum formae 4n + 1 esse summam duorum quadratorum" (*Novi Commentarii Academiae Petropolitanae ad annos 1754–1755*, vol. V), after years of fruitless attempts at its solution.

Today there are several proofs of the Fermat-Euler theorem. The following proof is distinguished by its great simplicity.

For the reader who is unfamiliar with problems of number theory we will provide several explanations that will be necessary for understanding this proof and will also be found useful for the problem dealt with in No. 22. At the same time, it is to be understood that the letters used here and in No. 22 represent *whole* numbers.

Two numbers a and b (according to Gauss), are called *congruent to the modulus m,*

written: $a \equiv b \bmod m$,   read a congruent to b modulo m,

when their difference is divisible by m. Every number, for example, in regard to the modulus (to the modulus, *modulo*) m, is congruent to the residue it leaves over when divided by m, for example $65 \equiv 2 \bmod 7$. And this is also true when the word residue is taken in its most general sense, in which it means the residue left after division when the quotient is *arbitrarily* chosen. If, for instance, we write $65/7 = 12$, we remain with a residue of -19.

Among the many possible residues two are of special importance: the *conventional* or *common* residue, which is positive and smaller than the divisor, and the *minimal* residue, the magnitude of which never exceeds half the divisor. A minimal residue of the division $89/13$ is, for example, -2, because $89/13 = 7 - \frac{2}{13}$, which can also be written $89 \equiv -2 \bmod 13$.

The following self-evident rules apply to congruences to the same modulus:

1. *If two numbers are congruent to a third, they are also congruent to each other.*

2. *Two congruences can be added, subtracted, and multiplied.*

From

$$A \equiv B \bmod m, \qquad a \equiv b \bmod m$$

it follows that

$$A \pm a \equiv B \pm b \bmod m$$

and

$$Aa \equiv Bb \bmod m.$$

[From $A = B + Gm$ and $a = b + gm$ it follows, for example, that $Aa = Bb + \mathfrak{g}m$ ($\mathfrak{g}$ integral), i.e., $Aa \equiv Bb \bmod m$.]

3. The congruence

$$a \equiv b \bmod m$$

may be multiplied by any whole number g:

$$ag \equiv bg \bmod m.$$

It can be divided by g only when g is a common divisor of a and b that has no common divisor with the modulus. If, for example, we divide $49 \equiv 14 \bmod 5$ by 7, we obtain a correct congruence $7 \equiv 2 \bmod 5$.

A system of m integral numbers no two of which are congruent to the modulus m is called a *complete residue system* to the modulus m. The simplest complete residue system is the system of the m common residues $0, 1, 2, \ldots, m - 1$, and the next simplest is the system of m minimal residues.

Every number z is congruent to the modulus m to one and only one number of a complete residue system mod m.

Of particular importance is the following theorem:

THEOREM: *If the numbers of a complete residue system are multiplied by a number possessing no common divisor with the modulus, there is obtained once again a complete residue system with respect to the modulus.*

PROOF. Let m be the modulus, a the multiplier possessing no common divisor with m. If then for two different numbers x and x' of the given residue system $ax \equiv ax' \bmod m$ were true, it would follow from congruence rule 3 that $x \equiv x' \bmod m$, which, however, is not the case.

From this theorem it follows directly that:

The congruence

$$ax \equiv b \bmod m,$$

in which a *and* m *possess no common divisor, possesses in each complete residue system mod* m *one and only one "root"* x.

QUADRATIC RESIDUES

Of two numbers possessing no common divisor one is called the *quadratic residue* of the other when it is congruent to a square number with respect to the other as modulus; if there is no such square number it is called a *quadratic nonresidue*. For example, 12 is a quadratic residue of 13, since $12 \equiv 8^2 \bmod 13$; -1 is a quadratic nonresidue of 3, since there exists no square number x^2 such that $x^2 \equiv -1 \bmod 3$.

The following theorems concerning quadratic residues and non-residues apply to odd *prime number modulus p :*

I. *There are a total of* $\mathfrak{p} = (p - 1)/2$ *mutually incongruent quadratic residues and just as many mutually incongruent nonresidues of* p. *The former are* $1^2, 2^2, 3^2, \ldots, \mathfrak{p}^2$, *or whichever numbers are congruent to them mod* p.

II. *The product of two residues is a residue, the product of a residue and a nonresidue is a nonresidue, and finally, the product of two nonresidues is a residue.*

PROOF OF I.   1. If two of the designated squares were congruent to each other, for example $x^2 \equiv y^2 \bmod p$, the product $(x + y)(x - y)$ [which is equal to $x^2 - y^2$] would be divisible by p, which is impossible, because both of its factors are smaller than p.

2. If we continue the series of squares beyond $\mathfrak{p}^2$, no new residues are obtained. The square $(\mathfrak{p} + h)^2$, for example, is congruent to $k^2 \bmod p$ if $k \leq \mathfrak{p}$ is so determined that $\mathfrak{p} + h + k$ is divisible by p, since then $\mathfrak{p} + h \equiv -k$ and moreover $(\mathfrak{p} + h)^2 \equiv k^2 \bmod p$. Since there are (aside from the number divisible by p, disregarded here) $2\mathfrak{p}$ numbers mutually incongruent mod p, there must be a total of $\mathfrak{p}$ mutually incongruent quadratic nonresidues of p.

PROOF OF II.   Let R and r be quadratic residues, N and n quadratic nonresidues of p.

1. From $A^2 \equiv R$, $a^2 \equiv r \bmod p$ we obtain by multiplication $(Aa)^2 \equiv Rr \bmod p$. Consequently, Rr is a residue.

2. The $2\mathfrak{p}$ numbers $1^2, 2^2, \ldots, \mathfrak{p}^2, N1^2, N2^2, \ldots, N\mathfrak{p}^2$ are mutually incongruent mod p. Since the first $\mathfrak{p}$ of these numbers are quadratic

residues of p, and since only $\mathfrak{p}$ residues exist, the $\mathfrak{p}$ numbers $N1^2$, $N2^2, \ldots, N\mathfrak{p}^2$ must be nonresidues, i.e., NR is a nonresidue.

3. The $2\mathfrak{p}$ numbers $n \cdot 1^2, n \cdot 2^2, n \cdot 3^2, \ldots, n \cdot \mathfrak{p}^2, n \cdot N1^2, n \cdot N2^2, \ldots,$ $n \cdot N\mathfrak{p}^2$ are mutually incongruent mod p. The first $\mathfrak{p}$ of these numbers are nonresidues in accordance with 2.; consequently, the others must be residues in accordance with 1.; however, among them is the product of the two nonresidues N and n.   Q.E.D.

Let us now consider the *bilinear congruence*

(0) $$xy \equiv D \bmod p,$$

in which the modulus p is once again an odd prime number, D a given number possessing no common divisor with p, and the "mutually conjugate" or "linked" magnitudes x and y are chosen in such a manner from the system Σ of the numbers $1, 2, 3, \ldots, p - 1$ that (0) is satisfied. For each x from Σ there is then only one conjugate y. [From $xy \equiv D \bmod p$ and $xy' \equiv D \bmod p$ it follows that $xy \equiv xy'$ mod p and from this $y \equiv y' \bmod p$ or $y - y' \equiv 0 \bmod p$. However, since both y and $y' \leq p - 1$, their difference is divisible by p only when $y' = y$.]

We select x_1 arbitrarily from Σ and determine y_1 such that

$$x_1 y_1 \equiv D \bmod p.$$

Then we select from Σ a number x_2 that differs from x_1 and y_1 and determine y_2 such that

$$x_2 y_2 \equiv D \bmod p.$$

y_2 then is different from x_1 as well as from y_1.

We continue in this manner until all the numbers of Σ have been arranged in the resulting congruences.

Here there are two cases to be distinguished:

1. y_ν never equals x_ν. In other words: the congruence $x_\nu^2 \equiv D \bmod p$ is impossible; D is a *quadratic nonresidue of p*. We then obtain exactly $\mathfrak{p} = (p - 1)/2$ pairs x_ν, y_ν of conjugate numbers, and multiplication of the $\mathfrak{p}$ congruences formed gives

(1) $$(p - 1)! \equiv D^\mathfrak{p} \bmod p.$$

2. For a certain index ν, $y_\nu = x_\nu$, thus $x_\nu^2 \equiv D \bmod p$; D is a *quadratic residue* of p. If aside from ν there is also an index μ for which the same occurs, then $x_\mu^2 \equiv D \bmod p$, and so $x_\mu^2 \equiv x_\nu^2 \bmod p$, i.e., $x_\mu^2 - x_\nu^2$ or $(x_\mu + x_\nu)(x_\mu - x_\nu)$ is divisible by p. Since $x_\mu - x_\nu$ is not

divisible by p, $x_\mu + x_\nu$ must be divisible by p, and consequently $x_\mu = p - x_\nu$. Actually, then $x_\mu^2 = p^2 - 2px_\nu + x_\nu^2 \equiv x_\nu^2 \equiv D \bmod p$. *Equal* linked magnitudes thus occur exactly twice if they occur at all. In our case $(y_\nu = x_\nu, y_\mu = x_\mu)$ we now have only $\mathfrak{p} - 1$ congruences $x_s y_s \equiv D \bmod p$, where y_s differs from x_s. To these $\mathfrak{p} - 1$ congruences we add the congruence

$$x_\nu x_\mu \equiv -D \bmod p,$$

multiply all $\mathfrak{p}$ congruences and obtain

(2) $(p - 1)! \equiv -D^\mathfrak{p} \bmod p.$

This is the case when, for example, $D = 1$, since then $1^2 \equiv D \bmod p$. Then we have the congruence

(2a) $(p - 1)! \equiv -1 \bmod p,$

which represents the so-called *Wilson theorem*.

Using Wilson's formula we write instead of (1) and (2)

(1a) $D^\mathfrak{p} \equiv -1 \,(\bmod\, p)$ (2a) $D^\mathfrak{p} \equiv 1 \,(\bmod\, p)$

and obtain

EULER'S THEOREM: *The number* D *that possesses no common divisor with the prime number* p *is either a quadratic residue or nonresidue of* p, *depending on whether* $D^\mathfrak{p}$ *is congruent mod* p *to the positive or negative unit.*

The introduction of the *Legendre symbol* makes it possible to express this criterion of the residue character of a number by a formula. The Legendre symbol $\left(\dfrac{D}{p}\right)$ represents the positive or negative unit, depending on whether or not D is a quadratic residue or nonresidue of p. Thus, for example, $\left(\dfrac{2}{7}\right) = 1$, since $3^2 - 2$ is divisible by 7, whereas $\left(\dfrac{2}{3}\right) = -1$, since there is no square number whose difference from 2 is divisible by 3.

When this symbol is used Euler's criterion assumes the simple form

(3) $\left(\dfrac{D}{p}\right) \equiv D^\mathfrak{p} \bmod p, \quad \text{with} \quad \mathfrak{p} = \dfrac{p - 1}{2}.$

In the simple case $D > -1$, congruence (3) is transformed into the equation

(4) $\left(\dfrac{-1}{p}\right) = (-1)^{(p-1)/2},$

since in this case both sides of (3) are units, and the difference between two units is divisible by the odd prime number p only when these units are equal.

Now $\dfrac{p-1}{2}$ is even or odd, depending on whether the prime number p is of the form $4n + 1$ or $4n + 3$. In the first case, then $\left(\dfrac{-1}{p}\right) = +1$, i.e., -1 is a quadratic residue of p, and in the second case $\left(\dfrac{-1}{p}\right) = -1$, i.e., -1 is a quadratic nonresidue of p. Consequently, the following is true:

THEOREM OF EULER: *The negative unit is a quadratic residue of the prime number* p, *when* p *has the form* 4n + 1 *and a quadratic nonresidue when* p *has the form* 4n + 3.

In other words: *The pure quadratic congruence*

$$x^2 + 1 \equiv 0 \bmod p$$

has integral solutions x *when* p *has the form* 4n + 1 *and has not when* p *has the form* 4n + 3.

Now for the *proof of the Fermat-Euler theorem*!

The following proof is based upon the above theorems and the

NORM THEOREM: *If a prime number goes into a norm but not into the bases of the norm, it is itself a norm.*

A *norm* is understood to mean the sum of the squares of two whole numbers, which are the "bases" of the norm.

PROOF OF THE NORM THEOREM. Let the prime number p go into the norm $a^2 + b^2$, but not into its bases a and b, so that

$$(5) \qquad a^2 + b^2 = pf,$$

it being assumed that the factor f is greater than 1 but smaller than $p/2$. This assumption does not represent a limitation of the theorem, since from $A^2 + B^2 = pF$, with $F > (P/2)$, we can immediately form the equation $a^2 + b^2 = pf$, with $f < (P/2)$, if the minimal residues $A - hp$ and $B - kp$ of the divisions A/p and B/p, respectively, are taken for a and b, respectively. On the one hand,

$$a^2 + b^2 = [A^2 + B^2] - 2(Ah + Bk)p + (h^2 + k^2)p^2$$

is divisible by p, and thus

$$a^2 + b^2 = pf;$$

while on the other hand, since $|a| < \frac{1}{2}p$ and $|b| < \frac{1}{2}p$, $a^2 + b^2$ is smaller than $\frac{1}{2}p^2$ or $pf < \frac{1}{2}p^2$ or $f < \frac{1}{2}p$. Moreover, p does not go into either a or b, because then (contrary to our assumption) it would go into $A = a + hp$ or into $B = b + kp$.

We determine the minimal residues $\alpha = a - mf$ and $\beta = b - nf$ of the divisions a/f and b/f and obtain similarly

(6) $$\alpha^2 + \beta^2 = ff', \quad \text{with} \quad f' \leq \tfrac{1}{2}f.$$

Multiplication of (5) and (6) gives us

$$(a^2 + b^2)(\alpha^2 + \beta^2) = pf^2f'$$

or

$$(a\alpha + b\beta)^2 + (a\beta - b\alpha)^2 = pf^2f'.$$

Since

$$a\alpha + b\beta = [a^2 + b^2] - (am + bn)f = a'f,$$

$$a\beta + b\alpha = (bm - an)f = b'f,$$

the equation obtained is written

(7) $$a'^2 + b'^2 = pf', \quad \text{where} \quad f' \leq \tfrac{1}{2}f.$$

Here f' cannot disappear. If $f' = 0$, then in accordance with (6) $\alpha = 0$ and $\beta = 0$, and from this it follows that $a = mf$ and $b = nf$; then according to (5) $p = (m^2 + n^2)f$. In this event p would have to be divisible by f, and then f would have to equal 1, which contradicts our premise.

If, then, $f' = 1$, (7) already gives us the norm expression of p.

If $f' > 1$, we obtain from (7)

(8) $$a''^2 + b''^2 = pf'' \quad \text{with} \quad 0 < f'' \leq \tfrac{1}{2}f',$$

just as (7) was obtained from (5). This method of constructing new equations with continuously *diminishing* factors f, f', f'', ... is continued until the factor 1 appears. The corresponding equation gives the prime number p represented as a norm.

Now we will prove

I. *A prime number* q *of the form* 4n + 3 *cannot be represented as a norm.*

II. *Every prime number* p *of the form* 4n + 1 *can be represented as a norm in only one way.*

PROOF OF I. If it were true that

$$a^2 + b^2 = q,$$

then it would follow that

$$b^2 \equiv -a^2 \bmod q$$

and the product $(-1)(a^2)$ of a quadratic nonresidue (-1) and a residue (a^2) of q would be a quadratic residue (b^2) of q, which according to the above is impossible.

PROOF OF II. According to Euler's theorem there is a whole number x such that the norm $x^2 + 1$ is divisible by p. According to the norm theorem, p is then itself a norm:

$$p = a^2 + b^2.$$

Here also there is only one possible norm representation.

If we assume a *second* such representation:

$$p = A^2 + B^2$$

(where a, b, A, B represent four different positive numbers), it follows that

$$p^2 = (a^2 + b^2)(A^2 + B^2) = (Aa + Bb)^2 + (Ab \mp Ba)^2,$$

where either the two upper signs or the two lower signs are possible. Then, since the product of the two factors $Aa + Bb$ and $Aa - Bb$:

$$A^2a^2 - B^2b^2 = A^2(a^2 + b^2) - b^2(A^2 + B^2)$$

is divisible by p, one of the factors must be divisible by p. Consequently, we select the upper or lower signs depending upon whether the first or second factor is divisible by p. Then either

$$Aa + Bb = p \quad \text{and at the same time} \quad Ab - Ba = 0$$

or

$$Ab + Ba = p \quad \text{and at the same time} \quad Aa - Bb = 0,$$

thus, either $A^2b^2 = B^2a^2$ or $A^2a^2 = B^2b^2$.

From the first of these equations it follows that

$$\frac{A^2}{a^2} = \frac{B^2}{b^2} = \frac{A^2 + B^2}{a^2 + b^2} = 1,$$

and from the second

$$\frac{A^2}{b^2} = \frac{B^2}{a^2} = \frac{A^2 + B^2}{b^2 + a^2} = 1,$$

thus, from the first $A = a$, while from the second $A = b$, both of which contradict the initial assumption, which requires that $A \neq a$ and $A \neq b$. There is therefore only one way of representing p as a norm, and the Fermat-Euler theorem is proved.

20   The Fermat Equation

Find the integral solutions of the equation

$$x^2 - dy^2 = 1,$$

in which d *is a nonquadratic positive whole number.*

This extremely important problem of number theory was posed by Pierre Fermat in 1657, first to his friend Frénicle and then to all contemporary mathematicians.

The first solution, a very complicated one, was obtained by the Englishmen Lord Brouncker and John Wallis.

The simplest and best solutions to this problem were discovered by Euler, Lagrange, and Gauss. [Euler: "De usu novi algorithmi . . .," *Novi Commentarii Academiae Petropolitanae ad annum 1765.* Lagrange: "Solution d'un problème d'arithmétique," *Miscellanea Taurinensia,* vol. IV, 1768. Gauss: *Disquisitiones arithmeticae,* 1801.] They are all based upon the properties of periodic continued fractions.

We will examine a somewhat modified form of this method with the more general equation

$$X^2 - DY^2 = 4,$$

which includes the original Fermat equation (with $X = 2x$, $Y = y$, $D = 4d$) as a special case, but includes as well the case in which D leaves a residue of 1 on being divided by 4.

For the sake of convenience we shall write the continued fraction

$$a + \frac{1}{b} + \frac{1}{c} + \frac{1}{d} + \cdots$$

in the abbreviated form $(a, b, c, d, \ldots)$.

A purely periodic continued fraction with an n-term period has the form

$$u = (g_1, g_2, \ldots, g_n, g_1, g_2, \ldots, g_n, \ldots),$$

so that we may write

$$u = (g_1, g_2, \ldots, g_N, u),$$

where N is an integral multiple of n, which we will assume to be even for reasons presently to be described. The terms (partial denominators) $g_1, g_2, \ldots$ are assumed to be positive whole numbers > 0. If we designate the numerator and denominator of the Nth approximation

$(g_1, g_2, \ldots, g_N)$ and of the $(N-1)$th approximation $(g_1, g_2, \ldots, g_{N-1})$ as P and Q and p and q, respectively, then according to continued fraction theory we obtain the two equations

(1) $$Pq - Qp = 1 \quad \text{and} \quad (2) \quad u = \frac{Pu + p}{Qu + q},$$

the second of which may also be expressed in the form

(2a) $$Qu^2 - Hu - p = 0 \quad \text{with} \quad H = P - q.$$

The discriminant $D = H^2 + 4Qp$ of the quadratic equation (2a) has, according to (1), the value $H^2 + 4Pq - 4 = (P + q)^2 - 4$; it is consequently smaller by 4 than a square number and therefore cannot itself be a square number. Its (positive) root $r = \sqrt{D}$ is therefore irrational. Moreover, since $r > H$ (because $r^2 = H^2 + 4Qp$), the second root $\bar{u} = (H - r)/2Q$ of the quadratic equation is negative, so that the first root $(H + r)/2Q$ represents our (improperly fractionated) continued fraction u. To obtain information about the magnitude of $\bar{u}$ we form the product of the roots $u\bar{u} = -p/Q$ and obtain

$$-\bar{u} = \frac{p/Q}{u}.$$

Since $P > p$ and $Q > q$, then

$$-\bar{u} < \frac{P/Q}{u} \quad \text{and} \quad -\bar{u} < \frac{p/q}{u}.$$

One of the right-hand fractions, however, is a proper fraction, since the value u of the continued fraction lies between the two successive approximations p/q and P/Q; therefore, $-\bar{u}$ must be a proper fraction.

A quadratic equation with integral coefficients and a nonquadratic discriminant whose first root is a positive and improper fraction while the second root is a negative proper fraction is called a *reduced equation*, and its first root is called a *reduced number*. Our conclusion therefore reads:

Every purely periodic, improperly fractionated, continued fraction is a reduced number.

We will now show conversely that the continued fraction of a reduced number is purely periodic.

First, we will solve the problem:

Obtain the first root $u = (r - b)/2a$ *of the quadratic equation*

(3) $$au^2 + bu + c = 0$$

with integral indivisible coefficients and the positive nonquadratic discriminant $D = r^2 = b^2 - 4ac$ *in the form of a continued fraction.*

We write

$$u = g + \frac{1}{u'},$$

where g is the largest whole number below u (in the following to be designated as $[u]$ and u' a positive improper fraction. We introduce three new magnitudes a', b', c' that are of the opposite sign and equal to the magnitudes $ag^2 + bg + c$, $2ag + b$, and a, and we obtain

$$u' = \frac{1}{u - g} = \frac{2a}{r + b'} = \frac{2a(r - b')}{r^2 - b'^2} = \frac{r - b'}{2a'}$$

with

$$b'^2 - 4a'c' = b^2 - 4ac = D.$$

Consequently, u' is the first root of the quadratic equation

(3') $a'u'^2 + b'u' + c' = 0,$

which likewise belongs to the discriminant D and possesses coefficients having no common divisor. (If a', b', c' possessed a common divisor, the latter because of the equations $-c' = a$, $-b' = 2ag + b$, $-a' = ag^2 + bg + c$ would go into a, b, c, which contradicts our assumption.) We call the new equation (3') the derivative of the initial equation (3) and its first root u' the derivative of u.

The new coefficients a', b', c' are calculated *in practice* in accordance with the following system:

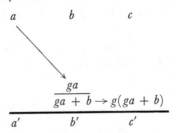

We add the two terms of the third column and change the sign of the sum, thus obtaining a'. We add the two *lower* terms of the second column, change the sign of the sum and get b'. We change the sign of a and get c'.

The derived quadratic equation (3') is treated in exactly this manner and the process continued as far as desired. The following example is presented to make the process completely clear.

Expand the positive root of the quadratic equation

$$3u^2 - 10u - 1 = 0$$

into a continued fraction. The discriminant is 112, thus $r = 10, \ldots$. In the scheme we will write in only the coefficients of the successive quadratic equations each of which is the derivative of the preceding one. In the last column we will write the first root of the appropriate equation and the highest integral contained in it that is at the same time the correct partial denominator of the continued fraction.

3	−10	−1		
	9		$\dfrac{10, \cdots + 10}{6} = 3 + \cdots$	
	−1	−3		
4	−8	−3		
	8		$\dfrac{10, \cdots + 8}{8} = 2 + \cdots$	
	0	0		
3	−8	−4		
	9		$\dfrac{10, \cdots + 8}{6} = 3 + \cdots$	
	1	3		
1	−10	−3		
	10		$\dfrac{10, \cdots + 10}{2} = 10 + \cdots$	
	0	0		
3	−10	−1		

Since we come back to the initial equation, the expansion is purely periodic, and we obtain

$$\frac{\sqrt{112} + 10}{6} = (3, 2, 3, 10, 3, 2, 3, 10, \ldots).$$

Now for the proof of the theorem that the expansion of a reduced number yields a purely periodic continued fraction!

Since the first root u of the reduced equation

$$au^2 + bu + c = 0$$

is a positive improper fraction, and the second one, $\bar{u}$, is a negative proper fraction, then according to the relations

$$u\bar{u} = \frac{c}{a}, \qquad u + \bar{u} = -\frac{b}{a}$$

between roots and coefficients, *both the free term* c *and the coefficient* b *of the linear term of a reduced equation are always negative* (the coefficient a is assumed to be always positive).

In accordance with the expansion examined above we write

(4) $$u = g + \frac{1}{u'}$$

with $g = [u]$ and $u' > 1$. From $u' = 1/(u - g)$ it follows initially that the first root u' of the derived equation is a positive improper fraction. If we then transform r into $-r$ in the equation $u' = 1/(u - g)$, the equation assumes the form $\bar{u}' = 1/(\bar{u} - g)$ and shows that the second root $\bar{u}'$ is a negative proper fraction. *The derivative of a reduced equation or number is consequently also reduced,* so that only reduced numbers occur in the continued fraction expansion of a reduced number.

If we write (4)

$$-\frac{1}{\bar{u}'} = g - \bar{u},$$

we see that *g can also be taken as the greatest integer that is contained in the reciprocal value of opposite sign of the second root of the* derived *equation.*

Now, the number of all the reduced numbers corresponding to a given discriminant D is finite. (From $D = b^2 - 4ac$ and $-ac > 0$ it follows first that the b's must be sought only among the numbers of the series $-1, -2, \ldots, -[r]$. Of these the only ones that need be considered are those for which $D - b^2$ is divisible by 4. We select these, and for each such b we determine the pairs of numbers a, c [with $a > 0, c < 0$] for which $-ac = (D - b^2)/4$, which in turn gives us a finite quantity of numbers a and c. Each number triplet a, b, c obtained in this way, however, leads to a reduced equation $au^2 + bu + c = 0$ and thus to a reduced number u only when $2a$ lies between $r + b$ and $r - b$.)

Consequently, in the continued fraction expansion of a reduced number U there must reappear after a finite number of steps a reduced number previously obtained, e.g., in such manner:

$$U = (K, L, u), \qquad u = (h, k, l, u).$$

But since, in accordance with the above, both l and L represent the greatest integer that is contained in the reciprocal value of $\bar{u}$ of opposite sign, $L = l$. Similarly, we find that $K = k$.

Consequently,

$$U = (k, l, h, k, l, h, \ldots),$$

i.e.: *The expansion of a reduced number yields a purely periodic continued fraction.*

After these preliminaries the solution of the Fermat equation becomes quite simple. We will show: I. that the continued fraction expansion of any reduced number belonging to the discriminant D possesses an infinite number of solutions of the Fermat equation; II. that *every* solution of the equation is obtained by this expansion.

I. Let

$$u = (g_1, g_2, \ldots, g_n, g_1, g_2, \ldots, g_n, \ldots)$$

be the positive root of the reduced equation

(5) $$au^2 + bu + c = 0$$

with the discriminant D and coefficients possessing no common divisor. Also, let

$$\frac{P}{Q} = (g_1, g_2, \ldots, g_N)$$

be an approximation of u and the index number N an even multiple of n, and let

$$\frac{p}{q} = (g_1, g_2, \ldots, g_{N-1})$$

be the preceding approximate fraction; then, according to (2a),

(5′) $$Qu^2 - Hu - p = 0 \qquad (H = P - q).$$

Since the roots of (5) and (5′) agree and the coefficients of (5) possess no common divisor, it must be possible to obtain (5′) from (5) by multiplication with a certain whole number y, such that

(6) $$y = \frac{Q}{a} = \frac{P - q}{-b} = \frac{p}{-c}.$$

If we then introduce the whole number

(7) $$x = P + q,$$

we obtain from (6) and (7)

$$x^2 - b^2 y^2 = (P + q)^2 - (P - q)^2$$

and

$$4acy^2 = -4Qp,$$

from which by addition we obtain

$$x^2 - Dy^2 = 4(Pq - Qp),$$

and, using (1),

$$x^2 - Dy^2 = 4.$$

II. Conversely, now let $x|y$ represent a solution of the Fermat equation

(8) $$x^2 - Dy^2 = 4$$

in nonevanescent positive integers x and y and let u represent the first root of a reduced equation

$$au^2 + bu + c = 0.$$

Making use of (6) and (7), we obtain the four nonevanescent positive integers

$$P = \frac{x - by}{2}, \qquad Q = ay, \qquad p = -cy, \qquad q = \frac{x + by}{2}.$$

(It is immediately obvious that Q and p are such numbers, whereas for P and q it follows from equation (8), if we make use also of the equation $D = b^2 - 4ac$ to write:

$$(x + by)(x - by) = x^2 - b^2y^2 = 4(1 - acy^2) = 4(1 + Qp).$$

We are then able to conclude from the appearance of the non-evanescent integer on the right, which is divisible by 4, that the two integral factors $2q$ and $2P$ of the product on the left-hand side have to be even and not equal to zero.)   According to (8) they satisfy the equation

(9) $$Pq - Qp = 1.$$

If we then replace the coefficients a, b, c in the reduced equation with Q/y, $-(P - q)/y$, $-p/y$, we get

(10) $$u = \frac{Pu + p}{Qu + q}.$$

Before we get from here to the continued fraction expansion, we still have to prove that $Q \geqq q$.

It is true that $2(Q - q) = [2a - b]y - x$.   Since the second root $\bar{u}$ of the reduced equation is a negative proper fraction, it follows that $r + b < 2a$ or $2a - b > r$.   Consequently,

$$2(Q - q) > ry - x = (r^2y^2 - x^2)/(ry + x) = -4/(ry + x)$$

or $(Q - q) > -2/(ry + x)$. However, since $D = r^2 = b^2 - 4ac$ is at least equal to 5, y is at least 1, and x at least 3, it follows that $ry + x > 5$ and from this $Q - q > -0.4$, i.e., $Q \geqq q$. Q.E.D.

We now expand P/Q into a continued fraction $(\gamma_1, \gamma_2, \ldots, \gamma_\nu)$ with the even number of terms ν in such a manner that between it and the last approximate fraction p'/q' there exists the relation

$$(9') \qquad\qquad Pq' - Qp' = 1.$$

From (9) and (9') it then follows by subtraction that

$$P(q' - q) = Q(p' - p).$$

However, since $q \leqq Q$, $q' < Q$, and $(q' - q)$ is divisible by Q, q' must equal q and therefore p' must also equal p. We then obtain

$$(\gamma_1, \gamma_2, \ldots, \gamma_\nu, u) = \frac{Pu + p}{Qu + q},$$

i.e., because of (10),

$$u = (\gamma_1, \gamma_2, \ldots, \gamma_\nu, u).$$

Every solution $x|y$ of the Fermat equation can therefore be obtained by the expansion of any reduced number u as a continued fraction.

FINAL RESULT: *The Fermat equation*

$$x^2 - Dy^2 = 4$$

has an infinite number of solutions; these can all be obtained in accordance with rules (6) and (7) from the approximation values, containing an even number of periods, obtained from the expansion as a continued fraction of any arbitrarily selected reduced number belonging to the discriminant D.

EXAMPLE. Find the smallest solution $x|y$ of the Fermat equation

$$x^2 - 112y^2 = 4.$$

A reduced equation applying to the discriminant 112 is the equation treated above

$$3u^2 - 10u - 1 = 0;$$

the expansion of the reduced number u reads

$$u = (3, 2, 3, 10, 3, 2, 3, 10, \ldots)$$

and has a four-termed period. The first four approximate fractions are

$$\frac{3}{1}, \quad \frac{7}{2}, \quad \frac{p}{q} = \frac{24}{7}, \quad \frac{P}{Q} = \frac{247}{72}.$$

Since here $a = 3$, $b = -10$, $c = -1$, we find, in accordance with (6) and (7), that
$$x = 254, \qquad y = 24.$$

It now remains to be shown that there is at least one reduced number corresponding to each discriminant D.

1. If $D = 4n$ and g is the maximum integer that is contained in $\sqrt{n}$, then
$$a = 1, \qquad b = -2g, \qquad c = g^2 - n$$
are the coefficients of a reduced equation.

PROOF. The discriminant of the equation is $b^2 - 4ac = 4n = D$. Moreover,
$$r + b < 2a < r - b,$$
since
$$2\sqrt{n} - 2g < 2 < 2\sqrt{n} + 2g.$$

2. If $D = 4n + 1$ and g is the largest integer for which $g^2 + g$ will be smaller than n (so that $(g + 1)^2 + (g + 1) > n$ or $g^2 + 3g + 2 > n$), then
$$a = 1, \qquad b = -(2g + 1), \qquad c = g^2 + g - n$$
are the coefficients of a reduced equation.

PROOF. The discriminant of the equation is $b^2 - 4ac = 4n + 1 = D$. Also,
$$r + b < 2a < r - b,$$
since
$$\sqrt{D} - (2g + 1) < 2 < \sqrt{D} + 2g + 1.$$

(That $\sqrt{D} - 2g - 1 < 2$ follows from the above condition $g^2 + 3g + 2 > n$. On multiplication by 4 this becomes
$$4g^2 + 12g + 9 > 4n + 1, \quad \text{i.e., it becomes} \quad (2g + 3)^2 > D.$$
From this it follows that
$$2g + 3 > \sqrt{D} \quad \text{or} \quad \sqrt{D} - 2g - 1 < 2.)$$

NOTE. If we have found the minimal solution of the Fermat equation (e.g., by the method just presented), we can find the other solutions (we will consider only positive solutions) in a simpler manner after Lagrange.

We assign to each solution $x \mid y$ the "Lagrange number"
$$z = \tfrac{1}{2}(x + yr)$$
and call x and y the *components* of the Lagrange number.

We will first prove the *auxiliary theorem*. *The product and the improperly fractionated quotient* $\zeta = \frac{1}{2}(\xi + \eta r)$ *of two Lagrange numbers* $Z = \frac{1}{2}(X + Yr)$ *and* $z = \frac{1}{2}(x + yr)$ *is also a Lagrange number.*

PROOF. We immediately find that

$$\zeta\bar{\zeta} = 1 \quad \text{or} \quad \xi^2 - D\eta^2 = 4$$

with

$$\xi = \frac{Xx \pm DYy}{2}, \qquad \eta = \frac{Yx \pm Xy}{2},$$

where the upper sign is used when we are concerned with the product and the lower when we are concerned with the quotient.

From $X > rY$ and $x > ry$ it follows that $Xx > DYy$, so that ξ is positive in *every* case. From

$$\left(\frac{X^2}{Y^2} - D\right)Y^2 = \left(\frac{x^2}{y^2} - D\right)y^2$$

it follows in the case of $\zeta = Z/z$, since then $Y > y$, that $X/Y < x/y$ or $Yx > Xy$, so that η is also positive in every case. Consequently, ζ is positive and improper because $\zeta\bar{\zeta} = 1$.

Now it merely remains to show that ξ and η are integers. Either D is divisible by 4 or D leaves a residue of 1 on division by 4. In the first case X and x are even. In the second case every solution of the Fermat equation consists either of two even or two odd numbers. In all cases ξ and η are consequently integers.

The method mentioned above is based upon the theorem:

Every Lagrange number is a power of the smallest Lagrange number with an integral exponent.

PROOF. Let $x|y$ be the minimal solution of the Fermat equation and thus $z = \frac{1}{2}(x + yr)$ the smallest Lagrange number. First it follows from the auxiliary theorem that every power of z is a Lagrange number.

Now let $Z = \frac{1}{2}(X + Yr)$ be a Lagrange number that is not a power of z. Then there must certainly exist two successive powers $\mathfrak{z} = z^n$ and $\mathfrak{z}' = z^{n+1}$ between which Z is situated. From

$$z^n < Z < z^{n+1}$$

it follows on division with z^n that

$$1 < Z/\mathfrak{z} < z.$$

Thus, the Lagrange number $\zeta = Z/3$ would be smaller than the smallest Lagrange number z, which is naturally absurd.

Consequently, the only Lagrange numbers are the powers

$$z, z^2, z^3, z^4, \ldots.$$

And the simplest way of finding the 2nd, 3rd, ... solution of the Fermat equation is to find them as components of the Lagrange numbers $z^2, z^3, \ldots.$

21 The Fermat-Gauss Impossibility Theorem

Prove that the sum of two cubic numbers cannot be a cubic number.

Thus, what must be proved is that the equation

$$x^3 + y^3 = z^3$$

cannot be composed of nonevanescent integers x, y, z.

The theorem that we have to prove is a special case of the famous *Fermat impossibility theorem*, which was expressed by Fermat in the following way in the arithmetic of Diophantus, edited by Fermat's son, and published in 1670:

"*It is impossible to divide a cube into two cubes, a fourth power into two fourth powers, and in general any power except the square into two powers with the same exponents.*"

Fermat added: "I have discovered a truly wonderful proof of this, but the margin (of the notebook) is too narrow to hold it." Unfortunately, Fermat neglected to disclose this "wonderful proof."

Fermat's impossibility theorem became very famous as a result of the fact that many of the greatest mathematicians since Fermat, including Euler, Legendre, Gauss, Dirichlet, Kummer, and others tried unsuccessfully to obtain the *general* proof of this theorem. To the present day a proof of the impossibility of the equation

$$x^n + y^n = z^n$$

is known only for special values of the exponent n, e.g., for the values from 3 to 100, and even this proof involves extraordinary complications and difficulties.

In the following we will limit ourselves to the simplest case, the case $n = 3$. The impossibility of the equation

$$x^3 + y^3 = z^3$$

was demonstrated by Euler in his algebra, which appeared in 1770, and later by Gauss (*Complete Works*, vol. II). This problem shows, as it often happens in mathematics, that the proof of a more general theorem is easier to obtain than that of a special case. To prove the impossibility of

$$(1) \qquad\qquad a^3 + b^3 = c^3$$

for the common integers a, b, c Euler had to resort to a relatively complicated method; Gauss, on the other hand, proved simply and clearly the impossibility of the more general equation

$$(2) \qquad\qquad \alpha^3 + \beta^3 = \gamma^3$$

for any numbers α, β, γ of the form $xJ + yO$, where x and y are any integers,

$$J = \frac{1 + i\sqrt{3}}{2} \quad \text{and} \quad O = \frac{1 - i\sqrt{3}}{2}$$

are cube roots of the (negative) unit.

For convenience in notation we will call numbers of the form $xJ + yO$ (in which x and y are integers) G-numbers.

That the case treated by Euler is simply a special case of (2) is apparent from the fact that every integer g is also a G-number: $g = gJ + gO$.

The G-numbers (which are the integers of the so-called group of the cubic unit roots) have many properties in common with common integers. Readers unfamiliar with these properties will find all the information necessary for an understanding of the Gauss proof in the supplement provided on p. 100.

Gauss' Proof of the Impossibility of the Equation

$$(2) \qquad\qquad \alpha^3 + \beta^3 = \gamma^3.$$

First, let Greek letters designate G-numbers and small Roman letters common integers.

We then replace α, β, γ with ξ, η, $-\zeta$, transforming (2) first into the symmetrical equation

$$(3) \qquad\qquad \xi^3 + \eta^3 + \zeta^3 = 0,$$

of which we assume that two of the three "bases" ξ, η, ζ will always have no common divisor; we will then refer to this equation as a Gauss equation. [The assumption we have just made in no way limits the proof. If, for example, ξ and η possessed a common prime factor δ, then, in accordance with (3), δ would also go into ζ^3 and consequently into ζ, so that division by δ^3 would eliminate the divisor δ from (3).]

The impossibility of (3) is obtained from the two following theorems, which we will derive from the assumption of the existence of (3).

I. *In every Gauss equation one and only one of the three bases—we will call it the special base—has the prime divisor $\pi = J - O$.*

II. *For every Gauss equation there is a second Gauss equation in which the special base contains the divisor π fewer times than the special base of the first equation.*

These two theorems, however, contradict each other. By continued application of II. it is possible to obtain a Gauss equation that no longer contains a special base, which contradicts theorem I.

PROOF OF I. If none of the three bases ξ, η, ζ were divisible by π, then

$$\xi^3 \equiv e,\ \eta^3 \equiv f,\ \zeta^3 \equiv g \bmod 9 \quad \text{with} \quad e^2 = f^2 = g^2 = 1$$

and consequently, because of (3), $e + f + g \equiv 0 \bmod 9$, which is, however, impossible. Therefore a situation such as the following must exist:

$$\zeta \equiv 0 \bmod \pi, \qquad \xi \not\equiv 0 \bmod \pi, \qquad \eta \not\equiv 0 \bmod \pi.$$

PROOF OF II. It follows from $\zeta^3 \equiv \bmod \pi^3$, according to (3), that $\xi^3 + \eta^3 \equiv 0 \bmod \pi^3$, and since $\xi^3 \equiv e \bmod 9$, $\eta^3 \equiv f \bmod 9$, $e + f \equiv 0 \bmod \pi^3$, then $e + f \equiv 0 \bmod 3$ must be true; from this it follows that $f = -e$. Now $\xi^3 + \eta^3 \equiv e + f \equiv 0 \bmod 9$, and consequently $\zeta^3 \equiv 0 \bmod \pi^4$ and

$$\zeta \equiv 0 \bmod \pi^2.$$

From $\xi^3 + \eta^3 \equiv 0 \bmod \pi^3$ and the identity

$$\xi^3 + \eta^3 = \varphi\psi\chi,$$

where

$$\varphi = \xi J + \eta O, \quad \psi = \xi O + \eta J, \quad \chi = \xi + \eta,$$

it follows that at least one of the factors φ, ψ, χ is divisible by π. From this and from $\varphi - \psi = (\xi - \eta)\pi$, $\varphi + \psi = \chi$ it follows that each one of the factors φ, ψ, χ is divisible by π, so that

$$\varphi = \pi\varphi', \qquad \psi = \pi\psi', \qquad \chi = \pi\chi'.$$

Thus no pair of the numbers φ', ψ', χ' possesses a common divisor.

[If, for example, φ' and ψ' possessed a common divisor δ, then also $\varphi' - \psi'$ would equal $\xi - \eta$ and $\pi(\varphi' + \psi') = \xi + \eta$, and then also 2ξ and 2η would be divisible by δ, so that δ would be equal to 2. Then we would either have $\xi = 2\lambda + \varepsilon$, $\eta = 2\mu + \varepsilon$, or $\xi = 2\lambda + \varepsilon$, $\eta = 2\mu - \varepsilon$, with $\varepsilon^3 = \pm 1$ and then $\varphi = 2\nu + \varepsilon$ or $\varphi = 2\nu + \varepsilon\pi$, which, however, is not divisible by $\delta = 2$.]

If we now set $\zeta/\pi = \omega$, then

$$\omega^3 = -\varphi'\psi'\chi' \quad \text{with} \quad \varphi' + \psi' = \chi'.$$

Since then no pair of φ', ψ', $-\chi'$ possesses a common divisor, these three magnitudes down to the possible unit factors α, β, γ must be cubes of the numbers ρ, σ, τ, no pair of which possesses a common divisor:

$$\varphi' = \alpha\rho^3, \quad \psi' = \beta\sigma^3, \quad -\chi' = \gamma\tau^3 \quad \text{with} \quad \alpha^6 = \beta^6 = \gamma^6 = 1,$$

so that

$$(4) \qquad \omega^3 = \alpha\beta\gamma\rho^3\sigma^3\tau^3, \qquad \alpha\rho^3 + \beta\sigma^3 + \gamma\tau^3 = 0.$$

However, if the cube of $\kappa = \omega/\rho\sigma\tau$ is the G-unit α, β, γ, then, since $\kappa^3 = E \bmod 9$, $\alpha\beta\gamma \equiv E \bmod 9$ also, and consequently

$$\alpha\beta\gamma = E \quad \text{with} \quad E^2 = 1.$$

From $\omega \equiv 0 \bmod \pi$ it follows, for example, that

$$\tau \equiv 0 \bmod \pi \quad \text{and} \quad \rho \not\equiv 0, \quad \sigma \not\equiv 0 \bmod \pi.$$

Then, however, $\rho^3 \equiv e$ and $\sigma^3 \equiv f \bmod 9$ $(e^2 = f^2 = 1)$, and consequently, according to (4), $e\alpha + f\beta \equiv 0 \bmod 3$, and from this $e\alpha + f\beta = 0$. Thus, we obtain

$$\beta = F\alpha, \quad F\alpha^2\gamma = E \quad (\text{with } F^2 = 1)$$

and from (4)

$$F\alpha^3\rho^3 + \alpha^3\sigma^3 + E\tau^3 = 0.$$

If we write here ξ', η', ζ' in place of $F\alpha\rho$, $\alpha\sigma$, $E\tau$, respectively, we finally obtain the Gauss equation

$$(3') \qquad\qquad \xi'^3 + \eta'^3 + \zeta'^3 = 0,$$

into the special base ζ' of which the factor π goes fewer times than into the special base ζ of (3).

Supplement. Properties of *G*-numbers

I. The magnitudes J and O satisfy the following equations:

$$J + O = 1, \quad JO = 1, \quad J^2 + O = 0,$$
$$O^2 + J = 0, \quad J^3 = -1, \quad O^3 = -1.$$

II. *The sum, difference, and products of G-numbers are also G-numbers.*

The product of the two numbers $aJ + bO$ and $a'J + b'O$ is, for example (according to I.), $pJ + qO$ with

$$p = ab' + ba' - bb' \quad \text{and} \quad q = ab' + ba' - aa'.$$

III. *Norm.* The norm of a complex number $\mathfrak{z} = \mathfrak{x} + i\mathfrak{y}$ is commonly understood to be the product

$$\mathfrak{z}_0 = N(\mathfrak{z}) = \mathfrak{z}\bar{\mathfrak{z}} = (\mathfrak{x} + i\mathfrak{y})(\mathfrak{x} - i\mathfrak{y}) = \mathfrak{x}^2 + \mathfrak{y}^2$$

of the two mutually conjugate numbers $\mathfrak{z}$ and $\bar{\mathfrak{z}} = \mathfrak{x} - i\mathfrak{y}$.

The norm of the G-number $aJ + bO$ *accordingly has the value* $a^2 + b^2 - ab$. *It is a positive integer* which disappears only when a and b are both zero. The smallest conceivable norms of *G*-numbers are 1, 2, 3.

From

$$a^2 + b^2 - ab = 1$$

we obtain one of the six following cases:

$a =$	1	0	-1	0	1	-1
$b =$	0	1	0	-1	1	-1

There are thus six *G*-numbers:

$$J, \ -J, \ O, \ -O, \ 1, \ -1$$

with the norm 1.

The equation

$$a^2 + b^2 - ab = 2$$

has no solution that is an integer. There is consequently no *G*-number whose norm is 2.

The equation

$$a^2 + b^2 - ab = 3$$

finally has six integral solutions

$$a = 1, \quad b = -1; \quad a = -1, \quad b = 1; \quad a = 1, \quad b = 2;$$
$$a = -1, \quad b = -2; \quad a = 2, \quad b = 1; \quad a = -2, \quad b = -1.$$

Accordingly, there are six G-numbers with the norm 3, the numbers $\pi = J - O = i\sqrt{3}$, πJ, πO, and their conjugates $\bar{\pi} = -\pi$, $-\pi O$, $-\pi J$.

The norm of the product of two numbers is equal to the product of the norms of these numbers.

PROOF. $N(\alpha\beta) = \alpha\beta \cdot \overline{\alpha\beta} = \alpha\beta \cdot \bar{\alpha} \cdot \bar{\beta} = \alpha\bar{\alpha} \cdot \beta\bar{\beta} = N(\alpha) \cdot N(\beta)$.

IV. *Units.* A G-number ε is called a *unit*, or more accurately a G-unit, when its reciprocal value η is also a G-number. From $\varepsilon\eta = 1$ it follows from norm formation that $\varepsilon_0\eta_0 = 1$, i.e., $\varepsilon_0 = 1$. According to III., there are consequently six G-units:

$$J, \ -J, \ O, \ -O, \ 1, \ -1.$$

These six units are the integral powers of J or O, e.g., J, J^2, J^3, J^4, J^5, and J^6.

V. *Associated numbers.* The six numbers that are obtained when a G-number ζ is multiplied by the six G-units are called the associated numbers of ζ.

The six associated numbers of $\pi = J - O$ are, for example,

$$\pi J = -1 - O, \qquad \pi J^2 = -1 - J, \qquad \pi J^3 = -\pi,$$
$$\pi J^4 = 1 + O, \qquad \pi J^5 = 1 + J, \qquad \pi J^6 = \pi.$$

VI. *Division.* The quotient $q = \alpha/\beta$ of two G-numbers α and β is not necessarily a G-number. If it is a G-number, however, β is called a divisor (G-divisor) of α or one says that β goes into α.

In order to divide any G-number α by any other β, we write

$$\frac{\alpha}{\beta} = \frac{\alpha\bar{\beta}}{\beta\bar{\beta}} = \frac{\alpha\bar{\beta}}{\beta_0} = \frac{hJ + kO}{\beta_0} = \frac{h}{\beta_0} J + \frac{k}{\beta_0} O.$$

Here we divide each rational fraction h/β_0 and k/β_0 into the integral components m and n, respectively, and the rational components $\mathfrak{r}$ and $\mathfrak{s}$, respectively, the absolute value of which never exceeds $\frac{1}{2}$ [Example: $\frac{19}{5} = 4 - 0.2$], we set $mJ + nO = \kappa$, $\mathfrak{r}J + \mathfrak{s}O = \Re$, and obtain

$$\frac{\alpha}{\beta} = \kappa + \Re \quad \text{or} \quad \alpha = \kappa\beta + \Re\beta.$$

From $\Re\beta = \alpha - \kappa\beta$ it follows that $\Re\beta$ is a G-number γ, and we have

$$\alpha = \kappa\beta + \gamma.$$

Here $\gamma_0 = \Re_0\beta_0 = (\mathfrak{r}^2 + \mathfrak{s}^2 - \mathfrak{r}\mathfrak{s})\beta_0$. Since, however, $|\mathfrak{r}| \leq \frac{1}{2}$ and $|\mathfrak{s}| \leq \frac{1}{2}$, then $\Re_0$ must certainly be $\leq \frac{3}{4}$, i.e., $\gamma_0 \leq \frac{3}{4}\beta_0$.

CONCLUSION. *The division of a G-number α by another G-number β results in a "quotient" κ and a "residue" γ such that*

$$\alpha = \kappa\beta + \gamma,$$

with the residue norm being at most equal to $\frac{3}{4}$ of the divisor norm.

VII. *The algorithm of the greatest common divisor.* We start with the division α/β and the related equation

$$(1) \qquad \alpha = \kappa\beta + \gamma \quad \text{with} \quad \gamma_0 \leqq \tfrac{3}{4}\beta_0,$$

and determine, as in VI., the quotient λ and the residue δ of the division β/γ; in this way we obtain the corresponding equation

$$(2) \qquad \beta = \lambda\gamma + \delta \quad \text{with} \quad \delta_0 \leqq \tfrac{3}{4}\gamma_0.$$

Then in a similar manner we obtain

$$(3) \qquad \gamma = \mu\delta + \varepsilon \quad \text{with} \quad \varepsilon_0 \leqq \tfrac{3}{4}\delta_0,$$

etc. Since the residue norms become progressively smaller, we must finally obtain a residue of zero. To avoid unnecessary writing we will assume that the division after (3) δ/ε leaves no residue, so that

$$(4) \qquad\qquad\qquad \delta = \nu\varepsilon.$$

Now it follows from (4) that every divisor τ of ε also goes into δ without residue, and, therefore, it follows from (3) that τ also goes into γ without residue; consequently, it follows from (2) that τ goes into β without residue, and, finally, from (1) it follows that τ goes into α without residue.

In reverse order: it follows from (1) that every common divisor τ of α and β is also a divisor of γ, then, from (2), that τ also goes into δ without residue, and, finally, from (3), that τ is also a divisor of ε.

Every common divisor of α and β consequently goes into ε without residue, and every divisor of ε goes into α and β without residue.

ε *is accordingly* (in terms of its absolute value) *the highest common divisor of α and β.*

If, in particular, ε is a *G*-unit, the numbers α and β are said to have *no common divisor* or to be *prime with respect to each other.*

The chain of equations (1,) (2), (3), . . . is nothing other than the *extension to G-numbers of the well-known algorithm for determination of the highest common divisor of common integers.*

VIII. *Unequivocal division of G-numbers into prime factors.* Just as with integers, the common theorems governing divisibility, indivisibility and unequivocal division into prime factors are derived from the divisional algorithm:

1. *If α and β possess no common divisor and $\alpha\mu$ is divisible by β, then μ is divisible by β.*

2. *If two G-numbers possess no common divisor with one and the same third G-number, their product also possesses no common divisor with this third G-number.*

3. *Every G-number can be divided into a product of prime factors* (i.e., *G*-primes) *in only one way.* [Divisions such as $\alpha\beta\gamma$ and $\alpha J \cdot \beta \cdot \gamma O$, in which one contains the associated numbers of the other rather than certain factors of it, are not considered different from each other.]

A G-*prime* is a *G*-number that possesses no divisor aside from its six associated numbers and the six units.

The numbers $\pi = J - O$ and 2 are, for example, primes.

If, for example, we assume that π is divisible: $\pi = \lambda\mu$, then $\pi_0 = \lambda_0\mu_0$ or $3 = \lambda_0\mu_0$. From this it follows that $\lambda_0 = 3$, $\mu_0 = 1$. μ is therefore a unit and the equation $\pi = \lambda\mu$ does not represent a division.

From $2 = \lambda\mu$ it follows that $2 = \lambda_0\mu_0$ or $4 = \lambda_0\mu_0$. The case of $\lambda_0 = 2$, $\mu_0 = 2$ is eliminated because, according to III., there is no *G*-number having a norm equal to 2.

Thus, we are left with $\lambda_0 = 4$, $\mu_0 = 1$. Once again μ is a unit and the equation $2 = \lambda\mu$ does not represent a division.

IX. *Congruence.* As in the theory of natural numbers, we say here also that two *G*-numbers α and β are congruent modulo μ—written $\alpha \equiv \beta \bmod \mu$—when their difference $\alpha - \beta$ is divisible by the *G*-number μ.

X. G-*numbers modulo* π. We will consider one more *G*-number $\kappa = aJ + bO$ in relation to the modulus $\pi = J - O$.

If κ is divisible by π:

$$aJ + bO = (mJ + nO)(J - O) = (2n - m)J + (n - 2m)O,$$

then $a = 2n - m$, $b = n - 2m$, thus

$$a + b = 3g \quad \text{with} \quad g = n - m.$$

Conversely, if $a + b = 3g$, m and n are determined from $n - m = g$ and $2n - m = a$, giving $\kappa = (mJ + nO)(J - O)$.

The G-number $\kappa = aJ + bO$ *is thus divisible by* π *only when* a + b *is divisible by* 3.

If κ is not divisible by π, then one of the three following formula pairs is valid:

$$a = 3h, \quad b = 3k + e; \qquad a = 3h + e, \quad b = 3k;$$

$$a = 3h + e, \quad b = 3k + e,$$

with $e^2 = 1$, and thus, if $hJ + kO$ is set equal to λ,

$$\kappa = 3\lambda + eO \quad \text{or} \quad \kappa = 3\lambda + eJ \quad \text{or} \quad \kappa = 3\lambda + e,$$

so that in every case κ has the form

$$\kappa = 3\lambda + \varepsilon,$$

where ε is a G-unit.

Let us now consider the cube of κ. It becomes

$$\kappa^3 = 9(3\lambda^3 + 3\lambda^2\varepsilon + \lambda\varepsilon^2) + \varepsilon^3,$$

and, because $\varepsilon^3 = \pm 1$, it has the form

$$\kappa^3 = 9\mu \pm 1.$$

If κ *is not divisible by* π *we then have the congruences* $\kappa \equiv \varepsilon \bmod 3$, $\kappa^3 \equiv \pm 1 \bmod 9$.

22   **The Quadratic Reciprocity Law**

(*The Euler-Legendre-Gauss theorem.*) *The reciprocal Legendre symbols of the odd prime numbers* p *and* q *are governed by the formula*

$$\left(\frac{p}{q}\right) \cdot \left(\frac{q}{p}\right) = (-1)^{[(p-1)/2] \cdot [(q-1)/2]}.$$

This law, the so-called quadratic reciprocity law, was formulated but not proved by Euler (*Opuscula analytica*, Petersburg, 1783). In 1785 Legendre discovered the same law (*Histoire de l'Académie des Sciences*) independently of Euler and proved it partially.

The first complete proof was presented by Karl Friedrich Gauss (1777–1855) in his famous *Disquisitiones arithmeticae* (published in 1801), a book that laid the foundations of contemporary number theory; this work, its five hundred quarto pages swarming with profound ideas, was written when Gauss was 20 years old. "It is really astonishing," says Kronecker, "to think that a single man of such young years was able to bring to light such a wealth of results, and above all to present such a profound and well organized treatment of an entirely new discipline."

Later Gauss discovered seven other proofs of the reciprocity theorem. (The Gauss proofs may be found in vol. 14 of Ostwald's *Klassiker der exakten Wissenschaften.*)

The quadratic reciprocity law is one of the most important theorems of number theory. Gauss called it the "*Theorema fundamentale.*" The American mathematician Dickson says in his *Theory of Numbers*: "The quadratic reciprocity law is doubtless the most important tool in the theory of numbers and occupies the central position in its history."

The importance of this law led other mathematicians like Jacobi, Cauchy, Liouville, Kronecker, Schering, and Frobenius to investigate it after Gauss and offer proofs of it. In his *Niedere Zahlentheorie*, P. Bachmann cites no fewer than 52 proofs and reports on the most important.

Probably the simplest of all the proofs is the following *arithmetic-geometric* proof, which arises from the combination of the so-called lemma of Gauss (Gauss' *Werke*, vol. II, p. 51) and a geometric idea of Cayley (Arthur Cayley [1821–1895], *Collected Mathematical Papers*, vol. II).

Before taking up the proof itself we will give the derivation of Gauss' lemma.

Let p be an odd prime number and D an integer that is not divisible by p. If x represents one of the numbers $1, 2, 3, \ldots, \mathfrak{p} = (p-1)/2$, R_x the common residue of the division Dx/p, g_x the corresponding integral quotient, then

$$(1) \qquad Dx = R_x + g_x p.$$

Accordingly as R_x is smaller or greater than $\frac{1}{2}p$, we set $R_x = \rho_x$ or $R_x = \rho_x + p$, where in the second case ρ_x represents the negative minimum residue of the division Dx/p, and we obtain

$$(1a) \qquad Dx = \rho_x + g_x p \qquad \text{or} \qquad (1b) \qquad Dx = \rho_x + p + g_x p.$$

If n is then the number of negative minimum residues occurring in the $\mathfrak{p}$ divisions Dx/p (for $x = 1, 2, 3, \ldots, \mathfrak{p}$), we have n equations of the form $(1b)$ and $m = \mathfrak{p} - n$ equations of the form $(1a)$.

We convert these equations into congruences mod p and obtain the $\mathfrak{p}$ congruences

(2) $Dx \equiv \rho_x \bmod p.$

Now the $\mathfrak{p}$ residues ρ_x agree, except with respect to sign and sequence, with the $\mathfrak{p}$ numbers 1 to $\mathfrak{p}$.

[If, for example, ρ_r were equal to ρ_s or $\rho_r = -\rho_s$ for two different values r and s of x, then $Dr \equiv \rho_r$ and $Ds \equiv \rho_s$ would yield by subtraction or addition, respectively, $D(r \mp s) \equiv 0 \bmod p$. This congruence is, however, impossible, because neither D nor $r \mp s$ is divisible by p.]

Multiplication of the $\mathfrak{p}$ congruences (2) results in

$$D^{\mathfrak{p}}\mathfrak{p}! \equiv (-1)^n \mathfrak{p}! \bmod p,$$

and from this we obtain

$$D^{\mathfrak{p}} \equiv (-1)^n \bmod p.$$

However, since, according to Euler's theorem (No. 19),

$$D^{\mathfrak{p}} \equiv \left(\frac{D}{p}\right) \bmod p,$$

we obtain

$$\left(\frac{D}{p}\right) \equiv (-1)^n \bmod p,$$

whence, since both sides of this congruence have the absolute value 1,

(3) $\left(\dfrac{D}{p}\right) = (-1)^n.$

This formula, in which n represents the number of negative minimum residues resulting from the $\mathfrak{p}$ divisions Dx/p $(x = 1, 2, 3, \ldots, \mathfrak{p})$, *is Gauss' lemma.*

Now let D be some odd prime number q that differs from p. We convert the $\mathfrak{p}$ equations $(1a)$ and $(1b)$ into congruences to the modulus 2, leave out all the excess multiples of 2, e.g., $(q - 1)x$, and obtain

$$x \equiv \rho_x + g_x \bmod 2 \quad \text{and} \quad x \equiv 1 + \rho_x + g_x \bmod 2.$$

Addition of these $\mathfrak{p}$ congruences yields

$$\sum x \equiv n + \sum \rho_x + \sum g_x \bmod 2.$$

However, since the absolute values of ρ_x are in agreement with the numbers 1 through $\mathfrak{p}$ and each summand can be replaced by its opposite value in a congruence mod 2, we will write $\sum x$ in the obtained congruence instead of $\sum \rho_x$ and $-n$ instead of n, thereby obtaining

$$\sum x + n \equiv \sum x + \sum g_x \bmod 2$$

or

(4) $$n \equiv \sum g_x \bmod 2.$$

In accordance with (4) we can now write (3) as

$$\left(\frac{q}{p}\right) = (-1)^{\Sigma g_x}.$$

Now g_x is the greatest integer contained in the quotient qx/p. If we designate this as $[qx/p]$, we obtain at last

(I) $$\left(\frac{q}{p}\right) = (-1)^{\Sigma[qx/p]},$$

where x passes through all the integers from 1 to $\mathfrak{p} = (p-1)/2$.

Accordingly,

(II) $$\left(\frac{p}{q}\right) = (-1)^{\Sigma[py/q]}$$

where y passes through all the integers from 1 to $\mathfrak{q} = (q-1)/2$.

Multiplication of (I) and (II) gives us

(III) $$\left(\frac{p}{q}\right) \cdot \left(\frac{q}{p}\right) = (-1)^{\Sigma[(q/p)x] + \Sigma[(p/q)y]}.$$

The exponent of the right-hand side is, however, easily found.

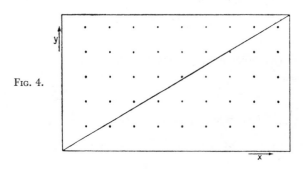

Fig. 4.

On a system of rectangular coordinates xy we draw the rectangle with the four angles

$$0\,|\,0, \qquad \frac{p}{2}\,\Big|\,0, \qquad \frac{p}{2}\,\Big|\,\frac{q}{2}, \qquad 0\,\Big|\,\frac{q}{2}$$

and bisect it with a diagonal d from the origin, possessing the equation $y = (qx/p)$; we then mark off all the lattice points* within the rectangle. (Cf. the figure, in which $p = 19$, $q = 11$.)

To begin with, it is clear that no marked lattice point $x\,|\,y$ lies on d, since here x would necessarily be $< \frac{1}{2}p$ and $y < \frac{1}{2}q$, which contradicts the condition $y/x = q/p$.

For an integral abscissa x the corresponding ordinate of d is $y = (qx/p)$ and the number of marked lattice points lying on this ordinate is $[qx/p]$. Consequently, the number of the marked lattice points lying in the lower half of the rectangle is $\sum[qx/p]$, where x passes through all the integers from 1 to $\mathfrak{p}$.

Similarly, the number of all the marked lattice points lying in the upper half of our rectangle is $\sum[py/q]$, where y passes through all the integers from 1 to $\mathfrak{q}$.

The exponent appearing in (III) is then the number of all the marked lattice points in our rectangle. This is a total of $\mathfrak{p} \cdot \mathfrak{q}$ elements. Consequently,

$$\left(\frac{p}{q}\right) \cdot \left(\frac{q}{p}\right) = (-1)^{\mathfrak{p}\mathfrak{q}}$$

or

$$\left(\frac{p}{q}\right) \cdot \left(\frac{q}{p}\right) = (-1)^{[(p-1)/2]\cdot[(q-1)/2]}. \qquad \text{Q.E.D.}$$

23 Gauss' Fundamental Theorem of Algebra

Every equation of the nth degree

$$z^n + C_1 z^{n-1} + C_2 z^{n-2} + \cdots + C_n = 0$$

has n *roots.*

Expressed more precisely, this theorem reads:
The polynomial

$$f(z) = z^n + C_1 z^{n-1} + C_2 z^{n-2} + \cdots + C_n$$

can always be divided into n *linear factors of the form* z $- \alpha_\nu$.

---

* A lattice point is a point whose coordinates are integers.

This famous theorem, the fundamental theorem of algebra, was first stated by d'Alembert in 1746, but only partially proved. The first rigorous proof was given in 1799 by Gauss, then twenty-one years old, in his doctoral dissertation *Demonstratio nova theorematis omnem functionem algebraicam rationalem integram unius variabilis in factores reales primi vel secundi gradus resolvi posso* (Helmstaedt, 1799). Subsequently, Gauss gave three other proofs of this theorem. All four are to be found in the third volume of his *Works*, as well as in vol. 14 of Ostwald's *Klassiker der exakten Wissenschaften*. Other authors after Gauss, including Argand, Cauchy, Ullherr, Weierstrass, and Kronecker also gave proofs of the fundamental theorem. The proof followed here (as modified by Cauchy) is Argand's (*Annales de Gergonne*, 1815), which is distinguished by its brevity and simplicity.

This proof (like most of the other proofs) falls into two steps. The first—and more difficult—step merely demonstrates that an equation of the nth degree will always contain *at least one* root; the second step shows that it has n roots and no more.

FIRST STEP

We set

$$z^n + C_1 z^{n-1} + C_2 z^{n-2} + \cdots + C_n = f(z) = w$$

and consider the different values that are assumed by the absolute magnitude $|w|$ when z is moved in the Gauss plane (the plane of complex numbers). Let the smallest of these values be μ and let it be attained, for example, at the site z_0, so that $|f(z_0)| = |w_0| = \mu$.

There are two possible cases:

1. The minimum μ is greater than zero.
2. The minimum μ is equal to zero.

We will begin by considering the first case. In the immediate vicinity of the point z_0, say, in the area defined by a small circle K of radius R with a center at z_0, $|w|$ is everywhere $\geq \mu$, since μ represents the smallest value of $|w|$; at z_0 itself $|w| = |w_0| = \mu$.

For any z in K, $z = z_0 + \zeta$, where $\zeta = \rho(\cos \vartheta + i \sin \vartheta)$ and ρ is the absolute magnitude of ζ, i.e., the line segment $z_0 z$, and ϑ the inclination of this segment toward the axis of the positive real numbers. We calculate

$$w = f(z) = f(z_0 + \zeta) = (z_0 + \zeta)^n + C_1(z_0 + \zeta)^{n-1} + \cdots + C_n,$$

eliminating the parentheses and arranging according to increasing powers of ζ. In this way we obtain

$$w = f(z) = z_0^n + C_1 z_0^{n-1} + C_2 z_0^{n-2} + \cdots + C_n$$
$$+ c_1 \zeta + c_2 \zeta^2 + \cdots c_n \zeta^n,$$

i.e.,

$$w = f(z_0) + c_1 \zeta + c_2 \zeta^2 + \cdots + c_n \zeta^n.$$

Since several coefficients c_r may be equal to zero, we call the first of the nonevanescent coefficients c, the second c', and so forth, so that

$$w = w_0 + c\zeta^\nu + c'\zeta^{\nu'} + c''\zeta^{\nu''} + \cdots,$$

with $\nu < \nu' < \nu'' \ldots$.

Division with w_0 and isolation of ζ^ν yields

$$\frac{w}{w_0} = 1 + q\zeta^\nu \cdot (1 + \zeta\xi),$$

where $q = c/w_0$ and ξ represents a sum of different powers of ζ with positive exponents and known coefficients.

We consider the product $q\zeta^\nu \cdot (1 + \zeta\xi)$. We write the *first* factor trigonometrically, abbreviating $\cos \varphi + i \sin \varphi$ to 1_φ, and, from $q = h(\cos \lambda + i \sin \lambda) = h \cdot 1_\lambda$ and $\zeta = \rho \cdot 1_\vartheta$, we obtain $q\zeta^\nu = h \cdot 1_\lambda \cdot \rho^\nu \cdot 1_{\nu\vartheta} = h\rho^\nu \cdot 1_{\lambda + \nu\vartheta}$. From now on we confine ourselves to z-values of K for which $\lambda + \nu\vartheta = \pi$, which consequently lie on the radius $z_0 H$ which forms the angle $\vartheta = (\pi - \lambda)/\nu$ with the real axis. For all these z's the number $1_{\lambda + \nu\vartheta} = 1_\pi$ has the value -1, and our product assumes the form $-h\rho^\nu \cdot (1 + \zeta\xi)$.

If we choose a sufficiently small radius R, the *second* factor $1 + \zeta\xi$ can be brought as close to unity as we desire, since $\rho = |\zeta| < R$. But this means that the product lies as close as desired to the value $-h\rho^\nu$, i.e., the fraction

$$\frac{w}{w_0} = 1 - h\rho^\nu \cdot (1 + \zeta\xi)$$

lies as close as we desire to the point $1 - h\rho^\nu$ of the Gauss plane, which shows that for *all* z's between z_0 and H the absolute magnitude $|w/w_0| < 1$. In other words, for *this* z, $|w| < \mu$, while for *all* z's in the vicinity of z_0, $|w|$ should be $\geqq \mu$. This is a contradiction, and consequently the first of the two possible cases given above ($\mu > 0$) is eliminated. This leaves only the second case: w_0 is equal to zero or

$$f(z_0) = 0.$$

Therefore: *Every equation* regardless of its degree, *has at least one root.*

Second Step

We begin with the demonstration of the *auxiliary theorem:* *If an algebraic equation* $f(z) = 0$ *has the root* α, *then the left side of the equation can be divided by* $z - \alpha$ *without a remainder.*

If we divide the polynomial $f(z)$ by $z - \alpha$ until the remainder R no longer contains any more z, we obtain

$$\frac{f(z)}{z - \alpha} = f_1(z) + \frac{R}{z - \alpha},$$

where R is a constant and $f_1(z)$ has the form

$$z^{n-1} + \mathfrak{C}_1 z^{n-2} + \mathfrak{C}_2 z^{n-3} + \cdots + \mathfrak{C}_{n-1}.$$

Multiplication with $z - \alpha$ gives

$$f(z) = (z - \alpha)f_1(z) + R.$$

If in this equation, which is valid for every z, we set $z = \alpha$, we obtain

$$R = f(\alpha) = 0$$

and thus for every z

$$f(z) = (z - \alpha)f_1(z). \qquad \text{Q.E.D.}$$

If we combine this auxiliary theorem with the theorem proved in the first step, which demonstrated the existence of one root, we obtain the new theorem: *Every polynomial of* z *can be represented as the product of a linear factor* z $- \alpha$ *with a polynomial one degree lower.*

We now write α_1 rather than α and obtain

$$f(z) = (z - \alpha_1)f_1(z).$$

We then apply the obtained theorem to the polynomial $f_1(z)$ and get

$$f_1(z) = (z - \alpha_2)f_2(z),$$

where $f_2(z)$ is of the $(n - 2)$th degree and α_2 is a root of the equation $f_1(z) = 0$. Also in similar fashion:

$$f_2(z) = (z - \alpha_3)f_3(z),$$

$$f_3(z) = (z - \alpha_4)f_4(z), \text{ etc.}$$

In this chain of equations, beginning with the next to last, if we replace every f on the right-hand side with its following value in the equation below, we finally obtain the theorem for the transformation of a polynomial of the nth degree into a product of n linear factors:

$$f(z) = (z - \alpha_1)(z - \alpha_2) \ldots (z - \alpha_n).$$

Expressed verbally: *Every integral rational function of the* n *th degree can be represented as the product of* n *linear factors.*

Thus, the previous equation $f(z) = 0$ allows us to write

$$(z - \alpha_1)(z - \alpha_2) \ldots (z - \alpha_n) = 0.$$

However, the product on the left becomes zero only when one factor is equal to zero. And since $z - \alpha_\nu = 0$ implies $z = \alpha_\nu$, we finally obtain:

The equation f$(z) = 0$ *possesses the* n *roots* $\alpha_1, \alpha_2, \ldots, \alpha_n$ *and no others.*

Thus we have proved the fundamental theorem.

NOTE. It is possible for several of the n roots $\alpha_1, \alpha_2, \ldots, \alpha_n$ to be equally great, for example, for α_2 and α_3 both to be equal to α_1, while $\alpha_4, \alpha_5, \ldots, \alpha_n$ may be different from α_1. In this case α_1 is called a multiple root, and specifically in the case we have assumed of *three* equal roots, a *triple* root.

24   Sturm's Problem of the Number of Roots

Find the number of real roots of an algebraic equation with real coefficients over a given interval.

This very important algebraic problem was solved in a surprisingly simple way in 1829 by the French mathematician Charles Sturm (1803–1855). The paper containing the famous Sturm theorem appeared in the eleventh volume of the *Bulletin des sciences de Férussac* and bears the title, "Mémoire sur la résolution des équations numériques."

"With this major discovery," says Liouville, "Sturm at once simplified and perfected the elements of algebra, enriching them with new results."

SOLUTION. We distinguish two cases:

I. The real roots of the equation in question are all simple over the given interval.

II. The equation also possesses multiple real roots over the interval.

We will first show that the second case leads us back to the first.

Let the prescribed equation $F(x) = 0$ have the distinct roots $\alpha, \beta, \gamma, \ldots$, and let the root α be a-fold, β b-fold, γ c-fold, $\ldots$, so that

$$F(x) = (x - \alpha)^a (x - \beta)^b (x - \gamma)^c \cdots.$$

For the derivative $F'(x)$ of $F(x)$ we obtain

$$\frac{F'(x)}{F(x)} = \frac{a}{x - \alpha} + \frac{b}{x - \beta} + \frac{c}{x - \gamma} + \cdots$$

$$= \frac{a(x - \beta)(x - \gamma)(x - \delta) \cdots + b(x - \alpha)(x - \gamma)(x - \delta) \cdots + \cdots}{(x - \alpha)(x - \beta)(x - \gamma) \cdots}.$$

If we then call the numerator of this fraction $p(x)$ and the denominator $q(x)$ and set the whole rational function $F(x)/q(x)$ equal to $G(x)$, then

$$F(x) = G(x) \cdot q(x) \quad \text{and} \quad F'(x) = G(x) \cdot p(x).$$

Now the functions $p(x)$ and $q(x)$ have no common divisor. (The factor $x - \beta$ of $q(x)$ may, for example, go into all the terms of $p(x)$ except the second with no remainder.) It follows from this that $G(x)$ is the greatest common divisor of $F(x)$ and $F'(x)$. This can be determined easily from the divisional algorithm and can therefore be considered known, as a result of which $q(x)$ is known also.

The equation $F(x) = 0$ then falls into the two equations

$$q(x) = 0 \quad \text{and} \quad G(x) = 0,$$

the first of which possesses only simple roots, while the second can be further reduced in the same way that $F(x) = 0$ was.

An equation with multiple roots can therefore always be transformed into equations (with known coefficients) possessing only simple roots.

Consequently, it is sufficient to solve the problem for the first case.

Let $f(x) = 0$ be an algebraic equation all of whose roots are simple. The derivative $f'(x)$ of $f(x)$ then vanishes for none of these roots and the highest common divisor of the functions $f(x)$ and $f'(x)$ is a constant K that differs from zero. We use the divisional algorithm to determine the highest common divisor of $f(x)$ and $f'(x)$, writing, for the sake of convenience in representation, $f_0(x')$ and $f_1(x)$ instead of $f(x)$ and $f'(x)$, and calling the quotients resulting from the successive divisions $q_0(x), q_1(x), q_2(x), \ldots$ and the remainders $-f_2(x), -f_3(x), \ldots$.

If we also drop the argument sign for the sake of brevity, we obtain the following scheme:

$$(0) \qquad\qquad f_0 = q_0 f_1 - f_2,$$

$$(1) \qquad\qquad f_1 = q_1 f_2 - f_3,$$

$$(2) \qquad\qquad f_2 = q_2 f_3 - f_4, \quad \text{etc.}$$

In this scheme there must at last appear—at the very latest with the remainder K—a remainder $-f_s(x)$ that does not vanish at any point of the interval and consequently possesses the same sign over the whole interval. Here we break off the algorithm. The functions involved

$$f_0, f_1, f_2, \ldots, f_s$$

form a "*Sturm chain*" and in this connection are called *Sturm functions*.

The Sturm functions possess the following three properties: 1. Two neighboring functions do not vanish simultaneously at any point of the interval. 2. At a null point of a Sturm function its two neighboring functions are of different sign. 3. Within a sufficiently small area surrounding a zero point of $f_0(x)$, $f_1(x)$ is everywhere greater than zero or everywhere smaller than zero.

PROOF OF 1. If, for example, f_2 and f_3 vanish at any point of an interval, f_4 [according to (2)] also vanishes at this point, and consequently f_5 also [according to (3)], and so forth, so that finally [according to the last line of the algorithm] f_s also vanishes, which, however, contradicts our assumption.

PROOF OF 2. If the function f_3 vanishes at the point σ, for example, of the interval, then it follows from (2) that

$$f_2(\sigma) = -f_4(\sigma).$$

PROOF OF 3. This proof follows from the known theorem: A function $[f_0(x)]$ rises or falls at a point depending on whether its derivative $[f_1(x)]$ at that point is greater or smaller than zero.

We now select any point x of the interval, note the sign of the values $f_0(x), f_1(x), \ldots, f_s(x)$, and obtain a *Sturm sign chain* (to obtain an unequivocal sign, however, it must be assumed that none of the designated $s + 1$ function values is zero). The sign chain will contain sign sequences ($++$ and $--$) and sign changes ($+-$ and $-+$).

We will consider the number $Z(x)$ of sign changes in the sign chain and the changes undergone by $Z(x)$ when x passes through the interval. A change can occur only if one or more of the Sturm functions changes sign, i.e., passes over from negative (positive) values through zero to positive (negative) values. We will accordingly study the effect produced on $Z(x)$ by the passage of a function $f_\nu(x)$ through zero.

Let k be a point at which f_ν disappears, h a point situated to the left, and l a point to the right of k and so close to k that over the interval h to l the following holds true: (1) $f_\nu(x)$ does not vanish except when $x = k$; (2) every neighbor $(f_{\nu+1}, f_{\nu-1})$ of f_ν does not change sign. We must distinguish between the cases $\nu > 0$ and $\nu = 0$; in the first case we are concerned with the *triplet* $f_{\nu-1}$, f_ν, $f_{\nu+1}$, in the second, with the *pair* f_0, f_1.

In the *triplet*, $f_{\nu-1}$ and $f_{\nu+1}$ possess either the $+$ and $-$ sign or the $-$ and $+$ sign at all three points h, k, l. Thus, whatever the sign of f_ν may be at these points, the triplet possesses one change of sign for each of the three arguments h, k, l. The passage through zero of the function f_ν does not change the number of sign changes in the chain!

In the *pair*, f_1 has either the $+$ or $-$ sign at all three points h, k, l. In the first case, f_0 is increasing and is thus negative at h and positive at l. In the second case, f_0 is decreasing and is positive at point h, and negative at l. In both cases a sign change is lost.

From our investigation we learn that: The Sturm sign chain undergoes a change in the number of sign changes $Z(x)$ only when x passes through a null point of $f(x)$; and specifically, the chain then loses (with an increasing x) exactly one sign change. Thus, if x passes through the interval (the ends of which do not represent roots of $f(x) = 0$) from left to right, the sign chain loses exactly as many sign changes as there are null points of $f(x)$ within the interval. Result:

STURM'S THEOREM: *The number of real roots of an algebraic equation with real coefficients whose real roots are simple over an interval the end points of which are not roots is equal to the difference between the numbers of sign changes of the Sturm sign chains formed for the interval ends.*

NOTE. The same considerations can also be applied unchanged to the series formed when we multiply $f_0, f_1, f_2, \ldots, f_s$ by any positive constants; this series is then likewise designated as a Sturm chain. In the formation of the Sturm function chain all fractional coefficients are accordingly avoided.

EXAMPLE 1. Determine the number and situation of the real roots of the equation $x^5 - 3x - 1 = 0$.

The Sturm chain is

$$f_0 = x^5 - 3x - 1, \quad f_1 = 5x^4 - 3, \quad f_2 = 12x + 5, \quad f_3 = 1.$$

The signs of f for $x = -2, -1, 0, +1, +2$ are

x	f_0	f_1	f_2	f_3
-2	$-$	$+$	$-$	$+$
-1	$+$	$+$	$-$	$+$
0	$-$	$-$	$+$	$+$
$+1$	$-$	$+$	$+$	$+$
$+2$	$+$	$+$	$+$	$+$

The equation thus has three real roots: one between -2 and -1, one between -1 and 0, one between $+1$ and $+2$. The other two roots are complex.

EXAMPLE 2. Determine the number of real roots of the equation $x^5 - ax - b = 0$ when a and b are positive magnitudes and $4^4 a^5 > 5^5 b^4$.

The Sturm chain reads

$$x^5 - ax - b, \quad 5x^4 - a, \quad 4ax + 5b, \quad 4^4 a^5 - 5^5 b^4.$$

For the values $x = -\infty$ and $+\infty$ it has the signs

$$- \quad + \quad - \quad +$$

and

$$+ \quad + \quad + \quad +, \quad \text{respectively.}$$

The equation has three real and two complex roots.

25   Abel's Impossibility Theorem

Equations of higher than the fourth degree are in general incapable of algebraic solution.

This famous theorem was first stated by the Italian physician Paolo Ruffini (1765–1822) in his book *Teoria generale delle equazioni*, published in Bologna in 1798. Ruffini's proof, however, is incomplete. The first rigorous proof was given in 1826 in the first volume of *Crelle's Journal für Mathematik* by the young Norwegian mathematician Niels Henrik Abel (1802–1829). His celebrated paper bore the title "Démonstration de l'impossibilité de la résolution algébraique des équations générales qui dépassent le quatrième degré."

The following proof of Abel's impossibility theorem is based on a theorem of Kronecker, published in 1856 in the *Monatsberichte der Berliner Akademie*.

We will begin by presenting in a short introduction the auxiliary algebraic theorems necessary for an understanding of the Kronecker proof.

A system $\Re$ of numbers is called a *number group* or *rational domain* when the addition, subtraction, multiplication, and division of two numbers of the system will also yield a number of the system. For brevity we will call the numbers of the system $\Re$-numbers. Two groups are called equal when every number of the one belongs also to the other. The simplest group is that composed of all rational numbers, the *group* $\Re$ of *rational numbers* or the *natural rationality domain*.

A group $\Re' = \Re(\alpha, \beta, \gamma, \ldots)$ created by the "*substitution* of the magnitudes $\alpha, \beta, \gamma, \ldots$ in a group $\Re$" is understood to mean the totality of all the numbers obtained from the $\Re$-numbers and the substituted magnitudes $\alpha, \beta, \gamma, \ldots$ by one or more applications of the four species, in other words, the totality of all the rational functions of $\alpha, \beta, \gamma, \ldots$ whose coefficients are $\Re$-numbers.

A function $f(x)$ *or an equation* $f(x) = 0$ *in a group* is a function or equation whose coefficients are numbers of the group. A *polynomial* in $\Re$ is understood to mean an integral rational function of the variable x whose coefficients are $\Re$-numbers.

A polynomial

$$F(x) = Ax^n + Bx^{n-1} + \cdots$$

or an equation

$$F(x) = 0$$

in a group $\Re$ is said to be reducible or irreducible in this group accordingly as $F(x)$ is divisible into a product of polynomials of lower degree in $\Re$ or not.

The function $x^2 - 10x + 7$, for example, is irreducible in the group $\Re$ whereas it is reducible in the group $\Re(\sqrt{2})$:

$$x^2 - 10x + 7 = (x - 5 - 3\sqrt{2})(x - 5 + 3\sqrt{2}).$$

ABEL'S LEMMA:* *The pure equation*

$$x^p = C$$

of the prime number degree p *is irreducible in a group* $\Re$ *when* C *is a number of the group but not the* p*th power of a group number.*

INDIRECT PROOF. Let $x^p - C = 0$ be reducible, so that

$$x^p - C = \psi(x)\varphi(x),$$

where ψ and φ are polynomials in $\Re$, whose free terms A and B are $\Re$-numbers. Since the roots of the equation $x^p = C$ are r, $r\varepsilon$, $r\varepsilon^2$, ..., $r\varepsilon^{p-1}$, where r is one of the roots and ε a complex pth unit root, and the free term of the equation $\psi(x) = 0$ or $\varphi(x) = 0$, independent of sign, represents the product of the equation's roots, then, for example,

$$A = r^\mu \varepsilon^M, \qquad B = r^\nu \varepsilon^N.$$

Since μ and ν possess no common divisor (because $\mu + \nu = p$), there are integers h, k such that

$$\mu h + \nu k = 1.$$

Thus, we obtain for the product K of the powers A^h and B^k the value $r\varepsilon^{hM+kN}$ and, consequently, the value $K^p = r^p = C$ for the pth power of the $\Re$-number K. It was assumed, however, that C must *not* be the pth power of a $\Re$-number. Consequently, $x^p = C$ cannot be reducible.

SCHOENEMANN'S THEOREM (*Crelle's Journal*, vol. XXXII, 1846): *If the integral coefficients* C_0, C_1, C_2, ..., C_{N-1} *of the polynomial*

$$f(x) = C_0 + C_1 x + C_2 x^2 + \cdots + C_{N-1} x^{N-1} + x^N$$

are divisible by a prime number p, *while the free term* C_0 *is not divisible by* p^2, *then* f(x) *is irreducible in the natural rationality domain.*

INDIRECT PROOF. Let f be reducible so that $f = \psi \cdot \varphi$, with

$$\psi = a_0 + a_1 x + a_2 x^2 + \cdots + a_{m-1} x^{m-1} + x^m,$$

$$\varphi = b_0 + b_1 x + b_2 x^2 + \cdots + b_{n-1} x^{n-1} + x^n.$$

---

* Abel, *Œuvres complètes*, vol. II, p. 196.

According to a theorem of Gauss* the coefficients a and b are here integers. We multiply the expressions for ψ and φ, obtaining, by comparison with f,

$$C_0 = a_0 b_0,$$
$$C_1 = a_0 b_1 + a_1 b_0,$$
$$C_2 = a_0 b_2 + a_1 b_1 + a_2 b_0, \quad \text{etc.}$$

Since C_0 is not divisible by p^2, let us say that a_0 is divisible by p, in which case b_0 is not. Since C_1 and a_0 are divisible by p, while b_0 is not, it follows from the second line of our scheme that a_1 is divisible by p. Then it follows according to the third line of our scheme, in which C_2, a_0, a_1 are divisible by p, that a_2 is also divisible by p, and so forth. Finally, we would be able to conclude that $a_m = 1$ is also divisible by p, which is naturally absurd. Consequently, f cannot be reducible.

Reducible and irreducible polynomials play the same role among polynomials that composite and prime numbers play among the integers. Thus, for example, every reducible polynomial can be divided in only one way into a product of irreducible polynomials. All of the theorems concerned here are based on the *fundamental theorem of irreducible functions.*

---

* GAUSS' THEOREM: *If a polynomial* $f = x^N + C_1 x^{N-1} + C_2 x^{N-2} + \cdots + C_N$ *with integral coefficients is divisible into a product of two polynomials* $\psi = x^m + \alpha_1 x^{m-1} + \cdots + \alpha_m$ *and* $\varphi = x^n + \beta_1 x^{n-1} + \cdots + \beta_n$ *with rational coefficients* $(f = \psi\varphi)$, *then the coefficients of this polynomial are integers.*

PROOF. We bring α_ν and β_ν to their highest common denominators a_0 and b_0, respectively, so that $\alpha_\nu = a_\nu/a_0$ and $\beta_\nu = b_\nu/b_0$, and the numbers a_0, a_1 $a_2, \ldots, a_m$, as well as the numbers $b_0, b_1, \ldots, b_n$, possess no common divisor, and we obtain

$$F = \Psi\Phi \quad \text{with} \quad F = a_0 b_0 f,$$
$$\Psi = a_0 x^m + a_1 x^{m-1} + \cdots + a_m, \qquad \Phi = b_0 x^n + b_1 x^{n-1} + \cdots + b_n.$$

Let p be a *prime* divisor of $a_0 b_0$.

Then all the coefficients of F are divisible by p, but not by Ψ and Φ. We combine these terms of Ψ and Φ, respectively, whose coefficients are divisible by p, to form the respective polynomials U and V, and similarly combine these terms whose coefficients are not divisible by p to form the polynomials u and v, so that $F = (U + u)(V + v)$, and consequently

$$uv = F - UV - Uv - Vu.$$

The right-hand side of this equation contains a polynomial in which, according to our assumptions for F, U, and V, every coefficient is divisible by p; the left side, however, does not, since the coefficient of the highest power of the left side, being the product of two factors a_r and b_s that are not divisible by p, is also not divisible by p.

This contradiction disappears only when $a_0 b_0$ has no prime divisor, i.e., when $a_0 = 1$ and $b_0 = 1$, in which case α_ν and β_ν are *integers.*

ABEL'S IRREDUCIBILITY THEOREM:* *If one root of the equation* $f(x) = 0$, *which is irreducible in $\Re$, is also a root of the equation* $F(x) = 0$ *in $\Re$, then all the roots of the irreducible equation are roots of* $F(x) = 0$. *At the same time* $F(x)$ *can be divided by* $f(x)$ *without a remainder:*

$$F(x) = f(x) \cdot F_1(x),$$

where $F_1(x)$ *is also a polynomial in $\Re$.*

The simple proof of this theorem is based on the familiar algorithm for finding the highest common divisor $g(x)$ of two arbitrary polynomials $F(x)$ and $f(x)$ in $\Re$. This algorithm leads through a chain of divisions, in which all the coefficients are $\Re$-numbers, to the pair of equations

$$F(x) = F_1(x) \cdot g(x), \qquad f(x) = f_1(x) \cdot g(x)$$

and to the equation

$$V(x)F(x) + v(x)f(x) = g(x),$$

where all the indicated functions are polynomials in $\Re$.

If the prescribed functions F and f have no common divisor, then $g(x)$ is a constant which is for convenience set equal to 1.

If f is *irreducible* and a root α of $f = 0$ is also a root of $F = 0$, then there exists a common divisor of at least the first degree $(x - \alpha)$. Since f is irreducible, $f_1(x)$ must equal 1 and $f(x) = g(x)$, and then

$$F(x)[= F_1(x) \cdot g(x)] = F_1(x) \cdot f(x).$$

$F(x)$ is thus divisible by $f(x)$ and vanishes for every zero point of $f(x)$. Q.E.D.

The fundamental theorem directly implies two important corollaries:

I. *If a root of an equation* $f(x) = 0$, *which is irreducible in $\Re$, is also a root of an equation* $F(x) = 0$ *in $\Re$ of lower degree than* f, *then all the coefficients of* F *are equal to zero.*

II. *If* $f(x) = 0$ *is an irreducible equation in a group $\Re$, then there is no other irreducible equation in $\Re$ that has a common root with* $f(x) = 0$.

The commonest case of substitution in a group $\Re$ consists of the substitution of a root α of an irreducible equation of the nth degree

$$f(x) = x^n + a_1 x^{n-1} + \cdots + a_n = 0$$

---

* N. H. Abel, "Mémoire sur une classe particulière d'équations résolubles algébriquement," *Crelle's Journal*, vol. IV, 1829.

into $\mathfrak{R}$. A number ζ of the group $\mathfrak{R}' = \mathfrak{R}(\alpha)$ defined by this substitution is a rational function of α with coefficients from $\mathfrak{R}$ and can be written $\zeta = \Psi(\alpha)/\Phi(\alpha)$, where Ψ and Φ are polynomials in $\mathfrak{R}$. Since $a^n = -a_1\alpha^{n-1} - a_2\alpha^{n-2} - \cdots - a_n$, every power of α with the exponent n or with a higher exponent can be expressed by the powers $\alpha^{n-1}, \alpha^{n-2}, \ldots, \alpha$, so that we may write $\zeta = \psi(\alpha)/\varphi(\alpha)$, where ψ and φ are polynomials in $\mathfrak{R}$ of no higher than the $(n-1)$th degree.

Since $f(x)$ and $\varphi(x)$ possess no common divisor, two polynomials $u(x)$ and $v(x)$ can be found (see above) in $\mathfrak{R}$, such that $u(x)\varphi(x) + v(x)f(x) = 1$. If in this equation we set $x = \alpha$, then [since $f(\alpha) = 0$] $u(\alpha)\cdot\varphi(\alpha) = 1$, i.e., $\zeta = \psi(\alpha)\cdot u(\alpha)$. We multiply this out and once again eliminate every power of α whose exponent $\geqq n$. This finally gives us

$$\zeta = c_0 + c_1\alpha + c_2\alpha^2 + \cdots + c_{n-1}\alpha^{n-1},$$

where the c_ν are $\mathfrak{R}$-numbers; i.e.,

III. *Every number of the group $\mathfrak{R}(\alpha)$, where α is a root of an irreducible equation of the nth degree in $\mathfrak{R}$, can be represented as a polynomial of the $(n-1)$th degree of α with coefficients that are $\mathfrak{R}$-numbers. There is only one such possible way of representing it.*

[From

$$c_0 + c_1\alpha + \cdots + c_{n-1}\alpha^{n-1} = C_0 + C_1\alpha + \cdots + C_{n-1}\alpha^{n-1}$$

it follows that

$$d_0 + d_1\alpha + \cdots + d_{n-1}\alpha^{n-1} = 0, \quad \text{with} \quad d_\nu = C_\nu - c_\nu.$$

Then the function of the $(n-1)$th degree

$$d_0 + d_1x + d_2x^2 + \cdots + d_{n-1}x^{n-1}$$

vanishes for a root of $f(x) = 0$ and, according to corollary I., must have nothing but evanescent coefficients. From $d_\nu = 0$, however, it follows that $C_\nu = c_\nu$.]

We have just seen a simple example of an irreducible function that became reducible by substitution of a root.

Let us consider the more general case in which an irreducible function $f(x)$ in $\mathfrak{R}$ of prime number degree p becomes reducible by substitution of a root α of an irreducible equation of the qth degree $g(x) = 0$ in $\mathfrak{R}$, in which, therefore, $f(x)$ can be divided into the product of the two polynomials $\psi(x, \alpha)$ and $\varphi(x, \alpha)$, which may be of the mth and nth degree of x, respectively.

Now the function in $\Re$

$$u(x) = f(r) - \psi(r, x)\varphi(r, x),$$

where r is some rational number, vanishes for $x = \alpha$. According to the fundamental theorem of irreducible functions, $u(x)$ is then evanescent for *all* roots $\alpha, \alpha', \alpha'', \ldots$ of the irreducible equation $g(x) = 0$.

Since, for example, the equation

$$f(x) - \psi(x, \alpha')\varphi(x, \alpha') = 0$$

is therefore valid for every *rational* x, it is valid for all the values of x, so that by identity

$$f(x) = \psi(x, \alpha')\varphi(x, \alpha')$$

and similarly for all other roots of $g(x) = 0$.

From the q equations

$$f(x) = \psi(x, \alpha)\varphi(x, \alpha),$$

$$f(x) = \psi(x, \alpha')\varphi(x, \alpha'), \quad \text{etc.,}$$

thus obtained, it follows by multiplication that

$$f(x)^q = \Psi(x) \cdot \Phi(x),$$

where $\Psi(x)$ and $\Phi(x)$ are the products of the q polynomials $\psi(x, \alpha)$, $\psi(x, \alpha'), \ldots$ and $\varphi(x, \alpha), \varphi(x, \alpha'), \ldots$, respectively. Since each of these products is a symmetrical function of the roots of $g(x) = 0$, each product can be expressed rationally according to the Waring theorem by the coefficients of $g(x) = 0$ [and naturally by x], so that $\Psi(x)$ and $\Phi(x)$ are polynomials in $\Re$.

Now $\Psi(x)$ certainly vanishes for at least one root of the irreducible equation $f(x) = 0$, as does $\Phi(x)$. Consequently both $\Psi(x)$ and $\Phi(x)$ can be divided without a remainder by $f(x)$, and since f is irreducible no other divisor than f is possible, as a result of which

$$\Psi(x) = f(x)^\mu, \qquad \Phi(x) = f(x)^\nu,$$

with $\mu + \nu = q$. Comparing the degree of the left and right sides, we obtain

$$mq = \mu p, \qquad nq = \nu p$$

and from these, since m and n are smaller than p, it follows that p is a divisor of q. We therefore obtain the theorem:

IV. *An irreducible equation of the prime number degree* p *in a group can become reducible through substitution of a root of another irreducible equation in this group only when* p *is a divisor of the degree of the latter equation.*

After this introduction we can turn to the proof of Abel's theorem. First, however, we will consider what is meant by an *algebraically soluble* equation.

An equation of the nth degree $f(x) = 0$ in a group $\Re$ is called *algebraically soluble* when it is soluble by a *series of radicals*, i.e., when a root w can be determined in the following manner:

1. Determination of the ath root $\alpha = \sqrt[a]{R}$ of an $\Re$-number R, which is not, however, an ath power of an $\Re$-number, and substitution of α into $\Re$, so that the group $\mathfrak{A} = \Re(\alpha)$ is formed;

2. Determination of the bth root $\beta = \sqrt[b]{A}$ of an $\mathfrak{A}$-number A, which, however, is not a bth power of an $\mathfrak{A}$-number, and substitution of β into $\mathfrak{A}$, so that the group $\mathfrak{B} = \mathfrak{A}(\beta) = \Re(\alpha, \beta)$ is formed;

3. Determination of the cth root $\gamma = \sqrt[c]{B}$ of a $\mathfrak{B}$-number B, which, however, is not a cth power of a $\mathfrak{B}$-number, and the substitution of γ into $\mathfrak{B}$, so that the group $\mathfrak{C} = \mathfrak{B}(\gamma) = \Re(\alpha, \beta, \gamma)$ is formed, etc., until these successive substitutions of radicals $\alpha, \beta, \gamma, \ldots$ at length result in a group to which ω, the sought-for root, belongs *and in which* f(x) [since it possesses the divisor $x - \omega$] *becomes reducible*. It is here assumed that all the radical exponents $a, b, c, \ldots$ are prime numbers. This does not represent a restriction since any extraction of roots with composite exponents can be reduced to successive extractions of roots with prime exponents (e.g., $\sqrt[15]{u} = \sqrt[5]{v}$ with $v = \sqrt[3]{u}$).

In order to shorten our task somewhat, we will limit ourselves to equations $f(x) = 0$ which possess rational coefficients, so that $\Re$ is the natural rationality domain, which are, moreover, irreducible in $\Re$, and which are of the degree n, which is an odd prime number.

Let the first substitution be that of the nth root of unity

$$\alpha = \eta = \sqrt[n]{1} = \cos\frac{2\pi}{n} + i\sin\frac{2\pi}{n}.$$

According to **IV.**, this substitution still does not make f reducible, since η is a root of the equation $x^{n-1} + x^{n-2} + \cdots + x + 1 = 0$, the degree of which is $< n$.

Also, with each substituted radical of our series, which still does not allow division of $f(x)$, we will also substitute at the same time the complex conjugate radical. Though this may be superfluous, it can certainly do no harm.

Let $\lambda = \sqrt[l]{K}$ be the radical the addition of which to the preceding radicals makes $f(x)$ reducible, so that $f(x)$ is still indivisible in the group $\Re$ (to which the number K belongs), but becomes divisible in $\mathfrak{L} = \Re(\lambda)$:

$$f(x) = \psi(x, \lambda) \cdot \varphi(x, \lambda) \cdot \chi(x, \lambda) \cdots.$$

Here the factors $\psi, \varphi, \chi, \ldots$ are irreducible polynomials in $\mathfrak{L}$ (but naturally not polynomials in $\Re$) whose coefficients are polynomials of λ in $\Re$.

Since, according to IV., the prime number n must be a divisor of the prime number l, l must be equal to n.

The l roots of the equation $x^l = K$, which is irreducible in $\Re$ according to Abel's lemma, are

$$\lambda_0 = \lambda,\ \lambda_1 = \lambda\eta,\ \lambda_2 = \lambda\eta^2, \ldots, \lambda_v = \lambda\eta^v, \ldots, \lambda_{n-1} = \lambda\eta^{n-1}.$$

Since $\psi(x, \lambda)$ is a divisor of $f(x)$, then $\psi(x, \lambda_v)$ also goes into $f(x)$ without a remainder (cf. the proof of IV.).

Every one of the n functions $\psi(x, \lambda_v)$ is irreducible in $\mathfrak{L}$.

[As in the proof of IV., it follows from $\psi(x, \lambda_v) = u(x, \lambda_v) \cdot v(x, \lambda_v)$ that $\psi(x, \lambda) = u(x, \lambda) \cdot v(x, \lambda)$, but this equation is impossible because $\psi(x, \lambda)$ is irreducible in $\mathfrak{L}$.]

No two of the n functions $\psi(x, \lambda_v)$ are equal. [In $\psi(x, \lambda\eta^\mu) = \psi(x, \lambda\eta^v)$, λ could, as before, be replaced by the root $\lambda\eta^{n-\mu}$, from which it would follow that

$$\psi(x, \lambda) = \psi(x, \lambda H),$$

where H represents the root of unity $\eta^{v-\mu}$. Here λ could in turn be replaced by λH, which would give

$$\psi(x, \lambda H) = \psi(x, \lambda H^2).$$

Similarly, it would follow that

$$\psi(x, \lambda H^2) = \psi(x, \lambda H^3),$$

etc. Thus, we would then have

$$\psi(x, \lambda) = \psi(x, \lambda H) = \psi(x, \lambda H^2) = \cdots,$$

i.e., also

$$\psi(x, \lambda) = \frac{\psi(x, \lambda) + \psi(x, \lambda H) + \cdots + \psi(x, \lambda H^{n-1})}{n}.$$

The right side of this equation, however, as a symmetrical function of the n roots λ, λH, λH^2, ... of $x^n = K$, is a polynomial of x in $\Re$, so that $\psi(x, \lambda)$ would also be a polynomial of x in $\Re$. This, however, contradicts what was stipulated above concerning $f(x)$.]

For these two reasons it follows that $f(x)$ is divisible by the product $\Psi(x)$ of the n different factors $\psi(x, \lambda)$, $\psi(x, \lambda\eta)$, ..., $\psi(x, \lambda\eta^{n-1})$ that are irreducible in $\mathfrak{L}$:

$$f(x) = \Psi(x) \cdot U(x),$$

where Ψ (as a symmetrical function of the roots of $x^n = K$), and consequently U as well, are polynomials of x in $\Re$. Now, since $f(x)$ is not reducible in $\Re$, $U(x)$ must equal 1 and necessarily

$$f(x) = \Psi(x) = \psi(x, \lambda)\psi(x, \lambda\eta) \ldots \psi(x, \lambda\eta^{n-1}).$$

The postulated divisibility of $f(x)$ for the group $\mathfrak{L}$ consequently reveals itself as a divisibility into *linear* factors. Thus, if ω, ω_1, ω_2, ..., ω_{n-1} are the roots and $x - \omega$, $x - \omega_1$, ..., $x - \omega_{n-1}$ are the linear factors of $f(x)$, then

$$x - \omega = \psi(x, \lambda), x - \omega_1 = \psi(x, \lambda\eta), \ldots x - \omega_{n-1} = \psi(x, \lambda\eta^{n-1}),$$

and consequently

$$\begin{aligned}
\omega &= K_0 + K_1\lambda &&+ K_2\lambda^2 + \cdots + K_{n-1}\lambda^{n-1}, \\
\omega_1 &= K_0 + K_1\lambda_1 &&+ K_2\lambda_1^2 + \cdots + K_{n-1}\lambda_1^{n-1}, \\
&\vdots \\
\omega_{n-1} &= K_0 + K_1\lambda_{n-1} &&+ K_2\lambda_{-1}^2 + \cdots + K_{n-1}\lambda_{n-1}^{n-1},
\end{aligned}$$

where all the K_ν are $\Re$-numbers.

Now the equation $f(x) = 0$ has at least one real root, since it is of an odd degree. Let this real root be

$$\omega = K_0 + K_1\lambda + \cdots + K_{n-1}\lambda^{n-1}.$$

We distinguish two cases:

I. *The base* K *of the reducible radical* λ *is real;*

II. *the base* K *is complex.*

CASE I. Here we can assume that λ is real, since the nth roots of unity belong to the group $\Re$. In that event the complex conjugate of ω is

$$\bar{\omega} = \overline{K}_0 + \overline{K}_1\lambda + \cdots + \overline{K}_{n-1}\lambda^{n-1},$$

where the complex conjugates $\overline{K}_\nu$ of K_ν are also $\mathfrak{R}$-numbers. From $\bar{\omega} = \omega$ it follows then that

$$(\overline{K}_0 - K_0) + (\overline{K}_1 - K_1)\lambda + \cdots + (\overline{K}_{n-1} - K_{n-1})\lambda^{n-1} = 0,$$

and from this, taking theorem I into consideration, it follows that $\overline{K}_\nu = K_\nu$ for every ν. The magnitudes $K_0, K_1, \ldots, K_{n-1}$ are therefore also real.

Furthermore,

$$\omega_\nu = K_0 + K_1\lambda_\nu \quad + \cdots + K_{n-1}\lambda_\nu^{n-1}$$

and

$$\omega_{n-\nu} = K_0 + K_1\lambda_{n-\nu} + \cdots + K_{n-1}\lambda_{n-\nu}^{n-1}.$$

However, since $\lambda_\nu = \lambda\eta^\nu$ and $\lambda_{n-\nu} = \lambda\eta^{n-\nu} = \lambda\eta^{-\nu}$ are complex conjugates, it follows that ω_ν and $\omega_{n-\nu}$ are also complex conjugates, i.e.:

The equation $f(x) = 0$ *possesses one real root and* $n - 1$ *paired conjugate complex roots* (ω_1 and ω_{n-1}, ω_2 and ω_{n-2}, etc.).

CASE II. In this case we substitute, in addition to the reducible radical $\lambda = \sqrt[n]{K}$, the complex conjugate $\bar{\lambda} = \sqrt[n]{\overline{K}}$ with the result that the real magnitude $\Lambda = \lambda\bar{\lambda}$ is also substituted.

If the substitution of $\Lambda = \sqrt[n]{K\overline{K}}$ alone (i.e., without λ) were sufficient to make $f(x)$ reducible, this would give us the situation of Case I. We may therefore assume that $f(x)$ is still irreducible in $\mathfrak{R}(\Lambda)$ and does not become reducible until the additional substitution of λ.

From

$$\omega = K_0 + K_1\lambda + \cdots + K_{n-1}\lambda^{n-1}$$

it follows that

$$\bar{\omega} = \overline{K}_0 + \overline{K}_1\bar{\lambda} + \cdots + \overline{K}_{n-1}\bar{\lambda}^{n-1}$$

$$= \overline{K}_0 + \overline{K}_1\left(\frac{\Lambda}{\lambda}\right) + \cdots + \overline{K}_{n-1}\left(\frac{\Lambda}{\lambda}\right)^{n-1},$$

and from this, since $\bar{\omega} = \omega$, that

$$K_0 + K_1\lambda + \cdots + K_{n-1}\lambda^{n-1}$$

$$= \overline{K}_0 + \overline{K}_1\left(\frac{\Lambda}{\lambda}\right) + \cdots + \overline{K}_{n-1}\left(\frac{\Lambda}{\lambda}\right)^{n-1}.$$

In this equation all of the magnitudes with the exception of λ belong to the group $\Re(\Lambda)$, and since the equation $x^n = K$ (according to Abel's lemma) is irreducible in this group, we are able to replace λ in the above equation by any root λ_ν of $x^n = K$.

If we do this and keep in mind that

$$\frac{\Lambda}{\lambda_\nu} = \frac{\Lambda}{\lambda\eta^\nu} = \frac{\bar{\lambda}}{\eta^\nu} = \bar{\lambda}\bar{\eta}^\nu = \overline{\lambda\eta^\nu} = \bar{\lambda}_\nu,$$

we obtain

$$K_0 + K_1\lambda_\nu + \cdots + K_{n-1}\lambda_\nu^{n-1} = \overline{K}_0 + \overline{K}_1\bar{\lambda}_\nu + \cdots + \overline{K}_{n-1}\bar{\lambda}_\nu^{n-1}$$

or

$$\omega_\nu = \bar{\omega}_\nu.$$

Thus, all the roots of $f(x) = 0$ *are real.*

The combination of the results of I. and II. yields the

KRONECKER* THEOREM: *An algebraically soluble equation of an odd degree that is a prime and which is irreducible in the natural rationality domain possesses either only one real root or only real roots.*

Kronecker's theorem proves at the same time that an equation of higher than the fourth degree cannot be solved generally by algebraic means.

The simple fifth-degree equation

$$x^5 - ax - b = 0,$$

for example, cannot be solved algebraically when a and b are positive integers that are divisible by a prime number p, b is indivisible by p^2, and when $4^4a^5 > 5^5b^4$.

According to Schoenemann's theorem the equation is irreducible. Sturm's theorem (No. 24) proves that it possesses *three* real roots and two complex roots. Consequently, the equation is algebraically insoluble according to Kronecker's theorem.

In exactly the same way it can be shown that

$$x^7 - ax - b = 0$$

is algebraically insoluble when $6^6a^7 > 7^7b^6$, etc.

---

* Leopold Kronecker (1823–1891), a German mathematician.

26   The Hermite-Lindemann Transcendence Theorem

The expression

$$A_1 e^{\alpha_1} + A_2 e^{\alpha_2} + A_3 e^{\alpha_3} + \cdots,$$

in which the coefficients A differ from zero and in which the exponents α are algebraic numbers differing from each other, cannot equal zero.

This extremely important theorem (see below) was proved in 1882 by the German mathematician Lindemann (in the *Berliner Sitzungsberichte*) after the French mathematician Hermite (1822–1901), in vol. 77 of the *Comptes rendus* in 1873, had proved the special case in which the coefficients and exponents were rational integers. Lindemann's proof, which required a great many higher mathematical tools, was simplified to such an extent, first (1885, *Berliner Sitzungsberichte*) by K. Weierstrass (1815–1897), then (1893, *Mathematische Annalen*, vol. 43) by P. Gordan (1837–1912), that the proof is now generally accessible. The proof is presented here essentially in the form given to it in his textbook of algebra by H. Weber (1842–1913).

The proof is indirect. We assume that there are l algebraic numbers $A_1, A_2, \ldots, A_l$ and l algebraic numbers $\alpha_1, \alpha_2, \ldots, \alpha_l$ differing from one another that satisfy the equation

$$(1) \qquad A_1 e^{\alpha_1} + A_2 e^{\alpha_2} + \cdots + A_l e^{\alpha_l} = 0,$$

and we show that this assumption leads to a contradiction. The demonstration is divided into four steps.

I. We consider the coefficients A as roots of a real equation $\mathfrak{A}(x) = 0$ with rational coefficients the degree of which, L, will generally be greater than l. Let the roots of this equation be $A_1, A_2, \ldots, A_l, \ldots, A_L$. We form all the possible l-termed expressions $A_r e^{\alpha_1} + A_s e^{\alpha_2} + \cdots$ [totaling $L(L-1)(L-2) \ldots (L-l+1)$ elements], where $A_r, A_s, \ldots$ are any l components of the series $A_1, A_2, \ldots, A_L$, and we multiply these expressions together, always combining each of the members with the *same* exponential factor e^*. The resulting product has the form

$$\Pi' = A_1' e^{\beta_1} + A_2' e^{\beta_2} + \cdots + A_m' e^{\beta_m},$$

where the A' are nonevanescent magnitudes.

[That the coefficients A' obtained by multiplying out and combining cannot *all* vanish is proved in the following manner. We call the first of the two complex numbers $x + iy$ and $X + iY$ the "smaller" when either $x < X$ or $x = X$ if y is at the same time $< Y$. Now the product Π' consists only of factors of the form $F_\nu = P_\nu e^{p_\nu} + Q_\nu e^{q_\nu} + R_\nu e^{r_\nu} + \cdots$, where none of the coefficients P, Q, R vanishes, and we can consider the terms as being arranged in such a manner that $p_\nu < q_\nu < r_\nu < \cdots$. On multiplying the factors F_ν the exponent $p_1 + p_2 + p_3 + \cdots$ of the first term obtained is then the *smallest* of all the exponents obtained and occurs *only once*. Consequently, at the very least the first term of the multiplied-out product differs from zero, which was what we set out to prove.]

The coefficients A' are not changed by transpositions of the magnitudes $A_1, A_2, \ldots, A_L$; in other words, they are symmetrical functions of the roots of $\mathfrak{A}(x) = 0$, and, therefore, according to the principal theorem concerning symmetrical functions, are *rational numbers*.

Since the left side of (1) is also among the factors of Π',

$$\Pi' = 0.$$

We multiply this equation by the common denominator of the A''s and obtain the new equation

$$(2) \qquad B_1 e^{\beta_1} + B_2 e^{\beta_2} + \cdots + B_m e^{\beta_m} = 0,$$

where the β different algebraic numbers and the coefficients B are nonevanescent *rational integers*.

II. Let us consider the exponents β as roots of an algebraic equation $\mathfrak{B}(x) = 0$ with rational coefficients of degree M, with M generally greater than m, and let us in the usual way think of the equation as being free of *identical* roots. We form the $M(M-1)(M-2)\ldots(M-m+1)$ m-termed sums

$$B_1 e^{v\beta_r} + B_2 e^{v\beta_s} + \cdots,$$

where v is a variable and $\beta_r, \beta_s, \ldots$ are any m roots of $\mathfrak{B}(x) = 0$, and multiply these sums by each other, once again combining terms with the same exponential factor e^*. The resulting product has the form

$$\Pi = C_1 e^{\gamma_1 v} + C_2 e^{\gamma_2 v} + \cdots + C_n e^{\gamma_n v},$$

where the coefficients C are nonevanescent rational integers and γ represents different algebraic numbers.

The product Π is a symmetrical function of the roots of $\mathfrak{B}(x) = 0$. Consequently, the coefficients of the expansion of Π according to the powers of v are also symmetrical functions of those roots; thus, for example, the coefficient k_v of v^v:

$$k_v = (C_1\gamma_1^v + C_2\gamma_2^v + \cdots + C_n\gamma_n^v)/v!.$$

Every coefficient k_v is therefore a *rational* number. Accordingly, if $\mathfrak{g}(x)$ is a rational function of x with coefficients that are rational integers, the sum $\displaystyle\sum_s^{1.n} C_s\mathfrak{g}(\gamma_s)$ is rationally composed of the coefficients k^v and is consequently a rational number.

Now since the product Π for $v = 1$ contains the factor $B_1e^{\beta_1}$, $B_2e^{\beta_2} + \cdots + B_me^{\beta_m}$, which is equal to zero according to (2), the product for $v = 1$ is also equal to zero, and we obtain the equation

(3) $$C_1e^{\gamma_1} + C_2e^{\gamma_2} + \cdots + C_ne^{\gamma_n} = 0,$$

in addition to which for every integral rational function $\mathfrak{g}(x)$ with integral rational coefficients

(3a) $$C_1\mathfrak{g}(\gamma_1) + C_2\mathfrak{g}(\gamma_2) + \cdots + C_n\mathfrak{g}(\gamma_n)$$

is a *rational number*.

III. We consider the exponents $\gamma_1, \gamma_2, \ldots, \gamma_n$ as roots of an algebraic equation

$$x^N + r_1x^{N-1} + r_2x^{N-2} + \cdots + r_N = 0$$

with rational coefficients of degree $N \geqq n$, possessing no identical roots.

We multiply this equation by the Nth power of the common denominator H of the coefficients $r_1, r_2, \ldots$ and obtain

$$(Hx)^N + Hr_1(Hx)^{N-1} + H^2r_2(Hx)^{N-2} + \cdots = 0$$

or, if we write X instead of Hx and call the integers $Hr_1, H^2r_2, H^3r_3, \ldots,$ $g_1, g_2, g_3, \ldots,$

$$f(X) = X^N + g_1X^{N-1} + g_2X^{N-2} + \cdots + g_N = 0.$$

If $\Gamma_1, \Gamma_2, \ldots, \Gamma_N$ are the roots of this equation, then

$$f(X) = (X - \Gamma_1)(X - \Gamma_2)\ldots(X - \Gamma_N).$$

The roots Γ possess the n values $\Gamma_1 = H\gamma_1, \Gamma_2 = H\gamma_2, \ldots,$ $\Gamma_n = H\gamma_n.$,

Since Γ represents *integral* algebraic numbers, then, as a result of $(3a)$,

$$(3b) \qquad C_1\mathfrak{g}(\Gamma_1) + C_2\mathfrak{g}(\Gamma_2) + \cdots + C_n\mathfrak{g}(\Gamma_n)$$

is a rational *integer*.

Besides $f(X)$ we will consider the function

$$\begin{aligned}
\varphi(X) &= \frac{f(X)}{X - \Gamma_1} + \frac{f(X)}{X - \Gamma_2} + \cdots + \frac{f(X)}{X - \Gamma_N} \\
&= (X - \Gamma_2)(X - \Gamma_3)\ldots(X - \Gamma_N) \\
&\quad + (X - \Gamma_1)(X - \Gamma_3)(X - \Gamma_4)\ldots(X - \Gamma_N) + \cdots \\
&= NX^{N-1} + N_1 X^{N-2} + \cdots,
\end{aligned}$$

which is not evanescent for any of the values $\Gamma_1, \Gamma_2, \ldots, \Gamma_N$, and the coefficients of which $N, N_1, \ldots$ (as symmetrical functions of the roots $\Gamma_1, \Gamma_2, \ldots, \Gamma_N$ of $f(X) = 0$) are rational integers.

If the sum

$$C_1\varphi(\Gamma_1) + C_2\varphi(\Gamma_2) + \cdots + C_n\varphi(\Gamma_n)$$

should by chance equal zero, we select the positive integral exponent $h(< n)$ in such a manner that the (integral) sum

$$G = C_1\Gamma_1^h\varphi(\Gamma_1) + C_2\Gamma_2^h\varphi(\Gamma_2) + \cdots + C_n\Gamma_n^h\varphi(\Gamma_n) \neq 0.$$

[Such an exponent must exist, because otherwise the n linear homogeneous equations

$$\begin{aligned}
1 \quad \cdot x_1 + 1 \quad \cdot x_2 + \cdots + 1 \quad \cdot x_n &= 0, \\
\Gamma_1 \quad \cdot x_1 + \Gamma_2 \quad \cdot x_2 + \cdots + \Gamma_n \quad \cdot x_n &= 0, \\
\Gamma_1^2 \quad \cdot x_1 + \Gamma_2^2 \quad \cdot x_2 + \cdots + \Gamma_n^2 \quad \cdot x_n &= 0, \\
&\vdots \\
\Gamma_1^{n-1}\cdot x_1 + \Gamma_2^{n-1}\cdot x_2 + \cdots + \Gamma_n^{n-1}\cdot x_n &= 0
\end{aligned}$$

would exist for the n nonevanescent "unknowns" $x_1 = C_1\varphi(\Gamma_1), \ldots,$ $x_n = C_n\varphi(\Gamma_n)$. This, however, is impossible, since then the determinant

$$\begin{vmatrix}
1 & 1 & \cdots & 1 \\
\Gamma_1 & \Gamma_2 & \cdots & \Gamma_n \\
\Gamma_1^2 & \Gamma_2^2 & \cdots & \Gamma_n^2 \\
\vdots & \vdots & \cdots & \vdots \\
\Gamma_1^{n-1} & \Gamma_2^{n-1} & \cdots & \Gamma_n^{n-1}
\end{vmatrix}$$

of the equation system would have to disappear; however, this determinant represents the product of all the differences $\Gamma_r - \Gamma_s$, in which $r > s$, and, in accordance with the above, *none* of which disappear.]

IV. Now we put the fundamental property of the exponential function—the series expansion for e^z—into the form most suited for our proof.

This is

$$e^x = 1 + x + \frac{x^2}{2!} + \cdots + \frac{x^\nu}{\nu!} + \cdots.$$

We multiply this equation by $H^\nu \nu!$ and obtain $(Hx = X)$

$$e^x \nu! H^\nu = H^\nu \nu! + \nu H^{\nu-1}(\nu - 1)!X + \nu_2 H^{\nu-2}(\nu - 2)!X^2 + \cdots$$
$$+ X^\nu + X^\nu \left[\frac{x}{\nu + 1} + \frac{x^2}{(\nu + 1)(\nu + 2)} + \cdots \right].$$

In order to write this formula more conveniently, we introduce the symbol $\mathfrak{S}$, which will be defined by the following direction:

A function $F(\mathfrak{S})$ shall be considered the expression obtained when $F(\mathfrak{S})$, on the assumption that $\mathfrak{S}$ is a number, is transformed in the usual way into a power series of $\mathfrak{S}$ and $\mathfrak{S}^\nu$ is replaced by $\nu! H^\nu$ at the end of expansion.

Our formula can then be written in the simple form:

$$e^x \mathfrak{S}^\nu = (\mathfrak{S} + X)^\nu + X^\nu \cdot [\ ].$$

If we then designate the absolute magnitude of x as ξ, the absolute magnitude of [ ] is smaller than

$$0 = \frac{\xi}{\nu + 1} + \frac{\xi^2}{(\nu + 1)(\nu + 2)} + \cdots,$$

and therefore certainly smaller than

$$1 + \xi + \frac{\xi^2}{2!} + \cdots = e^\xi.$$

If ε is understood to be a magnitude the absolute value of which is a proper fraction, we therefore obtain

(4) $e^x \mathfrak{S}^\nu = (X + \mathfrak{S})^\nu + e^\xi \cdot \varepsilon X^\nu.$

We will immediately extend this somewhat further. Let

$$V(X) = X^k + K_1 X^{k-1} + K_2 X^{k-2} + \cdots + K_k$$

represent an integral rational function of X with integral rational coefficients. We form (4) for $\nu = k, k - 1, k - 2, \ldots$, multiply the resulting equations by $1, K_1, K_2, \ldots$, and add. This gives us

$$(5) \qquad e^x V(\mathfrak{S}) = V(X + \mathfrak{S}) + e^{\xi} \overline{V}(X),$$

with

$$(5a) \qquad \overline{V}(X) = \varepsilon_0 X^k + \varepsilon_1 K_1 X^{k-1} + \varepsilon_2 K_2 X^{k-2} + \cdots,$$

where the absolute values of the magnitudes ε_κ are proper fractions.

If $\Delta_1, \Delta_2, \ldots$ represent the roots of $V(X) = 0$ and d represents the greatest of the k values $|X| + |\Delta_\kappa|$, it follows from

$$V(X) = (X - \Delta_1)(X - \Delta_2) \ldots$$

that the absolute magnitude of $\overline{V}(X)$ [like that of $V(X)$] is smaller than d^k:

$$(5b) \qquad |\overline{V}(X)| < d^k.$$

We apply the results (5), (5a), (5b) to the function

$$V(X) = F(X)^q \Phi(X),$$

in which

$$F(X) = X^h f(X), \qquad \Phi(X) = X^h \varphi(X),$$

$q = p - 1$, and p is a preliminarily selected, still undetermined prime number. Since the degree of $F(X)$ is $h + N$, and the degree of $\Phi(X)$ is $h + N - 1$, $V(X)$ is of the degree $k = (h + N)q + h + N - 1$.

Equation (5) is now transformed into

$$e^x V(\mathfrak{S}) = V(X + \mathfrak{S}) + \varepsilon e^{\xi} d^k,$$

where d is the greatest of the k values $|X| + |\Delta_\kappa|$ and ε is a number whose absolute magnitude is a proper fraction.

We now choose for x and X the values γ_ν and Γ_ν, respectively (ν is any one of the numbers 1 to n). Then ξ is the absolute magnitude ξ_ν of γ_ν and $d = d_\nu$ is the greatest of the k sums $|\Gamma_\nu| + |\Delta_\kappa|$.

If D then represents the greatest of the $2n$ numbers d_ν^{N+h} and $e^{\xi_\nu} d_\nu^{2(N+h)-1}$, then the improper fraction D/d_ν^{N+h} is $\geqq e^{\xi_\nu} d_\nu^{N+h-1}$, and consequently

$$(D/d_\nu^{N+h})^q \geqq e^{\xi_\nu} d_\nu^{N+h-1}$$

or

$$D^q \geqq e^{\xi_\nu} d_\nu^k$$

must be true, and we obtain the somewhat simpler formula

$$(6) \qquad e^{\gamma_\nu} V(\mathfrak{S}) = V(\Gamma_\nu + \mathfrak{S}) + \eta_\nu D^q,$$

where $|\eta_\nu| < 1$.

The expansion of $V(\Gamma_\nu + \mathfrak{S})$ according to the powers of $\mathfrak{S}$ gives us

$$V(\Gamma_\nu + \mathfrak{S}) = \psi_0 \mathfrak{S}^q + \psi_1 \mathfrak{S}^{q+1} + \psi_2 \mathfrak{S}^{q+2} + \cdots,$$

where the coefficients ψ are integral rational functions of Γ_ν with integral rational coefficients. In particular,

$$\psi_0 = \psi_0(\Gamma_\nu) = \Phi(\Gamma_\nu)^p.$$

[For $\nu = 1$, for example,

$$\begin{aligned}
F(\Gamma_1 + \mathfrak{S}) &= (\Gamma_1 + \mathfrak{S})^h [\mathfrak{S} \cdot (\mathfrak{S} + \Gamma_1 - \Gamma_2) \cdot (\mathfrak{S} + \Gamma_1 - \Gamma_3) \ldots] \\
&= \Gamma_1^h (\Gamma_1 - \Gamma_2)(\Gamma_1 - \Gamma_3) \ldots (\Gamma_1 - \Gamma_N) \cdot \mathfrak{S} + \cdots \\
&= \Gamma_1^h \varphi(\Gamma_1) \cdot \mathfrak{S} + \cdots
\end{aligned}$$

and

$$\Phi(\Gamma_1 + \mathfrak{S}) = (\Gamma_1 + \mathfrak{S})^h \varphi(\Gamma_1 + \mathfrak{S}) = \Gamma_1^h \varphi(\Gamma_1) + \cdots,$$

consequently

$$V(\Gamma_1 + \mathfrak{S}) = \Gamma_1^{hp} \varphi(\Gamma_1)^p \cdot \mathfrak{S}^q + \cdots = \Phi(\Gamma_1)^p \cdot \mathfrak{S}^q + \cdots.]$$

If we introduce this expansion into (6), we finally obtain

$$e^{\gamma_\nu} V(\mathfrak{S}) = \psi_0(\Gamma_\nu)\mathfrak{S}^q + \psi_1(\Gamma_\nu)\mathfrak{S}^p + \psi_2(\Gamma_\nu)\mathfrak{S}^{p+1} + \cdots + \eta_\nu D^q.$$

This formula, multiplied by C_ν, we then form for all ν from 1 through n, and we add the resulting n equations.

According to (3), we then obtain

$$(7) \quad 0 = G_0 \mathfrak{S}^q + G_1 \mathfrak{S}^p + G_2 \mathfrak{S}^{p+1} + \cdots + G_k \mathfrak{S}^{q+k} + \lambda D^q,$$

where

$$G_r = C_1 \psi_r(\Gamma_1) + C_2 \psi_r(\Gamma_2) + \cdots + C_n \psi_r(\Gamma_n)$$

is, according to (3b), a rational integer and λ is a number the absolute magnitude of which does not exceed the n-fold value of the maximum $|C|$-value.

We now replace $\mathfrak{S}^r$ with $H^r r!$, divide (7) by the then universally common factor H^q, abbreviate $D|H$ as E, and combine all the terms containing the factor $p!$, and we obtain

$$(8) \qquad\qquad G_0 q! + G' p! = \Lambda E^q,$$

where G' is an integer and $\Lambda = -\lambda$.

Now we compare

$$G_0 = C_1\Phi(\Gamma_1)^p + C_2\Phi(\Gamma_2)^p + \cdots + C_n\Phi(\Gamma_n)^p$$

with

$$G = C_1\Phi(\Gamma_1) + C_2\Phi(\Gamma_2) + \cdots + C_n\Phi(\Gamma_n),$$

the *latter* of which, according to our assumption concerning h, *differs from zero*.

If we expand G^p according to the polynomial theorem, every term of the expansion, with the exception of the n terms $C_\nu^p\Phi(\Gamma_\nu)^p$, is the p-multiple of an integral algebraic number, and, therefore,

$$(9) \qquad G^p = [C_1^p\Phi(\Gamma_1)^p + \cdots + C_n^p\Phi(\Gamma_n)^p] + \mu p,$$

where μ is an integral algebraic number (which is, in fact, integral and rational).

Now according to Fermat's theorem* every difference $C_\nu^p - C_\nu$, as well as $G^p - G$, is an integral multiple $c_\nu p$ and gp, respectively, of p. Accordingly, (9) is transformed into

$$G + gp = (C_1 + c_1 p)\Phi(\Gamma_1)^p + \cdots + (C_n + c_n p)\Phi(\Gamma_n)^p + \mu p$$
$$= C_1\Phi(\Gamma_1)^p + \cdots + C_n\Phi(\Gamma_n)^p + \mu'p = G_0 + \mu'p,$$

where μ' is also integral and algebraic.

This equation simplifies into

$$G_0 = G + g'p,$$

where $g' = g - \mu'$ is an *integral* algebraic number, and is also an integral *rational* number, as a result of $g' = (G_0 - G)/p$.

If we introduce this value into (8), we obtain

$$Gq! + g'p! + G'p! = \Lambda E^q$$

or, if the integer $G' + g'$ is designated as $\mathfrak{G}$,

$$(10) \qquad G + \mathfrak{G}p = \Lambda \frac{E^q}{q!}.$$

We now choose a prime number p so large that (1) $p > |G|$ and (2) the absolute magnitude of the right side of (10) is smaller than 1.

---

* FERMAT'S THEOREM: *For every integer* g *and every prime number* p *the difference* $g^p - g$ *is divisible by* p.

PROOF. The theorem is self-evident if g is divisible by p. For every g that is indivisible by p the theorem follows directly from the congruences (1a) and (2a) of No. 19, if g is substituted for D there and the congruences are squared. In both cases $g^{p-1} \equiv 1 \bmod p$ is obtained, and from this $g^p \equiv g \bmod p$.

Equation (10) then contains a contradiction. On the *left* side of the equation there is an *integer* that is indivisible by *p* (because $G \neq 0$) and is thus *not equal to zero*, while on the *right* there is a number whose absolute magnitude is *less than* 1. This is impossible. Consequently, the initial equation (1) is also impossible and Lindemann's theorem is proved.

The inferences that can be drawn from Lindemann's theorem are amazing. Here we present only a few:

1. THE TRANSCENDENCE OF *e*: *The Euler number* e *is transcendent*, i.e., it is not an algebraic number. (In other words, it cannot be a root of an algebraic equation with rational coefficients.)

2. THE TRANSCENDENCE OF π: *The Archimedes* (*Ludolph*) *number* π *is transcendent.*

According to Euler (No. 13), there exists the equation

$$e^{i\pi} + 1 = 0.$$

According to Lindemann's theorem the exponent $i\pi$ cannot, therefore, be an algebraic number. Consequently, it is also impossible for π to be an algebraic number. (If π were algebraic, then the product of the two algebraic numbers *i* and π would have to be algebraic.)

Thus, the ancient question of *squaring the circle* is answered, though the answer is negative:

It is impossible to draw with a compass and straight-edge a square that is equal in area to a given circle.

If, for example, we choose the radius of the given circle in such a manner that it is equal to the unit length, the area of the circle is π and the desired side of the square $\sqrt{\pi}$. If, however, $\sqrt{\pi}$ could be drawn with compass and straight-edge, then the square π of this segment could also be constructed, and, according to No. 36, π would have to be the root of an algebraic equation with rational coefficients (whose degree would be a power of 2). However, π is transcendent.

3. *The exponential curve* y $= $ e^x *passes through no algebraic point of the plane except the point* 0|1.

(An algebraic point is a point whose coordinates *x* and *y* are both algebraic numbers.) Since algebraic points are omnipresent in densely concentrated quantities within the plane, the exponential curve accomplishes the remarkably difficult feat of winding between all these points without touching any of them.

The same is, naturally, also true of the logarithmic curve $y = lx$.

4. *The sine curve* y = *sin* x *also passes through no algebraic points of the plane except the lattice point* 0|0.

If, for example, $\alpha|\beta$ were an algebraic point situated on the sine curve, β would be equal to sin α or, since $2i$ sin $\alpha = e^{i\alpha} - e^{-i\alpha}$, $e^{i\alpha} - e^{-i\alpha} - 2i\beta = 0$. However, according to Lindemann's theorem, this equation cannot exist for algebraic numbers α, β.

Planimetric Problems

27 Euler's Straight Line

In all triangles the center of the circumscribed circle, the point of intersection of the medians, and the point of intersection of the altitudes are situated in this order in a straight line—the Euler line—and are spaced in such a manner that the altitude intersection is twice as far from the median intersection as the center of the circumscribed circle is.

Leonhard Euler (1707–1783) was one of the greatest and most fertile mathematicians of all time. His writings comprise 45 volumes and over 700 papers, most of them long ones, published in periodicals.

The above theorem is among the results of the paper "Solutio facilis problematum quorundam geometricorum difficillimorum," which appeared in the journal *Novi commentarii Academiae Petropolitanae* (*ad annum 1765*).

The following proof of the Euler theorem is distinguished by its great simplicity.

In the triangle ABC let M be the midpoint of side AB, S the median intersection, which lies on CM, so that

$$(1) \qquad SC = 2 \cdot SM,$$

and U the center of the circle of circumscription, lying on the perpendicular bisector of AB.

We extend US by SO so that

$$(2) \qquad SO = 2 \cdot SU,$$

and join O to C.

According to (1) and (2) the triangles MUS and COS are similar. Consequently, $CO \parallel MU$, i.e., $CO \perp AB$, or expressed verbally, the line connecting the point O with a vertex of the triangle is perpendicular to the side of the triangle opposite the vertex; consequently, the connecting line is an altitude of the triangle.

The three altitudes consequently pass through point 0. This is, therefore, the altitude intersection, and Euler's theorem is proved.

NOTE. Our proof contains at the same time the solution to the interesting

PROBLEM OF SYLVESTER: *To find the resultant of the three vectors* UA, UB, UC *acting on the center of the circle of circumscription* U *of the triangle* ABC.

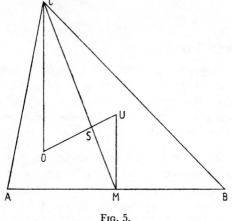

FIG. 5.

Since *UM* is *half* the resultant of the two vectors *UA* and *UB*, *CO* represents in magnitude and direction the *whole* resultant of these vectors. Now, since *UO* is the resultant of *UC* and *CO*, *UO* is the resultant we are seeking.

The resultant of the vectors represented by the three radii from the center of the circle of circumscription to the vertexes of the triangle is the segment extending from the center of the circle of circumscription to the altitude intersection.

James Joseph Sylvester (1814–1897) was an English jurist and mathematician.

28 The Feuerbach Circle

In every triangle the three midpoints of the sides, the three base points of the altitudes, and the midpoints of the three altitude sections touching the vertexes lie on a circle.

This circle was already known to Euler (1765), but is most commonly called the *Feuerbach circle* after Karl Feuerbach (1800–1834) [the uncle of the painter Anselm Feuerbach], who rediscovered it in 1822. It is also known as the *nine-point circle*, although it passes through many other significant points as well as those indicated above.

The proof consists of two steps. In the first we demonstrate that the circle circumscribing the triangle of the three midpoints of the sides passes through the base points of the altitudes; and in the second we show that the circle circumscribing the triangle of the altitude base points passes through the midpoints of altitude sections.

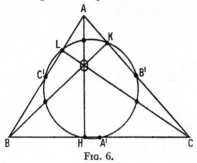

Fig. 6.

I. Let *ABC* represent the prescribed triangle, *A'*, *B'*, *C'* the midpoints, respectively, of sides *BC, CA, AB*. Let *H* be the base point of the altitude *AH*. Then the trapezoid *HA'B'C'* is *isosceles* (*A'B'*, as a midline of the triangle *ABC*, is equal to $\frac{1}{2}AB$; *HC'*, as the radius of the Thales circle having the diameter *AB*, is also equal to $\frac{1}{2}AB$.) The trapezoid is therefore a quadrilateral inscribed in a circle. All of the altitude base points consequently lie on the circle $\mathfrak{F}$ circumscribing the triangle *A'B'C'*.

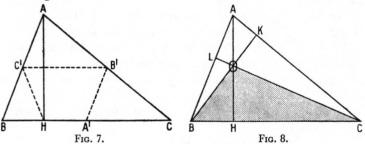

Fig. 7. Fig. 8.

II. Let the altitudes of the triangle *ABC* be *AH, BK, CL*, and *O* their point of intersection. We will now show that the center of each altitude section touching a vertex, let us say section *OC*, also lies on $\mathfrak{F}$. For this purpose we consider the triangle *OBC*, which also has the altitude bases *H, K, L*. According to I., the circle $\mathfrak{F}$ circumscribing the altitude base triangle (*HKL*) of this triangle passes

through the triangle at the side midpoints, e.g., through the center of
OB and *OC*, which completes the proof.

COROLLARY. *The midpoint F of the Feuerbach circle lies at the center of the
Euler line* OU, *and the radius* f *of the Feuerbach circle is equal to one half the
radius of the circle of circumscription of the triangle* ABC.

The first of these propositions follows from the fact that the perpendicular bisectors of the Feuerbach circle chords *HA'* and *KB'*, as
midlines of the trapezoids *UOHA'* and *UOKB'*, pass through the
center of *OU*, and the second, from the fact that the sides of the
triangle *A'B'C'* inscribed in the Feuerbach circle are one half the size
of the sides of the triangle *ABC*.

29 Castillon's Problem

*To inscribe in a given circle a triangle the sides of which pass through three
given points.*

This problem, posed by the Swiss mathematician Cramer, takes its
name from the Italian mathematician Castillon, who solved it in 1776.
(Gabriel Cramer, 1704–1752, in 1750 published his major work
Introduction à l'analyse des lignes courbes algébraiques, in which for the first
time, a system of linear equations was solved by means of determinants.
I. F. Salvemini, 1709–1791, took the name Castillon after his place of
birth Castiglione in Tuscany.)

The following simple, though not easily seen, solution of the
Castillon problem stems from the Italian Giordano.

We call the given circle $\mathfrak{K}$, the given points *A*, *B*, *C*, the desired
triangle *XYZ*, and let *YZ*, *ZX*, *XY* pass, respectively, through *A*, *B*, *C*.

Ottaiano in his solution makes use of four *auxiliary points*. These
are:

    I. the end point of the chord parallel to *AB* and beginning
      from *X*;

    II. the point of intersection of the lines *Y*I and *AB*;

    III. the end point of the chord beginning at *X* that is parallel to
      II*C*;

    IV. the point of intersection of the lines *C*II and I III.

The construction consists of the following five steps.

1. CONSTRUCTION OF AUXILIARY POINT II. The angles *A*II I and
*X*I*Y*, as alternate interior angles between parallels, are equal, and the

angles XZY and XIY are equal because they are inscribed in the same arc XY. Consequently,

$$\angle XZY = \angle AII\,I$$

and therefore $BZYII$ is a quadrilateral inscribed in a circle. It also follows from this that

$$AII \cdot AB = AY \cdot AZ.$$

Since, however, the right side of this equation is known to be the power P of the circle $\Re$ at A (see p. 152), it follows that

$$AII = P/AB$$

can be constructed, as a result of which II is known.

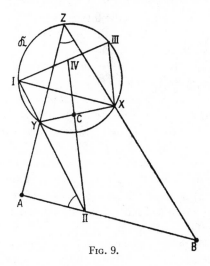

Fig. 9.

2. Construction of auxiliary point IV. The angles $YCIV$ and $YXIII$ are corresponding angles between parallels and are consequently equal, while angles $YI\,III$ and $YXIII$ are supplementary since they are opposite angles in the quadrilateral inscribed in the circle. Thus, $YI\,III$ and $YCIV$ are also supplementary, and $YCIV\,I$ is a quadrilateral inscribed in a circle. It follows from this that

$$IIC \cdot II\,IV = IIIY \cdot II\,I.$$

However, since the right side of this equation represents the power Π of circle $\mathfrak{K}$ at II, which, according to 1., is to be regarded as known, we find

$$\text{II IV} = \Pi/\text{II}C$$

and thus the auxiliary point IV.

3. DETERMINATION OF THE ANGLE IXIII $= \omega$. Since angle AII IV $= \kappa$ is known and since ω and κ, having pairwise parallel sides, are identical, it follows that

$$\omega = \kappa.$$

4. CONSTRUCTION OF THE CHORD I III. We draw through IV a chord subtending the angle $\omega = \kappa$. The points of intersection of this chord with $\mathfrak{K}$ are the remaining points I and III.

5. CONSTRUCTION OF THE TRIANGLE XYZ. We determine X as the point of intersection of $\mathfrak{K}$ with the line through III parallel to II IV; Y as the point of intersection of the line I II with $\mathfrak{K}$; and Z as the point of intersection of the line AY with $\mathfrak{K}$.

In comparison to this fairly intricate solution the following *projective solution* of the Castillon problem is very simple.

This solution is based upon Steiner's double element construction (No. 60) and the involution theorem: *If a ray is rotated about a fixed point, its two points of intersection with a circle describe on this circle* (involutional) *projective ranges of points* (No. 63).

We take any arbitrary point X_1 on the given circle $\mathfrak{K}$, determine the (second) point of intersection Z_1 of the circle with the secant BX_1, then the (second) point of intersection Y_1 of the circle with the secant AZ_1, and, finally, the (second) point of intersection X_1' of the circle with the secant CY_1. Only when X_1' happens to coincide with X_1 is $X_1Y_1Z_1$ the sought-for triangle. This favorable situation will, however, occur only rarely. We will consider the described construction as repeated with other starting points $X_2, X_3, \ldots$, giving us the points $Y_2, Y_3, \ldots, Z_2, Z_3, \ldots; X_2', X_3', \ldots$. According to the auxiliary theorem each of the fields of points $X_1, X_2, \ldots; Y_1, Y_2, \ldots; Z_1, Z_2, \ldots$, and X_1', X_2' is projective with respect to the following one; consequently,

$$(X_1, X_2, \ldots) \, \overline{\wedge} \, (X_1', X_2', \ldots).$$

The desired triangle is obtained from the described construction when the starting point X_ν coincides with the end point X_ν' and is accordingly determined by a double element of this projection. This gives us the following simple

CONSTRUCTION: We choose any three points X_1, X_2, X_3 on $\mathfrak{K}$, draw in the manner described the three corresponding points X'_1, X'_2, X'_3, and determine according to Steiner the double elements X_r and X_s of the projection on $\mathfrak{K}$ in which the points X'_1, X'_2, X'_3 correspond to X_1, X_2, X_3. Thus, each of the two triangles $X_rY_rZ_r$ and $X_sY_sZ_s$ satisfies the conditions of the Castillon problem.

NOTE. In a quite similar manner we are able to prove the *converse of the Castillon problem*:

To draw about a circle a triangle the angles of which lie on three given lines. The construction is based upon the auxiliary theorem:

If a point describes a straight line, the two tangents from the point to a circle determine upon this circle two (involutional) projective fields of tangents (No. 63).

We call the given circle $\mathfrak{K}$, the given lines a, b, c, the sides of the desired triangle x, y, z.

We draw any three tangents x_1, x_2, x_3 to $\mathfrak{K}$; through their points of intersection with b we draw three more tangents z_1, z_2, z_3; through the points of intersection of the latter with a we draw three new tangents y_1, y_2, y_3, and through their intersections with c three more tangents x'_1, x'_2, x'_3. We draw the double elements x_r and x_s of the projection defined on $\mathfrak{K}$ by the homologous triplets (x_1, x_2, x_3) and (x'_1, x'_2, x'_3). The triangles $x_ry_rz_r$ and $x_sy_sz_s$ obtained from these double elements are the ones we are seeking.

30   Malfatti's Problem

To draw within a given triangle three circles each of which is tangent to the other two and to two sides of the triangle.

This famous problem was posed by the Italian mathematician Malfatti (1731–1807) in 1803 and solved in the tenth volume of the *Memorie di Matematica e di Fisica della Società italiana delle Scienze.* This algebraic-geometric solution can be found, for example, in vol. 123 of Ostwald's *Klassiker der exakten Wissenschaften* (Supplement). The purely geometric solution of Malfatti's problem submitted by Jakob Steiner in 1826 without proof is also described and proved there. Here we will restrict ourselves to the exposition of the thoroughly simple solution published by Schellbach in volume 45 of *Crelle's Journal.*

Let ABC be the given triangle with sides a, b, c, the perimeter $2s$ and the angles α, β, γ. Let the Malfatti circles we are seeking (which are tangent to the arms of the angles α, β, γ) be $\mathfrak{P}$, $\mathfrak{Q}$, $\mathfrak{R}$, their midpoints P, Q, R, and their radii p, q, r. Let the tangents from the angles A, B, C to $\mathfrak{P}$, $\mathfrak{Q}$, $\mathfrak{R}$ be u, v, w.

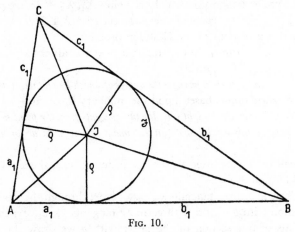

FIG. 10.

We introduce $\mathfrak{J}$, a circle inscribed in the triangle. Let its center be J and its radius ρ, and let the tangents to it from angles A, B, C be a_1, b_1, c_1, respectively. From the three equations

$$b_1 + c_1 = a, \qquad c_1 + a_1 = b, \qquad a_1 + b_1 = c$$

we obtain the values

$$a_1 = s - a, \qquad b_1 = s - b, \qquad c_1 = s - c.$$

Since the points P and J lie on the bisector of the angle α, it follows from the ray theorem that

$$p/\rho = u/a_1 \quad \text{or} \quad p = \frac{\rho}{a_1} u.$$

Similarly we find $q = \dfrac{\rho}{b_1} v$.

We call the points of tangency of $\mathfrak{P}$ and $\mathfrak{Q}$ with AB, U and V and calculate $UV = t$. Since PF, the perpendicular dropped from P to QV, is equal to t, it follows from the right triangle PQF that

$$PQ^2 = PF^2 + FQ^2 \quad \text{or} \quad (p + q)^2 = t^2 + (p - q)^2$$

and from this

$$UV = t = 2\sqrt{pq}.$$

If we then introduce here the values found above for p and q, we obtain

$$t = 2\sqrt{uv}\,\sqrt{\frac{\rho^2}{a_1 b_1}}.$$

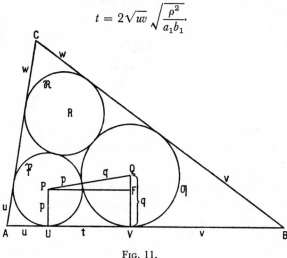

FIG. 11.

But it is known that

$$\rho^2 = a_1 b_1 c_1 / s.$$

This simplifies the value for t to

$$UV = t = 2\sqrt{\frac{c_1}{s}}\,\sqrt{uv}.$$

Since the side AB of the triangle is composed of the three segments AU, BV, and UV, we obtain the equation

$$u + v + 2\sqrt{\frac{c_1}{s}}\,\sqrt{uv} = c.$$

In the same way we obtain for the two other sides of the triangle BC and CA

$$v + w + 2\sqrt{\frac{a_1}{s}}\,\sqrt{vw} = a$$

and

$$w + u + 2\sqrt{\frac{b_1}{s}}\,\sqrt{wu} = b.$$

Taking half the perimeter as the unit length, we obtain somewhat more simply:

(1)
$$\begin{cases} v + w + 2\sqrt{a_1}\sqrt{vw} = a, \\ w + u + 2\sqrt{b_1}\sqrt{wu} = b, \\ u + v + 2\sqrt{c_1}\sqrt{uv} = c. \end{cases}$$

Now we take the proper fractions a, b, c, u, v, w as squares of the sines of six acute angles λ, μ, ν, ψ, φ, χ:

$$\sin^2 \lambda = a, \qquad \sin^2 \mu = b, \qquad \sin^2 \nu = c,$$
$$\sin^2 \psi = u, \qquad \sin^2 \varphi = v, \qquad \sin^2 \chi = w.$$

Then also (since $a + a_1 = s = 1$, $b + b_1 = 1$, $c + c_1 = 1$) $\cos^2 \lambda = a_1$, $\cos^2 \mu = b_1$, $\cos^2 \nu = c_1$, and the obtained equation triplet (1) assumes the form:

(2)
$$\begin{cases} \sin^2 \varphi + \sin^2 \chi + 2 \sin \varphi \sin \chi \cos \lambda = \sin^2 \lambda, \\ \sin^2 \chi + \sin^2 \psi + 2 \sin \chi \sin \psi \cos \mu = \sin^2 \mu, \\ \sin^2 \psi + \sin^2 \varphi + 2 \sin \psi \sin \varphi \cos \nu = \sin^2 \nu. \end{cases}$$

Now, for example, let us consider the first of these equations! It is nothing other than a trigonometric expression of the known relation $(\varphi + \chi = \lambda)$ between the angles φ and χ of the two vertexes of a triangle and the exterior angle λ of the third vertex. If, for example, we take such a triangle with a circle of circumscription of the diameter 1, then the three sides are $\sin \varphi$, $\sin \chi$, $\sin \lambda$, and the cosine theorem gives the equation

$$\sin^2 \lambda = \sin^2 \varphi + \sin^2 \chi + 2 \sin \varphi \sin \chi \cos \lambda.$$

It then follows from (2) that

$$\varphi + \chi = \lambda, \qquad \chi + \psi = \mu, \qquad \psi + \varphi = \nu$$

and from this

$$\psi = \sigma - \lambda, \qquad \varphi = \sigma - \mu, \qquad \chi = \sigma - \nu, \quad \text{with} \quad \sigma = \frac{\lambda + \mu + \nu}{2}.$$

Thus, we obtain the following simple

CONSTRUCTION:

1. We draw three angles λ, μ, ν whose sine squares are equal to the sides of the given triangle (where half the perimeter of the triangle is the unit length).

2. We draw the half sum

$$\sigma = \frac{\lambda + \mu + \nu}{2}$$

of the three angles λ, μ, ν and the three new angles

$$\psi = \sigma - \lambda, \qquad \varphi = \sigma - \mu, \qquad \chi = \sigma - \nu.$$

3. We draw the sine squares of the three angles ψ, φ, χ. These are the tangents from the triangle vertexes to the three Malfatti circles.

NOTE. If we are to draw the sine square $m = \sin^2 w$ for a given angle w, or to draw the angle w (whose sine square equals m) for a given segment m, we proceed in the following manner:

We draw a semicircle $\mathfrak{H}$ with the diameter $HK = 1$. We draw the given angle w at K on KH and from the intersection L of its free side with $\mathfrak{H}$ we drop the perpendicular LM to HK. Then $HM = m = \sin^2 w$.

Conversely, if m is given and we have to find w, we draw $HM = m$ on HK, erect at M a perpendicular on HK extending to the intersection L with $\mathfrak{H}$, and extend LK. Then $\angle HKL = w$.

PROOF. From the right triangle HML it follows that

$$m = HM = HL \cdot \sin HLM = HL \sin w,$$

and from the right triangle HKL

$$HL = HK \sin w = \sin w.$$

Consequently,

$$m = \sin^2 w.$$

31   Monge's Problem

To draw a circle that cuts three given circles perpendicularly.

The French mathematician Monge (1746–1818) was the founder of descriptive geometry.

In order to solve the problem, we seek the *locus of the centers of all the circles that are perpendicular to two given circles*.

[Two circles are said to intersect perpendicularly when the radii r and r' drawn to a single point of intersection are perpendicular to each other; in other words, when they form the base and altitude of a right triangle the hypotenuse z of which joins the centers of the circles, so that $r^2 + r'^2 = z^2$ or $z^2 - r^2 = r'^2$. Two circles are therefore perpendicular to each other when the power* of the one at the midpoint of the other is equal to the square of the radius of the other.]

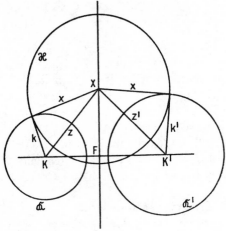

Fig. 12.

Let the given circles be $\mathfrak{K}$ and $\mathfrak{K}'$, their centers K and K', their radii k and k' ($> k$), the line joining their centers $KK' = l$. Let the circle $\mathfrak{X}$ with the midpoint X and the radius x be perpendicular to them. Let the center lines KX and $K'X$ be equal to z and z', respectively. Then $z^2 - k^2$ and $z'^2 - k'^2$ are each equal to x^2, so that

(1) $$z^2 - k^2 = z'^2 - k'^2.$$

Consequently, both circles $\mathfrak{K}$ and $\mathfrak{K}'$ have the *same power* at X. We therefore first attempt to find the locus of the point X at which the two given circles possess the same power. If X is a point possessing this locus and the perpendicular from X intercepts the center line KK' at the point F, and, moreover, if $KF = f$ and $K'F = f'$, then, according to the Pythagorean theorem, the square of the perpendicular is equal to $z^2 - f^2$ as well as to $z'^2 - f'^2$, so that

(2) $$z^2 - f^2 = z'^2 - f'^2.$$

---

* By the power of a circle at a point is meant the amount by which the square of the axis to the point exceeds the square of the radius of the circle. In accordance with the secant or chord theorem it can also be represented as the product of the two segments originating from the point that are generated by the circle through the point on any secant.

If we subtract (2) from (1) we obtain

$$(3) \qquad f^2 - k^2 = f'^2 - k'^2,$$

i.e., $\mathfrak{K}$ and $\mathfrak{K}'$ possess equal powers at F also.   If we figure the distances f and f' as positive in the directions KK' and $K'K$, respectively, then it is always true that

$$(4) \qquad f + f' = l.$$

Equations (3) and (4) give us fixed values for the unknowns f and f'. Consequently every locus point X lies on the perpendicular erected on the center line KK' at the fixed point F, and we obtain the

THEOREM OF THE CHORDAL: *The locus of the point at which two given circles possess the same powers is a straight line perpendicular to the line joining the midpoints of the circles and is known as the chordal or power line of the two circles.*

In the *construction of the chordal* we distinguish two different cases:

1. *The circles intersect.*   Since both circles have equal powers at each of their points of intersection, i.e., O, the points of intersection lie on the chordal.   *The chordal of two circles that intersect is the secant of intersection.*

2. *The circles do not intersect.*   Here the construction of the chordal is based upon the

THEOREM OF MONGE: *The three chordals of three circles pass through a point known as the power center of the three circles.*

[PROOF.   Let the circles be I, II, III.   We determine the point of intersection O of the chordals of the two pairs (II, III) and (III, I). At this point (1) II and III, (2) III and I possess equal powers; consequently II and I also have the same power at O, i.e., O lies on the chordal of I and II.]

Thus, to construct the chordal of two nonintersecting circles I and II, we draw an auxiliary circle III that intersects I and II and the chordals of the pairs (II, III) and (III, I).   The perpendicular from the intersection of these chordals to the line joining the centers of I and II is the chordal we are looking for.

From the theorem of the chordal it then follows:

The locus of the centers of all circles that are perpendicular to two given circles is the chordal of the given circles or, in the event that these circles intersect, the section of the chordal that lies outside the given circles. (The powers of the given circles at a single point must be positive!)

The *solution of Monge's problem* now becomes very simple.   We draw the power center O of the given circles.   If it lies outside the three circles, the circle with the midpoint O and the radius formed by the tangent from O to one of the given circles intersects perpendicularly with the given circles. If O is located inside even one of the given circles, the problem is insoluble.

32   The Tangency Problem of Apollonius

To draw a circle that is tangent to three given circles.

The circles may also comprise degenerate circles: points or straight lines.

This celebrated problem was put forth by the greatest mathematician of the ancient world after Euclid and Archimedes, Apollonius of Perga (ca. 260–170 B.C.), whose major work Κωνικά extended with an astonishing comprehensiveness the period's naturally slight knowledge of conic sections. His treatise *De Tactionibus*, which contained the solution of the tangency problem given above, has unfortunately been lost. François Viète, called Vieta, the greatest French mathematician of the sixteenth century (1540–1603), attempted about 1600 to restore the lost treatise of Apollonius and solved the tangency problem by treating each of its ten special cases individually, deriving each successive one from the preceding one. In contrast to this the solutions of Gauss (*Complete Works*, vol. IV, p. 399), Gergonne (*Annales de Mathématiques*, vol. IV), and Petersen (*Methoden und Theorien*) solve the general problem.

Here we will restrict ourselves to the exposition of the elegant *solution of Gergonne*. Since this proof presupposes, in addition to the chordal theorems proved in No. 31, a knowledge of the properties of similarity points and polars, we will begin with a brief discussion of these.

SIMILARITY POINTS

When we refer to the *external* or positive and *internal* or negative *similarity points*, respectively, of two circles $\Re$ and $\Re'$ with the centers M and M' and the radii r and r', we mean the points A and J, respectively, on the line MM' joining the centers for which

$$\frac{MA}{M'A} = +\frac{r}{r'} \quad \text{and} \quad \frac{MJ}{M'J} = -\frac{r}{r'}, \quad \text{respectively.}*$$

---

* The segment ratio $AX:BX$ is considered positive if X is situated outside AB and negative if X is inside AB.

It follows directly from the ray theorem that:

The line connecting the end points of two parallel (oppositely directed) radii of two circles passes through the external (internal) similarity point.

In particular, the external (internal) common tangents of the two circles pass through the external (internal) similarity point. We will further designate the external similarity point of the circles $\Re$ and $\Re'$ as $+\Re\Re'$, the internal one as $-\Re\Re'$, and, if the sign is not determined, we will indicate the similarity point as $\varepsilon\Re\Re'$. The symbol $\varepsilon\varepsilon'\varepsilon''\ldots$ is to be understood as meaning plus when the number of minus signs occurring among the symbols $\varepsilon,\varepsilon',\varepsilon'',\ldots$ is even and minus when it is odd.

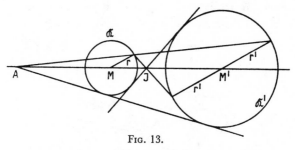

FIG. 13.

The similarity points of *three* circles are described by the

THEOREM OF D'ALEMBERT:* *If three circles $\mathfrak{A}$, $\mathfrak{B}$, $\mathfrak{C}$ are taken in pairs $(\mathfrak{B}, \mathfrak{C})$, $(\mathfrak{C}, \mathfrak{A})$, and $(\mathfrak{A}, \mathfrak{B})$, the external similarity points of the three pairs lie on a straight line; and, similarly, the external similarity point of one pair and the two internal similarity points of the other two pairs lie upon a straight line, a so-called similarity axis of the three circles.* More briefly:

If $\alpha\beta\gamma$ is plus, the three similarity points $\alpha\mathfrak{B}\mathfrak{C}$, $\beta\mathfrak{C}\mathfrak{A}$, and $\gamma\mathfrak{A}\mathfrak{B}$ lie on a straight line.

MONGE'S PROOF. Let the centers of the circles $\mathfrak{A}$, $\mathfrak{B}$, $\mathfrak{C}$ be A, B, C, and let the external similarity points of the pairs $(\mathfrak{B}, \mathfrak{C})$, $(\mathfrak{C}, \mathfrak{A})$, $(\mathfrak{A}, \mathfrak{B})$ be P, Q, R. If the circle pair $(\mathfrak{B}, \mathfrak{C})$ with its external tangents that pass through P is rotated about the axis PBC, we obtain the spheres $\mathfrak{B}_0$ and $\mathfrak{C}_0$ and their tangent cone with apex P. The case is similar for the other two circle pairs.

The planes E_1 and E_2 are tangent to the spheres $\mathfrak{A}_0$, $\mathfrak{B}_0$, $\mathfrak{C}_0$ in such a manner that the spheres always lie on one side of the plane, and both planes contain the point P, since this point lies on the external

---

* D'Alembert (1717–1783), a French mathematician.

tangent of $(\mathfrak{B}_0, \mathfrak{C}_0)$ within $E_1[E_2]$. They likewise contain the points Q and R.

The three points P, Q, R thus lie on the *line of intersection* of the planes E_1 and E_2.

If we are concerned with the internal similarity points of the pairs $(\mathfrak{B}, \mathfrak{C})$ and $(\mathfrak{A}, \mathfrak{C})$ and the external similarity point of $(\mathfrak{A}, \mathfrak{B})$, we must take the tangential planes so that $\mathfrak{A}_0$ and $\mathfrak{B}_0$ lie on one side of such a plane while $\mathfrak{C}_0$ lies on the other.

Let an arbitrary circle $\mathfrak{X}$ with the center X be homogeneously (nonhomogeneously) tangent to two fixed circles $\mathfrak{K}$ and $\mathfrak{K}'$, with centers K and K' and radii k and k', at P and Q'. Let the points of intersection of the straight line PQ' with the circles $\mathfrak{K}$ and $\mathfrak{K}'$ and the line KK' joining their centers be P, Q; P', Q', and S.

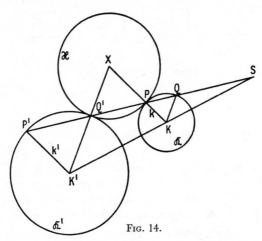

Fig. 14.

Since the base angles of the isosceles triangles KPQ, $K'P'Q'$, and XPQ' are also the opposite and coincident angles at P and Q', all six base angles are equal. Since the two base angles at P and P' are equal, the radii KP and $K'P'$ are parallel. Consequently, S is the external (internal) similarity point of $\mathfrak{K}$ and $\mathfrak{K}'$. From this it follows that

$$\frac{SP}{SP'} = \pm\frac{k}{k'}, \qquad \frac{SQ}{SQ'} = \pm\frac{k}{k'},$$

so that the two products $SP \cdot SQ'$ and $SQ \cdot SP'$ are equal. If we call their common value w, then

$$w^2 = SP \cdot SQ' \cdot SQ \cdot SP' = SP \cdot SQ \cdot SP' \cdot SQ',$$

i.e., w^2 is equal to the product of the powers Π and Π' of the two circles $\Re$ and $\Re'$ at S. Consequently,

$$SP \cdot SQ' = w = \sqrt{\Pi\Pi'}.$$

I.e.: The power $(SP \cdot SQ')$ of the circle $\mathfrak{X}$ at S is a constant $(\sqrt{\Pi\Pi'})$.

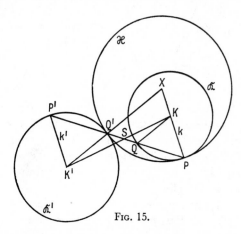

Fig. 15.

The result of our considerations is the following

TANGENCY THEOREM: *The external (internal) similarity point of two fixed circles is the point at which all the circles homogeneously (nonhomogeneously) tangent to the fixed circles have the same power and at which all the tangency secants* (which are determined by the points of tangency to the fixed circles) *intersect.*

POLE AND POLAR

Two points P and P' that lie on a ray originating at the center O of a circle $\Re$ with radius r in such manner that

$$OP \cdot OP' = r^2$$

are called conjugate with respect to each other in relation to the circle. Of two conjugate points one lies inside the circle and the other outside.

The conjugate of an external point A is the point of intersection J of the circle bisector from A with the tangency chord determined by the tangents AT_1 and AT_2 from A to the circle.

The conjugate of an internal point J is the point of intersection A of the tangents that pass through the end points T_1 and T_2 of the chord passing through J and perpendicular to the circle bisector from J.

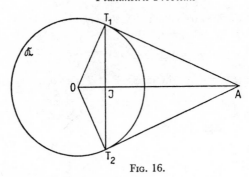

FIG. 16.

(From the right triangle OAT_1 it follows directly that $r^2 = OA \cdot OJ$.)

By the *polar of the point P* we mean the *line p* that is perpendicular to the circle bisector from P and passes through the conjugate of P.

Conversely, by the *pole of the line p* we mean the *point P* that is conjugate to the base point of the perpendicular dropped from the center of the circle to the line.

The relation between the pole and the polar is therefore reciprocal: *If* p *is the polar of* P, *then* P *is the pole of* p, and conversely.

Now let Q be any point on the polar p of P (that passes through the conjugate P' of P) and let Q' be the conjugate of Q. Then

$$OP \cdot OP' = OQ \cdot OQ' \ (= r^2),$$

and consequently $PP'QQ'$ is a quadrilateral inscribed in a circle. Since here the angle at P' is 90° the angle at Q' must also be 90°, i.e.,

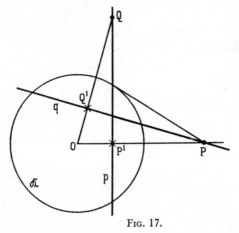

FIG. 17.

PQ' must be perpendicular to OQ. PQ' is therefore the polar q of Q, and we have the

THEOREM OF THE POLE AND POLAR: *If* Q *lies on the polar of* P, P *also lies on the polar of* Q. Or also: *If* p *passes through the pole of* q, q *also passes through the pole of* p.

Now for *Gergonne's solution of the tangency problem.*

In general, there are a number of circles that are tangent to three given circles $\mathfrak{A}$, $\mathfrak{B}$, $\mathfrak{C}$. Gergonne's solution is based upon the device of seeking the unknown circles in *pairs* rather than individually; in particular, one always seeks that pair $(\mathfrak{X}, \mathfrak{x})$ that is homogeneously or nonhomogeneously tangent to each of the given circles.

For the sake of convenience, we will call homogeneous tangencies positive $(+)$ and nonhomogeneous tangencies negative $(-)$ and combinations such as $\varepsilon\varepsilon'$ of the tangency signs ε and ε' will be treated in accordance with the rule that "like signs give plus and unlike minus."

Let the circles $\mathfrak{X}$ and $\mathfrak{x}$, respectively, be tangent to the circles $\mathfrak{A}$, $\mathfrak{B}$, $\mathfrak{C}$ at the points P, Q, R and p, q, r, respectively, and let the tangencies possess the signs A, B, C and a, b, c, respectively. Then

$$Aa = Bb = Cc = \varepsilon,$$

and

$$BC = bc = \alpha, \qquad CA = ca = \beta, \qquad AB = ab = \gamma$$

and

$$\alpha\beta\gamma = +.$$

Let us first consider $(\mathfrak{X}, \mathfrak{x})$ as the pair tangent to the circles $\mathfrak{A}$, $\mathfrak{B}$, $\mathfrak{C}$. According to the tangency theorem, *the similarity point* $\varepsilon\mathfrak{X}\mathfrak{x}$ *of* $\mathfrak{X}$ *and* $\mathfrak{x}$ *is the power center* O *of the three circles* $\mathfrak{A}$, $\mathfrak{B}$, $\mathfrak{C}$ *and the point of intersection of the three tangency chords* Pp, Qq, Rr.

We then take in succession $(\mathfrak{B}, \mathfrak{C})$, $(\mathfrak{C}, \mathfrak{A})$, $(\mathfrak{A}, \mathfrak{B})$ as the pair tangent to the circles $\mathfrak{X}$ and $\mathfrak{x}$. In accordance with the tangency theorem, the circles $\mathfrak{X}$ and $\mathfrak{x}$ then have the same powers at the similarity point $\alpha\mathfrak{B}\mathfrak{C} \equiv \mathrm{I}$, as well as at the similarity point $\beta\mathfrak{C}\mathfrak{A} \equiv \mathrm{II}$, and the similarity point $\gamma\mathfrak{A}\mathfrak{B} \equiv \mathrm{III}$. And since $\alpha\beta\gamma$ is $+$, the three points I, II, III, in accordance with d'Alembert's theorem, lie upon a similarity axis of $\mathfrak{A}$, $\mathfrak{B}$, $\mathfrak{C}$. *The similarity axis* I II III *is thus the chordal* χ *of the circles* $\mathfrak{X}$ *and* $\mathfrak{x}$.

Further, if S represents the point of intersection of the tangents to $\mathfrak{A}$ at P and p, then $SP = Sp$. Since these tangents also touch $\mathfrak{X}$ and $\mathfrak{x}$, S lies on the chordal χ of $\mathfrak{X}$ and $\mathfrak{x}$. Now S is also the pole of the tangency chord *Pp* with respect to circle 𝔄. Since χ therefore passes through the pole of *Pp*, it follows from the theorem of the pole and polar that *Pp* passes through the pole of χ. Since the same conclusions can be drawn with respect to the tangency chords *Qq* and *Rr*, we obtain the theorem: *The tangency chords* Pp, Qq, *and* Rr *pass respectively through the poles of the line* χ ≡ I II III *with respect to the circles* 𝔄, 𝔅, ℭ.

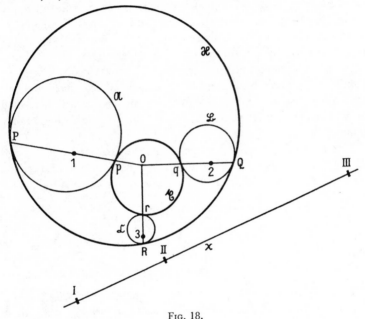

Fɪɢ. 18.

From the three theorems italicized in the last three paragraphs we obtain directly

Gᴇʀɢᴏɴɴᴇ's ᴄᴏɴsᴛʀᴜᴄᴛɪᴏɴ: *Draw the power center* O *of the given circles and the similarity axis* I II III ≡ χ. *Determine the poles* 1, 2, 3 *of* χ *in relation to the given circles and connect them with* O. *The connecting lines touch the given circles at the points at which they are tangent to the sought-for circles.*

Mascheroni's Compass Problem

To prove that any construction that can be carried out with a compass and straight-edge can be carried out with the compass alone.

The Italian L. Mascheroni (1750–1800) posed himself the problem of executing the geometric constructions with a compass alone (without the use of the straight-edge) and solved it in a masterly fashion in his book *La geometria del compasso*, which was published in Pavia in 1797.

If we examine the separate steps by which the circle and straight-edge constructions are carried out, we see that every step consists of one of the following three basic constructions:

   I. *Finding the point of intersection of two straight lines;*

   II. *finding the point of intersection of a straight line and a circle;*

   III. *finding the point of intersection of two circles.*

Consequently, we need only show that the two basic constructions I. and II. can be accomplished with a compass alone. (In Mascheroni's geometry of the compass a straight line is, naturally, regarded as given or determined if two of its points are known.)

First we must solve two preliminary problems.

PRELIMINARY PROBLEM 1.   *To draw the sum or difference of two given segments* a *and* b.

In other words: to lengthen or shorten a given segment $PQ = a$ by a segment $QX = b$.

SOLUTION.   1. We draw the arc $Q|b$,* take upon this arc any point H, draw the mirror image H' of H (the mirror image O' of a point O on a straight line AB is the point of intersection of the arcs $A|AO$ and $B|BO$) on the straight line $\mathfrak{g}$ determined by the points P and Q, and designate the segment HH' as h.   2. We draw the isosceles trapezoid $KHH'K'$ whose legs KH and $K'H'$ are equal to b and whose base $KK' = 2h$. (K is the point of intersection of the arcs $Q|h$ and $H|b$, K' is the mirror image of K on $\mathfrak{g}$.) Let the diagonal $KH' = HK'$ of the trapezoid be called d. Since the trapezoid is a quadrilateral that can be inscribed in a circle, according to Ptolemy the following equation is applicable:

$$d^2 = b^2 + 2h^2.$$

On the other hand, it follows from the right triangle $QK'X$, where $K'X$ will be designated as x, that

$$x^2 = b^2 + h^2.$$

---

* Let arc $Q|b$ mean the circle arc whose midpoint is Q and radius b.

From these two equations it follows that

$$d^2 = x^2 + h^2,$$

so that x is one of the legs of a right triangle with the hypotenuse d and the other leg h.　If we then find the point of intersection S of the

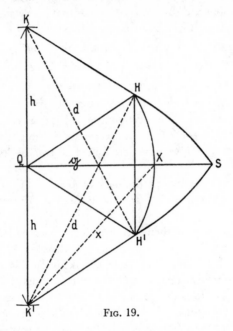

Fig. 19.

arcs $K|d$ and $K'|d$ on the straight line $\mathfrak{g}$, $QS = x$.　3. We draw the point of intersection of the arcs $K|x$ and $K'|x$; this is the point X that we have been trying to find.

PRELIMINARY PROBLEM 2.　*To find the fourth segment* x *that is in proportion to the three given segments* m, n, s.

In other words, draw the segment

$$x = \frac{n}{m}\, s.$$

The following solution that Mascheroni found for this fundamental problem is remarkable for its shortness and simplicity.

We draw two concentric circles $\mathfrak{M} \equiv Z|m$ and $\mathfrak{N} \equiv Z|n$, draw the chord $AB = s$ in $\mathfrak{M}$, lay off with the compass any length w from A

and from B on $\mathfrak{N}$, obtaining from the distance between the resulting points of intersection H and K the sought-for segment x. The proof follows directly from the similar triangles ZAB and ZHK.

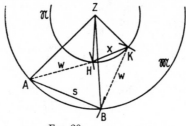

Fig. 20.

In this construction it is assumed that s falls within circle $\mathfrak{M}$. If this is not the case, we first transform the fraction n/m into N/M, where N and M, respectively, are sufficiently great integral multiples of n and m which can be drawn according to the first preliminary problem. (A comparatively simple method is the doubling that results, for example, when $PQ = m$, and the radius m of the circle $P|PQ$ is laid off three times in succession from Q. The end point after this laying off is separated from Q by the distance $2m$.)

After the solution of the preliminary problems, we go on to the solution of the two major problems.

I'. To find the point of intersection S of two straight lines AB and CD (each of which is given by two points) with the compass alone.

II'. To determine the point of intersection S of a given circle $\mathfrak{K}$ and a given straight line AB with the compass alone.

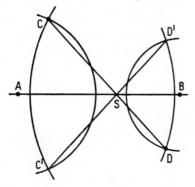

Fig. 21.

SOLUTION OF I'.   We draw the mirror images C' and D' of C and D with respect to AB.   The sought-for point of intersection S then also lies on $C'D'$.   According to the ray theorem, it follows that $CS/SD = CC'/DD'$, i.e., if we designate the segments CS, CD, CC', DD' as x, e, c, d, respectively, $x/(e - x) = c/d$ or

$$x = \frac{c}{c + d} \cdot e.$$

Now we begin by drawing $CH = c + d$ (H as the point of intersection of the arcs $C'|d$ and $D|e$); then we draw the segment x in accordance with preliminary problem 2; and finally we draw the sought-for point of intersection S as the intersection of the arcs $C|x$ and $C'|x$.

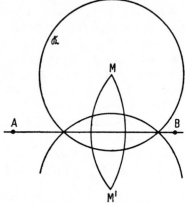

FIG. 22.

SOLUTION OF II'.   Let the center of the given circle be known as M, the radius as r.   We draw the mirror image M' of M with respect to the straight line AB and with the compass open to the radius r we strike off r on the circle $\Re$ from M'.   The resulting points of intersection are the sought-for points of intersection of the given straight line AB with the given circle $\Re$.

The construction cannot be carried out if the straight line AB happens to pass through M.   In this exceptional case we extend and shorten the segment AM by r in accordance with preliminary problem 1.   The end points of the extended and shortened segment are the sought-for points of intersection of $\Re$ and AB.

This completes the solution of Mascheroni's problem.

34 Steiner's Straight-edge Problem

To prove that every construction that can be executed with compass and straight-edge can be executed with a straight-edge alone in the event that within the picture plane there is also given a fixed circle.

As far back as 1759 Lambert had solved a whole series of geometric constructions with straight-edge alone in his book *Freie Perspektive*, which was published in Zurich that year. He is also the source of the term "straight-edge geometry." After Lambert the French mathematicians, primarily Poncelet and Brianchon, took up straight-edge geometry, particularly after the publication of Mascheroni's *Geometria del compasso* provided a new stimulus to these studies, and they attempted to execute as many constructions as possible with the straight-edge alone.

Now, with the use of a straight-edge alone it is possible to represent only those algebraic expressions whose algebraic form is rational (thus, for example, it is impossible to represent expressions such as $\sqrt{ab}$). This circumstance suggested to Poncelet that an additional fixed circle (as well as the center!) must be given inside the picture plane for it to be possible to draw with straight-edge alone all the algebraic expressions that can be constructed with a compass and straight-edge.

This suggestion was confirmed as a certainty by Jakob Steiner (1796–1863), the greatest geometer since the days of Apollonius, in his celebrated book *Die geometrischen Konstruktionen ausgeführt mittels der geraden Linie und Eines festen Kreises* (Geometrical Constructions Executed with a Straight Line and One Fixed Circle), published in Berlin, 1833.

The solution presented here is based upon that in Steiner's book, except that we have here eliminated everything that is not strictly essential for the purpose at hand, and we have also made it somewhat more elementary by dispensing with the theorems of homothety and chordals employed by Steiner.

Since in straight-edge geometry the intersection of two straight lines is known directly, we need only demonstrate that the two fundamental problems II. and III. of the previous section can be solved by means of a straight-edge and a fixed circle alone.

As in the solution of Mascheroni's problem, we must first solve several preliminary problems; in this case there are five rather than two.

Fɪɢ. 23.

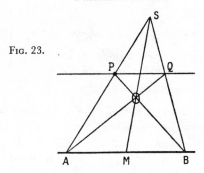

Preliminary problem 1: *To draw through a given point the parallel to a given line.*

Steiner distinguishes two cases: 1*a*. construction of the parallel to a directed straight line; 1*b*. construction of the parallel to an arbitrary straight line.

1*a*. A directed straight line is understood to mean a straight line in which two points A and B and the midpoint M of the segment joining them are known. In order to draw the parallel to such a line through a given point P, we draw AP, choose a point S on the extension of AP, connect this point with B and M, draw BP, and draw the straight line AO through the point of intersection O of BP and MS in such a manner that AO cuts BS at Q. PQ is then the desired parallel. A simple proof.

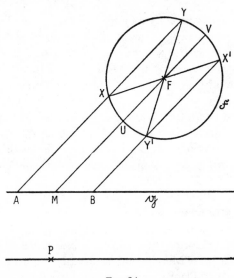

Fɪɢ. 24.

1*b*. We connect a given point M of the given straight line g with the midpoint F of the given fixed circle $\mathfrak{F}$ and designate the points of intersection of the connecting line and $\mathfrak{F}$ as U and V. The points U, F, V make the line FM a directed line. In accordance with 1*a*., we draw a parallel to FM in such a manner that it cuts $\mathfrak{F}$ at X and Y and g at A. If we then draw the diameters XFX' and YFY' and connect the end points X' and Y', the connecting line intersects the given line at a point B in such a manner that $MA = MB$ and g, defined by the three points A, M, B, is then a directed line. This makes it possible to determine the parallel to g in accordance with 1*a*.

Preliminary problem 1 gives us the solution to the problem: *shift a given segment* AB *parallel to itself in such a manner that one of its end points lies on a given point* P.

If P falls *outside* the straight line AB we find the point of intersection Q of the parallel through B to AP and the parallel through P to AB; PQ is then parallel to AB.

PRELIMINARY PROBLEM 2: *Draw a perpendicular through a given point* P *to a given straight line* g.

We draw g$'$ parallel to g in such a manner that it cuts $\mathfrak{F}$ at U and V. We then draw the diameter UFU' and the chord VU', which, according to Thales' theorem, is perpendicular to g$'$ and consequently also perpendicular to g. Finally, we draw the parallel to VU' through P in accordance with 1; this parallel is the desired perpendicular.

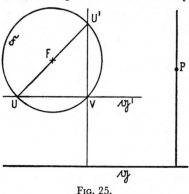

Fig. 25.

PRELIMINARY PROBLEM 3: *To lay off a given distance* PQ *from a given point* O *in a given direction.*

Let us consider the prescribed direction as given by the segment OH from O. First, in accordance with 1., we displace PQ parallel to itself to *OK*. Then from *F* we draw two radii *FU* and *FV* in the directions *OH* and *OK*. Finally, if we draw through *K* the parallel to *UV*, the point of intersection *S* of the parallel with the line *OH* gives the end point of the desired segment.

PRELIMINARY PROBLEM 4: *If three distances* m, n, s *are given, draw the fourth proportional.*

From any point *O* we draw two rays I and II, mark off the two distances $OM = m$ and $ON = n$ on I and the distance $OS = s$ on II; we draw the parallel to *MS* through *N* and designate its point of intersection with II as *X*. Then

$$OX = \frac{n}{m} s$$

is the desired fourth proportional.

PRELIMINARY PROBLEM 5: *If two segments* a *and* b *are given, draw the mean proportional.*

We designate the sought-for mean proportional ($\sqrt{ab}$) as *x*, the diameter of the fixed circle as *d*, the sum $a + b$ that can be constructed according to 3. as *c*, and we write

$$x \frac{c}{d} s, \quad \text{with} \quad s = \sqrt{hk}, \; h = \frac{d}{c} a, \; k = \frac{d}{c} b$$

(so that $h + k = d$).

First, in accordance with 4., we draw the segments *h* and *k*, and in accordance with 3., we make $HO = h$ on a diameter *HK* of the fixed circle, so that *KO* will necessarily equal *k*. Then, according to 2., we construct through *O* the perpendicular to *HK* and call the intersection of the perpendicular with the fixed circle *S*. Then $OS = \sqrt{hk} = s$. Finally, we draw the desired segment $x(= (c/d)s)$ according to 4.

Now that we have solved these five preliminary problems, the solution of the two basic problems II and III is simple.

BASIC PROBLEM II: *To draw the points of intersection of a given line and a given circle.*

In straight-edge geometry a circle is considered determined if its center and radius are known. Let us designate the given circle as $\Re$, its center as *C*, its diameter as *r*, the given straight line as $\mathfrak{g}$, the points of intersection of $\mathfrak{g}$ with circle $\Re$ as *X* and *Y*, the chord of intersection as 2s, the midpoint of the chord as *M*, its distance from the center *C* as *l*. From the right triangle *CMX* we obtain the equation

$$s^2 = r^2 - l^2 \quad \text{or} \quad s = \sqrt{(r + l)(r - l)}.$$

Then, in accordance with 2., we drop the perpendicular $CM = l$ to g; we draw the segments $a = r + l$ and $b = r - l$ in accordance with 3.; then, according to 5., we draw the segment $s = \sqrt{ab}$; and finally, according to 3., we lay off s from M on g in both directions. The end points of the laid-off segments are the desired points of intersection X and Y.

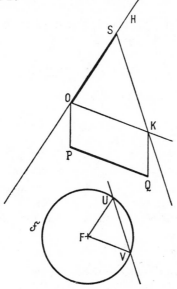

Fig. 26.

BASIC PROBLEM III: *Find the points of intersection of two given circles.*

Let us designate the circles as $\mathfrak{A}$ and $\mathfrak{B}$, their midpoints as A and B, their radii as a and b, the line AB joining their centers as c, the sought-for points of intersection as X and Y, the point of intersection of the chord XY with the center line AB as O, and, finally, the unknown segments AO and OX as q and x.

FINDING q. From the triangle ABX it may be inferred, in accordance with the expanded Pythagorean theorem, $b^2 = c^2 + a^2 - 2cq$; thus, if we set $c^2 + a^2$ equal to d^2,

$$q = \frac{(d + b)(d - b)}{2c}.$$

Consequently, we draw, in accordance with 2. and 3., a right triangle with the short legs a and c and obtain as the hypotenuse d.

Then, according to 3., we draw the segments

$$n = d + b, \qquad m = 2c, \qquad s = d - b$$

and finally, according to 4.,

$$q = \frac{n}{m}\, s.$$

FINDING x.   From $\triangle OAX$ it follows, according to the Pythagorean theorem, that $x^2 = a^2 - q^2$; thus

$$x = \sqrt{(a + q)(a - q)}.$$

According to 3., we draw $h = a + q$, $k = a - q$ and, according to 5.,

$$x = \sqrt{hk}.$$

CONSTRUCTION OF X AND Y.   According to 3., we lay off q from A on AB.   At O, the end of the segment laid off, we erect the perpendicular to AB in accordance with 2. and (according to 3.) we lay off x on it in both directions.   The end points of the laid-off segments are the points of intersection that we are looking for.

35   The Delian Cube-doubling Problem

To construct the edge of a cube that is double the size of a given cube.

The name "Delian problem," according to an account given by the mathematician and historian Eutocius (sixth century A.D.), goes back to an old legend according to which the Delphic oracle in one of its utterances demanded that the Delian altar block be doubled.

If k is the edge of the given cube and x the edge of the cube we are seeking, the respective volumes of the two cubes are k^3 and x^3. Consequently we are confronted with the problem of finding, when the segment k is given, a second segment x such that

$$x^3 = 2k^3.$$

This problem is not capable of solution with compass and straight-edge. (See the Supplement to No. 36.)

The numerous solutions to this problem, some of which were found in antiquity, consequently make use of more advanced means.

Thus, the solution of the Greek mathematician Menaechmus (ca. 375–325 B.C.) is based upon finding the point of intersection of the two parabolas

(1) $\qquad x^2 = ky \qquad$ and $\qquad$ (2) $\qquad y^2 = 2kx$

with the parameters k and $2k$. The abscissa x of the point of intersection satisfies the condition $x^3 = 2k^3$ as a result of the fact that $x^4 = k^2 y^2 = 2k^3 x$, and the sought-for edge x is thereby obtained.

Descartes (1596–1650) showed that one of the two parabolas (1) and (2) was sufficient. For their point of intersection $x|y$ the following equation is also true:

$$x^2 + y^2 = ky + 2kx;$$

and this is the equation of a circle with the midpoint coordinates k and $k/2$ which passes through the common apex of the two parabolas. Thus, it is only necessary to find the intersection of this circle with one of the two parabolas to find the sought-for point of intersection.

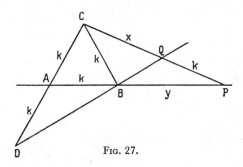

Fig. 27.

The simplest and most accurate method of constructing

$$x = k\sqrt[3]{2}$$

is by *paper strip construction*. 1. We draw an equilateral triangle ABC with the side k, extend CA by $AD = k$, and draw the line DB. 2. We mark off on the sharp edge of a paper strip the distance k. 3. We place the paper strip in such a way that the edge passes through C and the end points of the marked-off distance fall upon two points P and Q of the extensions of AB and DB.

Then

$$CQ = x = k\sqrt[3]{2}.$$

PROOF. Let $CQ = x$, $BP = y$. According to the leg transversal theorem used in figure $CABP$, $(x + k)^2 - k^2 = y(k + y)$ or

(I) $x^2 + 2kx = y^2 + ky.$

According to the theorem applied by Menelaus to the triangle ACP with the transversal DBQ, $AD \cdot CQ \cdot BP = PQ \cdot AB \cdot CD$ or

(II) $xy = 2k^2.$

A glance at equations (I) and (II) shows that they are satisfied by the roots x and y of equations (1) and (2). The unknowns x and y, which are determined by (I) and (II), are therefore at the same time the coordinates of the point of intersection of Menaechmus' parabolas. In particular, $x = k\sqrt[3]{2}$.

Naturally, this result can also be obtained without reference to these parabolas.

NOTE. The doubled cube can also be constructed by means of the so-called *conchoid* of Nicomedes, a Greek mathematician who lived at the beginning of the second century B.C.; we cannot, however, present this construction here.

36   Trisection of an Angle

To divide an angle into three equal angles.

This famous problem cannot be solved with compass and straight-edge (see the supplement).

The simplest solution is by means of the following *paper strip construction* of Archimedes.

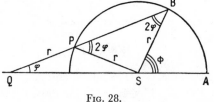

FIG. 28.

Taking as the center the apex S of the angle Φ to be trisected, we draw a circle of radius r that intersects the legs of the angle at A and B. We mark off a segment of length r on the edge of a paper strip. We place the edge on the figure in such a way that it passes through B and that one end point of the marked-off segment coincides with a point P on the circle, while the other end point coincides with a point Q (outside the circle) of the extension of AS. Then $\angle PQS = \varphi$ is one third of the given angle Φ.

PROOF. Since $PS = PQ \ (= r)$, $\triangle PQS$ is isosceles and $\angle PSQ$ is therefore also equal to φ, while the external angle $\angle SPB$ is equal to 2φ. Since $\triangle SPB$ is also isosceles, $\angle SBP = \angle SPB = 2\varphi$. Finally, since the external angle Φ at S of the triangle SBQ is equal to the sum of the two nonadjacent internal angles SQB and SBQ, we find that $\Phi = \varphi + 2\varphi$ or

$$\varphi = \tfrac{1}{3}\Phi. \qquad \text{Q.E.D.}$$

The problem of the trisection of an angle can also be solved by means of a *fixed* hyperbola, as the Greek mathematician Pappus (ca. 300 A.D.) demonstrated in his ingenious masterwork Συναγωγαὶ μαθηματικαί (*Collectiones mathematicae*).

In order to understand the construction we must first solve the problem: *Find the locus of the vertex* P *of a triangle* ABP *with fixed base* AB *when the base angles* α *and* β *are to each other in the proportion of* 2 *to* 1.

Let $AB = 3k$, $AP = u$. We lay off the angle β at P on PB and designate the point of intersection of the free leg with segment AB as Q. The triangles BPQ and APQ are then isosceles ($\angle AQP$ as the external angle of BPQ is equal to $2\beta = \alpha$); consequently, $AP = QP = BQ = u$. We then extend AB by $BC = k$ and set CP equal to v. From figure $AQCP$ it then follows, according to the apex transversal theorem, that

$$v^2 - u^2 = CA \cdot CQ = 4k(k + u)$$

or

$$v^2 = (u + 2k)^2,$$

more simply

$$v = u + 2k$$

or also

$$v - u = 2k.$$

This is the equation for the locus in bipolar coordinates u, v.

The locus of the point P *is thus a hyperbola with the foci* A *and* C *and the major axis* BD = 2k. (*D* lies between A and B in such a way that, according to the locus equation $w - u = 2k$, $CD = 3k$, and AD is equal to k.)

Let us now consider this hyperbola as having been drawn once and for all for any k. (The half of the branch belonging to the focus A, lying above the major axis, is sufficient.)

In order to trisect the prescribed angle ω we draw about AB as chord the arc subtending the angle $180° - \omega$ and call its intersection with the hyperbola P.   Then

$$\measuredangle ABP = \beta = \tfrac{1}{3}\omega.$$

PROOF.   From $\measuredangle APB = 180° - \omega$ it follows that $\alpha + \beta = \omega$, i.e., (because $\alpha = 2\beta$), $3\beta = \omega$.

NOTE.   It is also possible to trisect an angle by means of Nicomedes' conchoid; this method, however, now possesses only historical interest.

SUPPLEMENT TO NOS. 35, 36, AND 37

On the degree of irreducible equations that can be solved by quadratic roots:

Let a rational function of one or more magnitudes be known as an $\Re$-function and an algebraic equation with rational coefficients as an $\Re$-equation; in particular, let us designate an integral rational function of several magnitudes with rational coefficients as an $\Re$-polynomial.   We will also call a quadratic root of a rational number or an $\Re$-function of such quadratic roots an expression of the first order, and a quadratic root of an expression of the first order or an $\Re$-function of such quadratic roots an expression of the second order, etc.

In every expression of the mth order we assume that none of its roots of the mth order can be expressed rationally by the remaining ones or even by expressions of lower than the mth order; we assume as well that the expression (by elimination of irrational denominators and powers higher than the first of the relevant quadratic roots) has been put into its simplest form—the normal form.   An expression of the mth order that contains the root of the mth order $\sqrt{\alpha}$ will thus appear in the form $\mathfrak{a} + a\sqrt{\alpha}$, where $\mathfrak{a}$ and a are expressions of the mth order (or lower) in which the $\sqrt{\alpha}$ does not recur.

Now let x_1 be an expression of the mth order which contains the mth-order roots $\sqrt{\alpha}$, $\sqrt{\beta}$, $\sqrt{\gamma}$, ... and in which a total of n different roots [of mth and lower order] occur.   If we change the signs of these n roots in every possible way, we obtain a total of $2^n = N$ similarly constructed root expressions $x_1, x_2, x_3, \ldots, x_N$.

We form the function

$$F(x) = (x - x_1)(x - x_2) \ldots (x - x_N).$$

If everywhere in this expression we change the sign of any of the above n roots contained in it, the value of the expression is not changed. Thus, if we multiply out the parentheses, the resulting polynomial of x—as we know from computations with root expressions—will merely contain the squares of the roots and is consequently an $\Re$-function of x. The equation

$$(1) \qquad\qquad F(x) = 0$$

is thus an $\Re$-equation with the roots $x_1, x_2, \ldots, x_N$, which moreover need not all be different.

We now postulate:

If an $\Re$-polynomial $f(x)$ vanishes for a null value, such as x_1, of $F(x)$, then $f(x)$ will vanish for all the roots of $F(x) = 0$.

PROOF. We write $x_1 = \mathfrak{a} + a\sqrt{\alpha}$ (see above) and introduce this value into $f(x)$, and on computation we obtain

$$0 = f(x_1) = \mathfrak{A} + A\sqrt{\alpha},$$

where $\mathfrak{A}$ and A contain expressions of the mth degree and lower with the exception of $\sqrt{\alpha}$. Now, since it is assumed that $\sqrt{\alpha}$ is independent of these expressions, A cannot differ from zero (for otherwise it would follow that $\sqrt{\alpha} = -\mathfrak{A}/A$ and thus $\sqrt{\alpha}$ would be a function of $\sqrt{\beta}, \sqrt{\gamma}, \ldots$) and, therefore, necessarily

$$A = 0 \quad \text{and} \quad \mathfrak{A} = 0.$$

We will write the expressions A and $\mathfrak{A}$ as $\mathfrak{b} + b\sqrt{\beta}$ and $\mathfrak{B} + B\sqrt{\beta}$, where $\mathfrak{b}, b, \mathfrak{B}, B$ are no longer dependent upon $\sqrt{\alpha}$ and $\sqrt{\beta}$. From

$$\mathfrak{b} + b\sqrt{\beta} = 0 \quad \text{and} \quad \mathfrak{B} + B\sqrt{\beta} = 0$$

it follows as above that

$$\mathfrak{b} = 0, \qquad b = 0, \qquad \mathfrak{B} = 0, \qquad B = 0,$$

etc. From these values we finally obtain equations that possess no roots but only rational numbers and which are, in other words, independent of the signs of the n roots occurring in x_1 and consequently are unchanged when the signs are changed in any way. Now, since this change of sign transforms x_1 into one of the values $x_2, x_3, \ldots, x_N, f(x)$ must therefore also vanish for $x_2, x_3, \ldots, x_N$, which is what we set out to prove.

Among all the $\Re$-polynomials $f(x)$ that vanish for $x = x_1$ there is one possessing the lowest possible degree ν; let this be called $\varphi(x)$.

The polynomial $\varphi(x)$ is irreducible in the natural rationality domain (cf. No. 24).

[If φ were divisible: $\varphi(x) = u(x) \cdot v(x)$, then when $\varphi(x_1) = 0$ it would necessarily follow that one of the factors such as $v(x_1)$ must equal zero: this would contradict our assumption in that there would be a polynomial v of lower degree than φ with the null value x_1.]

Since the $\Re$-polynomial $F(x)$ vanishes for a null value x_1 of the irreducible polynomial $\varphi(x)$, $F(x)$, according to Abel's irreducibility theorem (No. 25), is divisible by $\varphi(x)$:

$$F(x) = F_1(x)\varphi(x).$$

Since, moreover, the $\Re$-polynomial $F_1(x)$ vanishes for a null value of F, thus also for φ, F_1 is also divisible by φ and $F_1(x) = F_2(x)\varphi(x)$; consequently

$$F(x) = F_2(x)\varphi(x)^2,$$

etc.   Finally we obtain

$$F(x) = \varphi(x)^\mu$$

(assuming that the first coefficient of F and φ has the value 1).

If we compare the degree of the polynomial on the right-hand side of this equation with that of the polynomial on the left, we find that

$$N = \mu\nu.$$

Since, however, $N = 2^n$, ν must also be a power of 2.

CONCLUSION: *The degree of an irreducible equation with rational coefficients for which a single expression formed from quadratic roots will suffice must be a power of 2.*   From this the two following theorems are easily obtained:

I. *It is impossible to double a cube with compass and straight-edge.*

II. *It is in general impossible to trisect an angle with compass and straight-edge.*

In *both* problems the specific magnitude x to be constructed is a root of an irreducible equation of the *third* degree, and according to our conclusion it is impossible for such an equation to be constructed from quadratic roots, and therefore with compass and straight-edge. [As is well known, all expressions that can be represented by compass and straight-edge constructions are either rational or built up from quadratic roots.]

Thus it merely remains to show that the equations for doubling a cube and trisecting an angle are cubic and irreducible.

The edge x of the cube that is twice the size of a cube with an edge equal to 1 satisfies the equation

$$x^3 - 2 = 0.$$

If this equation were reducible, then it would necessarily follow that

$$x^3 - 2 = (x^2 + hx + k)(x - l),$$

where h, k, l are rational numbers. Accordingly, the equation $x^3 = 2$ would have to possess the rational root $l = p/q$, where we may assume that p and q have no common divisor, and consequently $(p/q)^3$ would have to be equal to 2 or p^3 equal to $2q^3$. Consequently, p^3 would have to be divisible by q^3 and therefore p would also have to be divisible by q, which is not the case.

In the trisection of an angle we can consider the given angle α and the angle we are looking for φ as peripheral angles of a unit circle, so that the subtended arcs are $a = 2 \sin \alpha$ and $x = 2 \sin \varphi$, respectively. From $\alpha = 3\varphi$ and $\sin 3\varphi = 3 \sin \varphi - 4 \sin^3 \varphi$ it follows that

$$\sin \alpha = 3 \sin \varphi - 4 \sin^3 \varphi$$

or

$$x^3 - 3x + a = 0.$$

If we assume an arc a of length $3m/n$, where m and n possess no common divisors and are integers that cannot be divided by 3, and if we multiply the equation by n^3 and set $nx = X$, the equation assumes the form

$$X^3 - 3n^2 X + 3 mn^2 = 0.$$

But according to Schoenemann's theorem (No. 25) this equation is irreducible, since the coefficient of X is divisible by the prime number 3 and the free term is divisible by 3, but not by 3^2.

37   The Regular Heptadecagon

To construct a regular heptadecagon.

In other words: *To divide the perimeter of a circle into 17 equal parts.*

This celebrated problem was solved by Gauss in his major work *Disquisitiones arithmeticae*, published in 1801. In the section of this work dealing with the solution of the binomial equations $x^n = 1$ Gauss proved the important theorem:

A regular polygon can be constructed with compass and straight-edge when and only when the number of its sides has the form $2^m p_1 p_2 \ldots p_\nu$, *where* $p_1, p_2, \ldots, p_\nu$, *are all different prime numbers of the form* $2^n + 1$.

For $m = 0$, $\nu = 1$, and $p_1 = 3$ and $p_1 = 5$, we obtain the cases of the regular triangle and pentagon, respectively, which had already been solved in antiquity.

In the conclusion to his investigations Gauss said, "The division of a circle into three and into five equal parts was already known in Euclid's time; it is amazing that nothing new was added to these discoveries in the next two thousand years, that the geometers considered it as confirmed that, except for these cases and those that could be derived from them, regular polygons could not be constructed with compass and straight-edge."

The great advances made in the division of the circle by Gauss were possible only because Gauss transformed the originally purely geometrical problem into an algebraic one. He arrived at this transformation in the course of his representation of complex numbers in the Gauss plane, which was named after him.

An arbitrary complex number $c = a + bi$ is conventionally represented in this plane by a point with the coordinates $a|b$; this point itself is designated as "the complex number c." Another common method is the trigonometric representation

$$c = r(\cos \vartheta + i \sin \vartheta)$$

of the complex number c, where r represents the so-called *magnitude* (modulus) of the number, the distance of the number c from the null point O of the number plane and ϑ, the so-called *angle* of the number, which is the angle formed by the distance r and the axis of the positive real numbers.

The points of the unit circle $\Re$ drawn about the center O represent the so-called Gauss numbers, i.e., numbers of the form

$$\gamma = \cos \varphi + i \sin \varphi,$$

where φ is the angle of the number γ.

We will write for short

$$\cos \varphi + i \sin \varphi = 1\varphi.$$

The fundamental property of the Gauss numbers is described by the relation

$$1_\varphi \cdot 1_\psi = 1_{\varphi + \psi},$$

i.e., *the product of two Gauss numbers is also a Gauss number; the angle of the product is the sum of the angles of the factors.*

It is easily confirmed that the theorem also holds for products of more than two Gauss numbers.

For example,

$$1_\varphi^n = 1_\varphi \cdot 1_\varphi \cdot 1_\varphi \cdots = 1_{n\varphi},$$

or, written out fully,

$$(\cos \varphi + i \sin \varphi)^n = \cos n\varphi + i \sin n\varphi.$$

This is Demoivre's formula (Abraham Demoivre, 1667–1754).

To obtain a regular polygon of n angles we mark off the angle $\varphi = (2\pi/n)$ n times in succession from point 1 on $\Re$. The resulting points representing the divisions are

$$\varepsilon_1 = \varepsilon = \cos \varphi + i \sin \varphi, \qquad \varepsilon_2 = \cos 2\varphi + i \sin 2\varphi, \ldots$$

$$\varepsilon_n = \cos n\varphi + i \sin n\varphi = 1.$$

Then

$$\varepsilon_\nu = \varepsilon_1^\nu = \varepsilon^\nu \quad \text{and} \quad \varepsilon_\nu^n = \varepsilon^{\nu n} = (\varepsilon^n)^\nu = 1.$$

The n *angles* $\varepsilon_1, \varepsilon_2, \ldots, \varepsilon_n$ *of a regular polygon of* n *angles are therefore the roots of the equation*

$$z^n = 1.$$

Thus the geometric problem of "constructing a regular polygon of n angles," following Gauss, turns out to be the problem "of finding the roots of the equation $z^n = 1$."

Since one of the n roots of this equation has the value 1, we need only find the other $(n - 1)$ roots. These satisfy the equation

$$\frac{z^n - 1}{z - 1} = z^{n-1} + z^{n-2} + \cdots + z^2 + z + 1 = 0,$$

the so-called *circle partition equation.* In the case of $n = 3$, for example, the equation reads

$$z^2 + z + 1 = 0$$

and has the roots

$$\varepsilon_1 = \frac{-1 + i\sqrt{3}}{2}, \qquad \varepsilon_2 = \frac{-1 - i\sqrt{3}}{2}.$$

Since the complex numbers ε_1 and ε_2 both possess the real component $-\frac{1}{2}$, the angles ε_1 and ε_2 of the regular triangle are the points of intersection of $\mathfrak{K}$ with the parallel to the imaginary number axis that passes through the point $-\frac{1}{2}$.

A proof of the general theorem of Gauss would take us too far, so that we will restrict ourselves here to a brief exposition of the basic idea and the elements that are necessary for an understanding of the construction of the regular heptadecagon.

Let us first take note of the fact that the construction of the regular $2^m N$-gon, where N is the product of the odd prime numbers $p, q, r, \ldots$, is equivalent to drawing the regular p-gon, q-gon, r-gon, etc. If we have these polygons, we determine the integral numbers x, y, z in such manner that

$$\frac{N}{p} \cdot x + \frac{N}{q} \cdot y + \frac{N}{r} \cdot z + \cdots = 1.$$

This can be done because the numbers

$$\frac{N}{p}, \frac{N}{q}, \frac{N}{r}, \ldots$$

have no common divisor. Then

$$\frac{1}{N} = \frac{x}{p} + \frac{y}{q} + \frac{z}{r} + \cdots,$$

so that the Nth part of $\mathfrak{K}$ is obtained by joining the x $p$ths, y $q$ths, z $r$ths, ... of the circle perimeter.

Consequently, we need only be concerned with the solution of the circle partition equation

(1) $$z^{p-1} + z^{p-2} + \cdots + z^2 + z + 1 = 0,$$

in which p is a prime number of the form $2^n + 1$.

The brilliant idea underlying Gauss' method of solution consists in grouping the roots $\varepsilon_1, \varepsilon_2, \ldots, \varepsilon_{p-1}$ of (1) (where $\varepsilon_v = \varepsilon_1^v = \varepsilon^v$, $\varepsilon = \cos \varphi + i \sin \varphi$, $\varphi = 2\pi/p$) into so-called *periods*. The Gauss periods are root sums in which each successive term is the gth power of the preceding term, and the gth power of the last sum term results once again in the first term (hence the name period). The exponent g is here a so-called *primitive root of the prime number p*, i.e., an integer such that g^{p-1} is the *smallest* of its integral powers that leaves a residue of 1 on division by p. In other words, g is an integer such that the roots of (1) can be expressed in the form

$$z_0 = \varepsilon, \; z_1 = \varepsilon^g, \; z_2 = \varepsilon^{g^2}, \ldots, z_{p-2} = \varepsilon^{g^{p-2}}.$$

The next period is

$$z_0 + z_1 + z_2 + \cdots + z_{p-2}.$$

In fact,

$$z_{v+1} = z_v^g \text{ and } z_{p-2}^g = \varepsilon^{g^{p-1}} = \varepsilon^{sp+1} \text{ (where } s \text{ is an integer) } = \varepsilon.$$

The following period contains only $a = (p-1)/2$ terms and reads

$$z_0 + z_2 + z_4 + \cdots + z_r \qquad (r = 2a - 2).$$

In this period each term is the Gth power of the preceding term and $z_r^G = z_0$, where $G = g^2$ is similarly a primitive root of p.

Let

$$b = \tfrac{1}{2}a, \qquad c = \tfrac{1}{2}b, \qquad d = \tfrac{1}{2}c, \quad \text{etc.}$$

Gauss' method for solving the circle partition equation consists of reducing (1) to a chain of groups of quadratic equations. The first group contains one, the second group two, the third group four, the fourth group eight, etc., and the last group a quadratic equations. The roots of the first group form periods of a terms, those of the second group periods of b terms, those of the third periods of c terms, those of the last periods of a single term, i.e., the roots of (1) itself. The coefficients of the equations of one group can be determined from the coefficients of the preceding group, so that the equations of the last group give us the roots of (1) directly.

In the successive determination of coefficients the formula

$$(2) \qquad \qquad \varepsilon^E = \varepsilon^r,$$

in which r represents the residue remaining when the integral exponent E is divided by p, plays a predominant role.

We will now use the Gauss method to solve the equation for the heptadecagon ($p = 17$).

$$z^{16} + z^{15} + \cdots + z^2 + z + 1 = 0.$$

Let $\varphi = 2\pi/17$, $\varepsilon = \varepsilon_1 = \cos\varphi + i\sin\varphi$, $\varepsilon_v = \varepsilon^v$, and accordingly, let $\varepsilon_1, \varepsilon_2, \varepsilon_3, \ldots, \varepsilon_{17}$ be the corners of the heptadecagon, for which $z_v = \varepsilon^{g^v}$, where g represents the (smallest) primitive root 3 of 17. The powers $3^1, 3^2, 3^3, \ldots, 3^{16}$ on division by 17 leave the residues

$$3, 9, 10, 13, 5, 15, 11, 16, 14, 8, 7, 4, 12, 2, 6, 1.$$

Consequently, according to (2),

$$z_0 = \varepsilon, \quad z_2 = \varepsilon^9, \quad z_4 = \varepsilon^{13}, \quad z_6 = \varepsilon^{15}, \quad z_8 = \varepsilon^{16}, \quad z_{10} = \varepsilon^8,$$
$$z_{12} = \varepsilon^4, \quad z_{14} = \varepsilon^2, \quad z_1 = \varepsilon^3, \quad z_3 = \varepsilon^{10}, \quad z_5 = \varepsilon^5, \quad z_7 = \varepsilon^{11},$$
$$z_9 = \varepsilon^{14}, \quad z_{11} = \varepsilon^7, \quad z_{13} = \varepsilon^{12}, \quad z_{15} = \varepsilon^6.$$

Each root in the series $z_0, z_1, z_2, \ldots$ is the cube of the preceding one.

The first group in the chain contains a quadratic equation the roots of which are the periods

$$X = z_0 + z_2 + z_4 + z_6 + z_8 + z_{10} + z_{12} + z_{14}$$
$$= \varepsilon + \varepsilon^9 + \varepsilon^{13} + \varepsilon^{15} + \varepsilon^{16} + \varepsilon^8 + \varepsilon^4 + \varepsilon^2$$

and

$$x = z_1 + z_3 + z_5 + z_7 + z_9 + z_{11} + z_{13} + z_{15}$$
$$= \varepsilon^3 + \varepsilon^{10} + \varepsilon^5 + \varepsilon^{11} + \varepsilon^{14} + \varepsilon^7 + \varepsilon^{12} + \varepsilon^6.$$

Since the sum of the roots of (1) possesses the value -1, we obtain the relation

$$X + x = -1.$$

Making use of (2), we find on computation that Xx is equal to four times the sum of all the roots of (1), and consequently

$$Xx = -4.$$

The quadratic equation for the periods X and x consequently reads

(I) $$t^2 + t - 4 = 0.$$

Its roots are

$$X = \frac{-1 + \sqrt{17}}{2} \quad \text{and} \quad x = \frac{-1 - \sqrt{17}}{2}.$$

That $X > x$ is shown in the following manner. If we designate the real component of the complex number c as $\Re c$, then (cf. Fig. 29)

(3) $$\Re \varepsilon^\mu = \Re \varepsilon^\nu \quad \text{if} \quad \mu + \nu = 17,$$

since the corners ε^μ and ε^ν of the heptadecagon are symmetrical to the real axis. Applying this rule, we obtain

$$\Re X = 2[\Re \varepsilon_1 + \Re \varepsilon_2 + \Re \varepsilon_4 + \Re \varepsilon_8],$$
$$\Re x = 2(\Re \varepsilon_3 + \Re \varepsilon_5 + \Re \varepsilon_6 + \Re \varepsilon_7).$$

A glance at the figure shows that the bracket is positive and the parenthesis negative.

The four four-term periods are

$$U = z_0 + z_4 + z_8 + z_{12} = \varepsilon + \varepsilon^{13} + \varepsilon^{16} + \varepsilon^4,$$
$$u = z_2 + z_6 + z_{10} + z_{14} = \varepsilon^9 + \varepsilon^{15} + \varepsilon^8 + \varepsilon^2,$$
$$V = z_1 + z_5 + z_9 + z_{13} = \varepsilon^3 + \varepsilon^5 + \varepsilon^{14} + \varepsilon^{12},$$
$$v = z_3 + z_7 + z_{11} + z_{15} = \varepsilon^{10} + \varepsilon^{11} + \varepsilon^7 + \varepsilon^6.$$

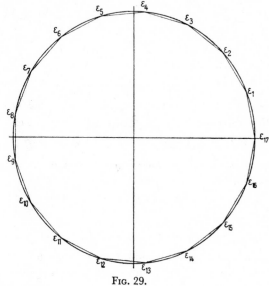

Fig. 29.

Here we obtain

$$U + u = X \qquad \qquad V + v = x$$

and, applying rule (2),

$$Uu = \varepsilon^1 + \varepsilon^2 + \cdots + \varepsilon^{16} = -1 \mid Vv = \varepsilon^1 + \varepsilon^2 + \cdots + \varepsilon^{16} = -1.$$

The respective quadratic equations are

(II) $\qquad t^2 - Xt - 1 = 0 \qquad \qquad t^2 - xt - 1 = 0.$

Their roots are

$$U = \frac{X + \sqrt{X^2 + 4}}{2}, \qquad V = \frac{x + \sqrt{x^2 + 4}}{2}.$$
$$u = \frac{X - \sqrt{X^2 + 4}}{2}, \qquad v = \frac{x - \sqrt{x^2 + 4}}{2}.$$

It follows from rule (3) that $U > u$ and $V > v$.   Consequently,

$$\Re U = 2[\Re\varepsilon_1 + \Re\varepsilon_4], \qquad\qquad \Re V = 2[\Re\varepsilon_3 + \Re\varepsilon_5],$$
$$\Re u = 2(\Re\varepsilon_2 + \Re\varepsilon_8), \qquad\qquad \Re v = 2(\Re\varepsilon_6 + \Re\varepsilon_7).$$

A look at the heptadecagon shows that the brackets are larger than the parentheses immediately below them.

Of the two-membered periods obtained we need only the two

$$W = z_0 + z_8 = \varepsilon + \varepsilon^{16} \quad\text{and}\quad w = z_4 + z_{12} = \varepsilon^{13} + \varepsilon^4.$$

Here we find

$$W + w = U$$

and, according to (2),

$$Ww = \varepsilon^5 + \varepsilon^{14} + \varepsilon^3 + \varepsilon^{12} = V.$$

Here also $W > w$, since $\Re W = 2\Re\varepsilon_1$ and $\Re w = 2\Re\varepsilon_4$, but $\Re\varepsilon_1 > \Re\varepsilon_4$.

The quadratic equation with the roots W and w reads

(III) $$t^2 - Ut + V = 0.$$

The *construction of the heptadecagon* accordingly consists of the following four steps:

   I. Construction of X and x;

  II. construction of U and V;

 III. construction of W and w according to (III);

 IV. finding the points W and w on the real number axis.   The perpendicular bisectors of the lines joining them to the null point cut the circle $\Re$ at the corners ε_1, ε_{16} and ε_4, ε_{13} of the regular heptadecagon (thus all the other corners are also determined).

38   Archimedes' Determination of the Number π

Archimedes of Syracuse (287?–212 B.C.) was the greatest mathematician of the ancient world.

The most famous of his achievements is the measurement of the circle.   The crux of this problem is the calculation of the number π, i.e., the number by which the diameter and the square of the radius must be multiplied to determine the circumference and area, respectively, of a circle.*

---

* The proposal that this number be designated as π came from Leonhard Euler (*Commentarii Academiae Petropolitanae ad annum 1739*, vol. IX).

The idea upon which Archimedes' method was based is the following. The circumference of a circle lies between the perimeters of a circumscribed and inscribed n-gon, and in particular, the greater n is, the smaller is the deviation of the circumference of the circle from the perimeters of the two n-gons. Then the object is to calculate the perimeters of a circumscribed and inscribed regular polygon with so great a number of sides that their difference is equal to a very negligible magnitude ε. Then if the circumference of the circle is set equal to the perimeter of one of these polygons, the resulting deviation from the true circumference of the circle is smaller than ε, with the result that when ε is sufficiently small the circumference of the circle is determined with sufficient accuracy.

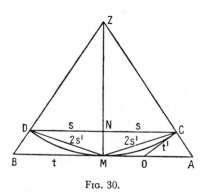

Fig. 30.

The particular achievement of Archimedes was to indicate a method by which the perimeters of such *many-sided* polygons could be calculated.

This method, the so-called *Archimedes algorithm*, is based upon the two *Archimedes recurrence formulas* which we will now derive.

In Figure 30, let Z be the center of the circle, let $AB = 2t$ be the side of the circumscribed and $CD = 2s$ the side of the inscribed regular n-gon. Let M be the midpoint of AB and N the midpoint of CD, let O be the point of intersection with MA of the tangent to the circle passing through C. Accordingly, $OM = OC = t'$ is half the side of the circumscribed $2n$-gon and $MC = MD = 2s'$ is the side of the inscribed regular $2n$-gon.

Since ACO and AMZ are similar right triangles,

$$t'/(t - t') = OC/OA = MZ/AZ,$$

and from the ray theorem,

$$s/t = NC/MA = CZ/AZ.$$

Since the right sides of these proportions are equal, we obtain $t'/(t - t') = s/t$ or

$$t' = \frac{ts}{t + s}.$$

Since the isosceles triangles CMD and COM are similar, $2s'/2s = t'/2s'$, i.e.,

$$2s'^2 = st'.$$

If a is the perimeter of the circumscribed n-gon and b the perimeter of the inscribed n-gon, and a' and b' are the perimeters, respectively, of the circumscribed and inscribed $2n$-gons, we then have

$$a = 2nt, \qquad b = 2ns, \qquad a' = 4nt', \qquad b' = 4ns'.$$

If we then introduce the values obtained for t, s, t', s' from these equations into the two formulas we have found, they are transformed into the *Archimedes recurrence formulas:*

(I) $\qquad a' = \dfrac{2ab}{a + b},$ $\qquad$ (II) $\qquad b' = \sqrt{ba'}.$

Thus, a′ is the harmonic mean of a and b, b′ the geometric mean of b and a′.

Now let us consider in succession by the regular n-gon, $2n$-gon, $4n$-gon, $8n$-gon, etc., and let us designate the perimeters of the circumscribed and inscribed $2^{\nu}n$-gons as a_{ν} and b_{ν}, respectively. We then obtain the *Archimedes series*

$$a_0, b_0, a_1, b_1, a_2, b_2, \ldots$$

of the successive perimeters. Here the recurrence formulas (I) and (II) read

(1) $\qquad a_{\nu+1} = \dfrac{2a_{\nu}b_{\nu}}{a_{\nu} + b_{\nu}},$ $\qquad$ (2) $\qquad b_{\nu+1} = \sqrt{b_{\nu}a_{\nu+1}}.$

That is: *Each term of the Archimedes series is alternately the harmonic and geometric mean of the two preceding terms.*

Using this rule, we are able to calculate all the terms of the series if the first two terms are known. The *Archimedes algorithm* consists of this calculation of the successive perimeters of the polygons.

Archimedes chose as his initial polygon the regular hexagon, the perimeters of which are $a_0 = 4\sqrt{3}r$ and $b_0 = 6r$, respectively, and worked out the series a_1, b_1, a_2, b_2, a_3, b_3, a_4, b_4 up to the perimeters a_4 and b_4 of the circumscribed and inscribed regular 96-cornered polygon.   He found that

$$a_4 = 3\tfrac{10}{70}d, \qquad b_4 = 3\tfrac{10}{71}d,$$

where d is the diameter of the circle.   The *Archimedes approximation* for the value of π is consequently

$$\pi = 3\tfrac{1}{7} = 3.14.$$

NOTE.   The calculations involved in the Archimedes method are very laborious.   For this reason Christian Huygens, in his treatise published in Leyden in 1654, *De circuli magnitudine inventa*, replaced the limits a_ν and b_ν of the circumference u of the Archimedes method by the limits α_ν and β_ν, which gave a closer approximation of u, since it made it possible to obtain π correctly to two decimal places for $\nu = 1$. Huygens' method, however, involves rather complicated considerations.   The following method supplied by the author is faster and more convenient; it is based on the known theorem: *The harmonic mean of two numbers is smaller than the geometric mean of the numbers.* This can be expressed as

$$\frac{2xy}{x + y} < \sqrt{xy}.$$

[Since $(\sqrt{x} - \sqrt{y})^2 > 0$, it follows that $2\sqrt{xy} < x + y$, and from this, multiplication with $\sqrt{xy}/(x + y)$ gives the designated inequality.]

According to this theorem, we obtain from (1) $a_{\nu+1} < \sqrt{a_\nu b_\nu}$.   If we multiply the square of this inequality by the square of (2), we obtain

$$a_{\nu+1} b_{\nu+1}^2 < a_\nu b_\nu^2$$

or, if we set

$$\sqrt[3]{a_\nu b_\nu^2} = A_\nu$$

then

(3) $$A_{\nu+1} < A_\nu.$$

According to the same theorem, it follows from (2) that

$$b_{\nu+1} > \frac{2b_\nu a_{\nu+1}}{b_\nu + a_{\nu+1}} \quad \text{or} \quad \frac{2}{b_{\nu+1}} < \frac{1}{b_\nu} + \frac{1}{a_{\nu+1}}.$$

If we then add to this inequality the equation

$$\frac{2}{a_{\nu+1}} = \frac{1}{a_\nu} + \frac{1}{b_\nu},$$

which is only a different manner of writing (1), we obtain

$$\frac{1}{a_{v+1}} + \frac{2}{b_{v+1}} < \frac{1}{a_v} + \frac{2}{b_v}$$

or

$$\frac{3a_{v+1}b_{v+1}}{2a_{v+1} + b_{v+1}} > \frac{3a_v b_v}{2a_v + b_v},$$

or, in abbreviated form, if we set

$$\frac{3a_v b_v}{2a_v + b_v} = B_v,$$

then

(4) $$B_{v+1} > B_v.$$

The inequalities (3) and (4) imply that as v increases, A_v grows continuously smaller, B_v continuously larger.

Since for infinitely great v, both A_v and B_v become the circumference u of the circle, for every finite v it must be true that

$$B_v < u < A_v.$$

The limits A_v and B_v of this inequality are much narrower than the Archimedes limits a_v and b_v. If we take the hexagon, for example, as our initial polygon and $d = 1$, then $a_0 = 2\sqrt{3}$, $b_0 = 3$, $u = \pi$, and we obtain $A_1 = 3.1423$ and $B_0 = 3.1402$; thus we are able to obtain the correct value of π to two accurate decimal places by using only the inscribed hexagon and the circumscribed dodecagon, whereas the same precision is achieved by the Archimedes method only with the use of the polygon of 96 sides.

39   Fuss' Problem of the Chord-Tangent Quadrilateral

To find the relation between the radii and the line joining the centers of the circles of circumscription and inscription of a bicentric quadrilateral.

A *bicentric* or *chord-tangent quadrilateral* is defined as a quadrilateral that is simultaneously inscribed in one circle and circumscribed about another. Let $PQRS$ be such a quadrilateral, $\mathfrak{C}$ the circumscribed circle, Γ the inscribed circle. Let the points of tangency of the opposite sides PQ and RS with circle Γ be X and X', let the points of tangency of the opposite sides QR and SP be Y and Y', and let the

point of intersection of the tangency chords XX' and YY' be O.   If
we then apply the theorem of the sum of the angles of a quadrilateral

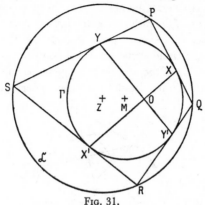

Fig. 31.

to the two quadrilaterals $OXPY$ and $OX'RY'$, designating the
quadrilateral angles by means of a line over the letter representing
the corner, we obtain the two equations

$$\bar{O} + \bar{X} + \bar{P} + \bar{Y} = 360°, \qquad \bar{O} + \bar{X}' + \bar{R} + \bar{Y}' = 360°.$$

Since the angles $\bar{X}$ and $\bar{X}'$ ($\bar{Y}$ and $\bar{Y}'$) situated at opposite sides of the
chord XX' (YY') add up to 180°, addition of the two equations gives
the following relation

(1) $$2\bar{O} + \bar{P} + \bar{R} = 360°.$$

Now the sum of the two opposite angles $\bar{P}$ and $\bar{R}$ of the chord
quadrilateral $PQRS$ is 180°; consequently, $\bar{O} = 90°$.

The tangency chords of the two pairs of opposite sides of a bicentric
quadrilateral are therefore perpendicular to each other.

This condition is also sufficient: *A bicentric quadrilateral* PQRS *is
obtained if the tangents* PQ, RS, SP, QR *are drawn through the end points*
X, X', Y, Y' *of two perpendicular chords* XX' *and* YY' *of an arbitrary
circle* Γ.   In fact, it now follows from (1), since $\bar{O} = 90°$, that the sum
of the opposite angles $\bar{P}$ and $\bar{R}$ is 180°, i.e., that *PQRS* is also a *chord
quadrilateral.*

The simplest way of obtaining the desired relation between the
radii and the axis of the centers of the circumscribed and inscribed
circles is by means of the following locus problem.   *A right angle is
rotated about its fixed vertex, which is located inside a circle; find the locus of*

the point of intersection of the two circle tangents that pass through the point of intersection of the legs of the angle with the circle.

SOLUTION OF THE LOCUS PROBLEM.   Let the given circle be known as Γ, its midpoint as M, its radius as ρ, the fixed vertex of the right angle as O, the distance of the vertex from M as e.   Let the legs of the right angle intersect the circle at the (moving) points X and Y; and let the point of intersection of the two circle tangents passing through X and Y be known as P and its distance from the center of the circle as p.

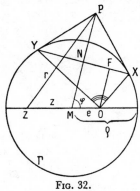

FIG. 32.

We will first determine the relation between p and its angle φ ($= \angle OMP$) with the fixed line MO.

Since OXY is a right triangle,

$$OF^2 = FX \cdot FY,$$

where F represents the base point of the altitude to the hypotenuse. If we introduce the projections $\rho' = MN$ and $e' = e \cos \varphi$ and $\rho'' = NX$ and $e'' = e \sin \varphi$ ($= NF$) on the lines MP and XY, respectively, the equation can be written

$$(\rho' - e')^2 = (\rho'' - e'')(\rho'' + e'')$$

or

$$2\rho'^2 - 2\rho'e' + e'^2 + e''^2 = \rho'^2 + \rho''^2$$

or

(2) $$2\rho'^2 - 2\rho'e \cos \varphi + e^2 = \rho^2.$$

Since MXP is a right triangle,

$$MX^2 = MP \cdot MN$$

or

(3) $$\rho^2 = p\rho'.$$

If we introduce the value of ρ' from (3) into (2), we obtain the relation we are looking for:

$$(4) \qquad p^2 + 2\,\frac{\rho^2 e}{\rho^2 - e^2}\,p\cos\varphi = \frac{2\rho^4}{\rho^2 - e^2}.$$

The distance $r = ZP$ of a point Z from P on the extension of OM at a distance of $MZ = z$ from M is obtained by the cosine theorem

$$(5) \qquad r^2 = z^2 + p^2 + 2zp\cos\varphi.$$

If for z, which up to this point has been arbitrary, we now choose the value

$$(\text{I}) \qquad MZ = z = \frac{\rho^2}{\rho^2 - e^2}\cdot e,$$

we obtain, in accordance with (4),

$$(\text{II}) \qquad r^2 = z^2 + \frac{2\rho^4}{\rho^2 - e^2},$$

and consequently r has a *constant* value!

The desired locus of the point of intersection P *is thus a circle* $\mathfrak{C}$ whose center Z, which is situated on the extension of OM, is determined by (I) and whose radius r is determined by (II).

Naturally, also belonging to this locus are the points of intersection Q, R, S of the tangents, which are obtained when we draw the tangents through the points of intersection of the circle Γ with the extensions of XO and YO.

The quadrilateral $PQRS$ is simultaneously a tangent and chord quadrilateral, in that it circumscribes circle Γ and is inscribed in circle $\mathfrak{C}$. If the right angle XOY is rotated about O so that the points X, Y describe the circle Γ, the quadrilateral $PQRS$ continuously assumes different positions but always circumscribes circle Γ and is always inscribed in circle $\mathfrak{C}$. Similarly, we see that in this way *all* the bicentric quadrilaterals belonging to the two circles Γ and $\mathfrak{C}$ are obtained. The obtained formulas (I) and (II) contain the solution to the problem posed.

We substitute the value obtained from (II) for $\rho^2 - e^2$ in (I) and obtain $e = 2z\rho^2/(r^2 - z^2)$. From this there follows $\rho^2 - e^2 = \rho^2[(r^2 - z^2)^2 - 4\rho^2 z^2]/(r^2 - z^2)^2$. When this value is introduced into (II) we finally obtain the sought-for *relation between the radii* r *and* ρ *and the axis* z *connecting the centers of the circumscribed and inscribed circles of the bicentric quadrilateral:*

$$2\rho^2(r^2 + z^2) = (r^2 - z^2)^2.$$

The developed formula comes from Nicolaus Fuss (1755–1826), a student and friend of Leonhard Euler. Fuss also found the corresponding formulas for the bicentric pentagon, hexagon, heptagon, and octagon (*Nova Acta Petropol.*, XIII, 1798).

The corresponding formula for the triangle had already been given by Euler. It is

$$r^2 - z^2 = 2r\rho$$

and is easily obtained in the following manner. Let *ABC* be any triangle, let *Z* and *M* be the respective centers, *r* and *ρ* the radii of the circles of circumscription and inscription, respectively; thus, *ZM* = *z* is the axis connecting the centers; further, let *D* be the point at which the extension of *CM* meets the circumscribed circle, so that *DM* = *DA* = *DB*. The power of the circumscribed circle at *M* is

$$MC \cdot MD = r^2 - z^2.$$

However, since we can replace sin ($\gamma/2$) by the ratio ρ/MC as well as by $AD/2r$ or $MD/2r$, $\rho/MC = MD/2r$, i.e.,

$$MC \cdot MD = 2r\rho.$$

When the two values found for the product $MC \cdot MD$ are set equal to each other we obtain Euler's formula.

Note. Much more remarkable than the Fuss formula is a theorem concerning bicentric quadrilaterals that follows directly from the preceding locus consideration. For convenience in expression we will make a prefatory observation.

Let a circle Γ lie completely inside another circle $\mathfrak{C}$. If from any point on $\mathfrak{C}$ we draw a tangent to Γ, extend the tangent line so that it intersects $\mathfrak{C}$, and draw from the point of intersection a new tangent to Γ, extend this tangent similarly to intersect $\mathfrak{C}$, and continue in this manner, we obtain a so-called *Poncelet traverse* which, when it consists of *n* chords of the larger circle, is called *n*-sided.

The theorem concerning bicentric quadrilaterals now reads:

If on the circle of circumscription there is one *point of origin for which a four-sided Poncelet traverse is closed, then the four-sided traverse will also close for* any *other point of origin on the circle.*

The French mathematician Poncelet (1788–1867) demonstrated that this theorem is not limited to four-sided traverses only, but is generally true for *n*-sided traverses, and not only for circles, but for any type of conic section. The general theorem reads:

PONCELET'S CLOSURE THEOREM: *If an* n-*sided Poncelet traverse constructed for two given conic sections is closed for* one *position of the point of origin, it is closed for* any *position of the point of origin.*

40 Annex to a Survey

To determine the position of unknown but accessible points of the earth's surface by taking the bearings of known points.

(A point on the earth's surface is considered as known when its geographic coordinates [length and width] are known.)

This problem is of great importance in the incorporation of new points of the earth's surface into a survey and consequently in the preparation of accurate maps.

Land surveyors and sailors are specifically confronted with the following two cases:

I. THE SNELLIUS-POTHENOT PROBLEM; THE PROBLEM OF THREE INACCESSIBLE POINTS: *Determine the position of an unknown accessible point* P *by its bearings from three inaccessible known points* A, B, C.

This most famous of all land surveying problems was posed and solved by the Dutchman Willebrord Snellius (1581–1626) in his 1617 work, *Eratosthenes Batavus*, but attracted no attention among his contemporaries. It was not commonly known until it was solved once again by the Frenchman Pothenot (died 1732) in a paper submitted in 1692 to the French Academy. Since then it has been known as the Pothenot problem.

II. HANSEN'S PROBLEM; THE PROBLEM OF THE INACCESSIBLE DISTANCE: *From the position of two known but inaccessible points* A *and* B, *determine the position of two unknown accessible points* P *and* P' *by bearings from* A, B, P' *to* P *and* A, B, P *to* P'.

This problem was solved by the German astronomer Hansen (1795–1874), but was solved as well by other authors before him.

TRIGONOMETRIC SOLUTION

This type of solution is required when accuracy is important, as in land surveying. For both problems this type of solution is based upon the *sine tangent theorem:*

If

$$\sin \alpha / \sin \beta = m/n,$$

then also

$$\tan \frac{\alpha - \beta}{2} \Big/ \tan \frac{\alpha + \beta}{2} = (m - n)/(m + n).$$

[From $\sin \alpha/\sin \beta = m/n$ it first follows that

$$(\sin \alpha - \sin \beta)/(\sin \alpha + \sin \beta) = (m - n)/(m + n).$$

If the numerator and denominator of the fraction on the left of the equation are converted into products, we obtain

$$\cos \frac{\alpha + \beta}{2} \sin \frac{\alpha - \beta}{2} \Big/ \sin \frac{\alpha + \beta}{2} \cos \frac{\alpha - \beta}{2} = (m - n)/(m + n)$$

or

$$\tan \frac{\alpha - \beta}{2} \Big/ \tan \frac{\alpha + \beta}{2} = (m - n)/(m + n).]$$

SOLUTION OF THE POTHENOT PROBLEM

Known are the five elements $AC = a$, $BC = b$, $\angle ACB = \gamma$, $\angle APC = \alpha$, $\angle BPC = \beta$; to be found are the five elements $AP = x$, $BP = y$, $CP = z$, $\angle CAP = \psi$, $\angle CBP = \varphi$. If the sine theorem is applied to the triangles ACP and BCP,

$$\frac{\sin \psi}{\sin \alpha} = \frac{z}{a} \quad \text{and} \quad \frac{\sin \varphi}{\sin \beta} = \frac{z}{b}.$$

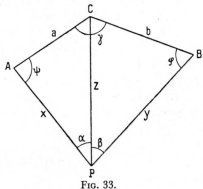

FIG. 33.

On division it follows from this that

$$\sin \psi/\sin \varphi = b \sin \alpha/a \sin \beta.$$

We determine the auxiliary angle μ whose tangent is $b \sin \alpha/a \sin \beta$, and obtain

$$\sin \psi/\sin \varphi = \tan \mu.$$

From this it follows according to the sine tangent theorem that

$$\frac{\tan\dfrac{\psi-\varphi}{2}}{\tan\dfrac{\psi+\varphi}{2}} = \frac{\tan\mu-1}{1+\tan\mu} = \tan(\mu-45°),$$

i.e.,

$$\tan\frac{\psi-\varphi}{2} = \tan\frac{\psi+\varphi}{2}\cdot\tan(\mu-45°).$$

Since $\psi+\varphi\;(=360°-\alpha-\beta-\gamma)$ is known, this equation gives us

$$\frac{\psi-\varphi}{2}.$$

From

$$\frac{\psi+\varphi}{2}\quad\text{and}\quad\frac{\psi-\varphi}{2},$$

addition and subtraction give us ψ and φ.

The unknowns x, y, z are obtained from the following formulas derived from the sine theorem:

$$\frac{x}{a} = \frac{\sin(\alpha+\psi)}{\sin\alpha},\qquad \frac{y}{b} = \frac{\sin(\beta+\varphi)}{\sin\beta},\qquad \frac{z}{a} = \frac{\sin\psi}{\sin\alpha}.$$

The position of the point P is determined from the magnitudes ψ, φ, x, y, z.

Solution of Hansen's Problem

Known are the five elements $AB = c$, $\angle APB = \gamma$, $\angle AP'B = \gamma'$, $\angle BPP' = \delta$, $\angle AP'P = \delta'$, and consequently also the angles $PAP' = \alpha$ and $PBP' = \beta$; we do not know the seven elements $AP = x$, $AP' = x'$, $BP = y$, $BP' = y'$, $\angle BAP = \psi$, $\angle ABP = \varphi$, and $PP' = s$.

We now represent the four ratios of the adjacent sides of the quadrilateral as sine ratios in accordance with the sine theorem:

$$\frac{c}{x} = \frac{\sin\gamma}{\sin\varphi},\qquad \frac{x}{s} = \frac{\sin\delta'}{\sin\alpha},\qquad \frac{s}{y'} = \frac{\sin\beta}{\sin\delta},\qquad \frac{y'}{c} = \frac{\sin\psi}{\sin\gamma}.$$

Multiplication of these equations gives us

$$\frac{\sin\psi\sin\beta\sin\gamma\sin\delta'}{\sin\varphi\sin\alpha\sin\gamma'\sin\delta} = 1\quad\text{or}\quad \frac{\sin\psi}{\sin\varphi} = \frac{\sin\alpha\sin\gamma'\sin\delta}{\sin\beta\sin\gamma\sin\delta'}.$$

We then determine an auxiliary angle μ whose tangent is equal to the right side of this equation, and we obtain

$$\frac{\sin \psi}{\sin \varphi} = \tan \mu,$$

i.e., according to the sine tangent theorem as above,

$$\tan \frac{\psi - \varphi}{2} = \tan \frac{\psi + \varphi}{2} \tan (\mu - 45°).$$

As above, we find from this

$$\frac{\psi - \varphi}{2} \quad \text{(since } \psi + \varphi = \delta + \delta' \text{ is known)}$$

and then ψ and φ.   Now the remaining unknowns are easily obtained by the sine theorem.

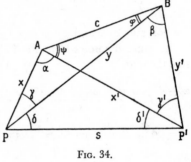

Fig. 34.

The positions of P and P' are determined by the values found for the six unknowns.

The Drawing Solution

This is adequate when great accuracy is not requisite, for example, in sailing along a coast where A, B, C are known landmarks, P and P' unknown positions of a ship with a bearing on these landmarks.

The solution of Pothenot's problem is extremely simple. The ship's position P is the point of intersection of the two circles to be drawn on the ship's chart with the chords AC and BC and the corresponding peripheral angles α and β.

Hansen's problem is solved in the following way. We draw a quadrilateral $abp'p$ having the same form as $ABP'P$ (beginning with an arbitrary distance pp') and lay this off on the chart so that b falls on B

and *a* on *AB*. The ship's position *P* is the point of intersection of *Bp* with the parallel to *ap* passing through *A*, the ship's position *P'* is the point of intersection of *Bp'* with the parallel to *pp'* passing through *P*.

41   Alhazen's Billiard Problem

To describe in a given circle an isosceles triangle whose legs pass through two given points inside the circle.

This problem stems from the Arabic mathematician Abu Ali al Hassan ibn al Hassan ibn Alhaitham (ca. 965 – ca. 1039), whose name was transformed into Alhazen by the translators of his *Optics*. In his *Optics* the above problem has the following form: "*Find the point on a spherical concave mirror at which a ray of light coming from a given point must strike in order to be reflected to another given point.*"

This problem can be posed in various other forms, e.g.: "*On a circular billiard table there are two balls; in what manner must one be struck in order for it to strike the other after rebounding from the cushion?*" or "*On the circumference of a circle find a point the sum of whose distances from two given points within the circle is equal to a minimum (or maximum).*"

A whole series of famous mathematicians took up this problem after Alhazen, among them Huygens, Barrow, de L'Hôpital, Riccati, and Quételet.

SOLUTION.   Let us call the given circle $\Re$, its center *M*, its radius *r*, the given points *P* and *p*, and let us make *M* the origin of a mutually perpendicular coordinate system *xy* in which *P* and *p* have the coordinates *A*|*B* and *a*|*b*.

If *OS* and *Os*, which pass through *P* and *p*, are the legs of the isosceles triangle *OSs* that we are looking for, the angles Φ and φ, which these legs form with the radius *OM*, must be equal.

If we designate the angles that the lines *PO*, *MO*, *pO* form with the *x*-axis as Λ, μ, λ, then, on the one hand, $\Phi = \Lambda - \mu$ and $\varphi = \mu - \lambda$ or

$$\tan \Phi = \frac{\tan \Lambda - \tan \mu}{1 + \tan \mu \tan \Lambda} \quad \text{and} \quad \tan \varphi = \frac{\tan \mu - \tan \lambda}{1 + \tan \mu \tan \lambda},$$

while, on the other hand, if *x*|*y* are the coordinates of *O*,

$$\tan \Lambda = \frac{y - B}{x - A}, \qquad \tan \mu = \frac{y}{x}, \qquad \tan \lambda = \frac{y - b}{x - a},$$

and consequently, since $\tan \Phi = \tan \varphi$,

$$\frac{\dfrac{y-B}{x-A} - \dfrac{y}{x}}{1 + \dfrac{y}{x}\dfrac{y-B}{x-A}} = \frac{\dfrac{y}{x} - \dfrac{y-b}{x-a}}{1 + \dfrac{y}{x}\dfrac{y-b}{x-a}}$$

or

$$\frac{Ay - Bx}{x^2 + y^2 - Ax - By} = \frac{bx - ay}{x^2 + y^2 - ax - by},$$

or finally, if we set

$$Ab + Ba = H, \qquad Aa - Bb = K, \qquad A + a = h, \qquad B + b = k,$$

then

$$H(x^2 - y^2) - 2Kxy + (x^2 + y^2)[hy - kx] = 0.$$

Since the point $O(x|y)$ has to lie upon the circle $\mathfrak{K}$, the circle equation

(1) $$x^2 + y^2 = r^2$$

consequently applies here, and our condition assumes the form

(2) $$H(x^2 - y^2) - 2Kxy + r^2[hy - kx] = 0.$$

Since equation (2) represents a hyperbola, our conclusion reads as follows:

The point O *that we are looking for is the point of intersection of the circle* (1) *with hyperbola* (2).

Since there are in general four points of intersection for a circle and a hyperbola, there are in general four solutions to our problem.

Possessing particular interest is the special case in which the distances C and c of the given points P and p from the center M are equally great. In this case we naturally take the perpendicular bisector of Pp as the x-axis, and then we have

$$A = a, \qquad B = -b, \qquad H = 0, \qquad K = c^2, \qquad h = 2a, \qquad k = 0$$

and, according to (2)

$$-2c^2xy + 2ar^2y = 0.$$

This equation is satisfied by each of the conditions

(3) $$y = 0 \qquad \text{and} \quad (4) \qquad x = a\frac{r^2}{c^2}.$$

From (3) follows the corresponding $x = \pm r$. Consequently, the points of intersection of $\mathfrak{K}$ with the x-axis satisfy the condition for the point O we are looking for.

From (4) it follows that

$$\frac{c^2}{a} = \frac{r^2}{x}.$$

If we then draw through M a circle $\mathfrak{k}$ whose diameter $MN = d = c^2/a$ lies on the x-axis, and if $Q(X \mid Y)$ is a point of intersection of this circle with $\mathfrak{K}$, it follows, since MNQ is a right triangle, that

$$MQ^2 = MN \cdot X \quad \text{or} \quad r^2 = dX.$$

However, since $r^2/x = d$, we obtain

$$X = x.$$

Consequently, the points of intersection of the circles $\mathfrak{K}$ and $\mathfrak{k}$ also satisfy the condition for the point O we are looking for.

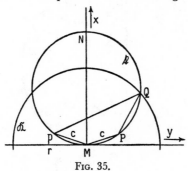

Fig. 35.

For these points of intersection to exist, d must be $> r$ or $c^2 > ar$. We will assume that this condition is satisfied.

Now the quadrilateral $MPpQ$ in circle $\mathfrak{k}$ is a chord quadrilateral, and therefore, according to Ptolemy's theorem, the sum of the products of the opposite sides must be equal to the product of the diagonals:

$$PQ \cdot Mp + pQ \cdot MP = MQ \cdot Pp$$

or

(5) $$(PQ + pQ)c = 2br.$$

For any other point Q' of $\mathfrak{K}$, $MPpQ'$ is not a chord quadrilateral, and therefore the sum of the products of the opposite sides must be greater than the product of the diagonals:

(6) $$(PQ' + pQ')c > 2br.$$

From (5) and (6) we obtain

$$PQ + pQ < PQ' + pQ'.$$

The problem: "*On a given circle find a point the sum of whose distances from two given points located in the circle at an equal distance from the midpoint of the circle is a minimum*" has the following striking solution:

The point we are looking for is the point of intersection of the given circle with the circle that passes through the given points and the center of the given circle.

NOTE. In connection with the above problem Alhazen also solved the problem: "*How to strike a ball lying on a circular billiard table in such a way that after twice striking the cushion the ball will return to its original position.*"

SOLUTION. Let the billiard table possess the radius r and the center M. Let the initial position of the ball be P, so that $MP = c$ is known. Let the ball first strike the circle at U, cross the extension of

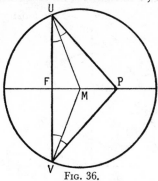

FIG. 36.

PM at a right angle at F, then strike the circle at V and return from here to P. UM and VM are then angle bisectors of the triangle PUV. We set

$$MF = x, \qquad FU = y, \qquad UP = z.$$

Applying the angle bisector theorem to the triangle FUP,

$$y/z = x/c,$$

and according to the Pythagorean theorem

$$r^2 = x^2 + y^2 \quad \text{and} \quad z^2 = y^2 + (x + c)^2.$$

If we eliminate y and z from these three equations, we obtain the quadratic equation

$$2cx^2 + r^2x = cr^2$$

for the unknown x. From this, x is easily constructed.

Problems Concerning Conic Sections and Cycloids

42   An Ellipse from Conjugate Radii

To draw an ellipse for which the magnitude and position of two conjugate radii are given.

SOLUTION.   Let the ellipse have the center equation

$$(1) \qquad \left(\frac{x}{a}\right)^2 + \left(\frac{y}{b}\right)^2 = 1.$$

Let the prescribed conjugate radii be OP and OQ such that the coordinates $x|y$ and $x'|y'$ of their end points satisfy the conditions

$$(2) \qquad \frac{x'}{a} = -\frac{y}{b}, \qquad \frac{y'}{b} = \frac{x}{a}.$$

(The conditions (2) give us directly for the product of the slopes y/x and y'/x' of the two radii the known value $-b^2/a^2$ for the product of the slopes of the conjugate radii.)

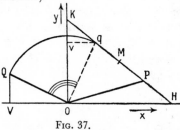

FIG. 37.

Let the base point of the ordinate from Q be V. We rotate the right triangle OQV clockwise about O by $90°$ to the position Oqv and extend the straight line Pq to intersect with the axes of the ellipse at H and K. According to (2), the distances of the points q and P from the x-axis and the distances of the points P and q from the y-axis are in the ratio of a/b. Consequently (according to the ray theorem),

$$\frac{Hq}{HP} = \frac{a}{b} \quad \text{and} \quad \frac{KP}{Kq} = \frac{a}{b}.$$

It then follows from this that

$$\frac{HP + Pq}{HP} = \frac{Kq + qP}{Kq}, \quad \text{i.e.,} \quad HP = Kq,$$

so that the center M of Pq is also the center of HK.

If we substitute HP for Kq, one of our proportions becomes

(3) $KP/HP = a/b.$

In order to obtain a second equation for the unknowns KP and HP, we obtain the cosine and sine of the angle v from HK to the x-axis:

$$\cos v = x/KP, \quad \sin v = y/HP;$$

squaring and adding, we obtain

(4) $\dfrac{x^2}{KP^2} + \dfrac{y^2}{HP^2} = 1.$

From (1), (3), and (4) it immediately follows that

$$KP = a, \quad HP = b.$$

This gives us the following simple

CONSTRUCTION. 1. We rotate OQ about O 90° through the interior of the obtuse angle POQ to the position Oq. 2. We determine the center M of Pq and the points of intersection H and K of the line Pq with the circle of center M and radius MO.

KP and HP *are then equal to half the length of the axes of the ellipse, while* OH *and* OK *represent the positions of the axes of the ellipse.*

The rest is simple.

43   An Ellipse in a Parallelogram

To inscribe in a prescribed parallelogram an ellipse that is tangent to the parallelogram at a boundary point.

The solution of this problem is based upon the theorem: Every ellipse can be considered as a normal projection of a circle.

Let $ABCD$ be the given quadrilateral, N the given boundary point lying on AB. Let the other points at which the ellipse touches the boundary of the parallelogram be K on BC, M on CD, and H on DA.

In the normal projection, in which the ellipse has the image of a circle, the parallelogram $ABCD$ and the tangency points N, K, M, H

appear as projections of a parallelogram circumscribing a circle, and specifically of a rhombus *abcd* with the tangency points *n, k, m, h*.

Since $nk \parallel hm \parallel ac$ and $nh \parallel km \parallel bd$ and since parallelism is preserved in a normal projection, $NK \parallel HM \parallel AC$ and $NH \parallel KM \parallel BD$. Thus, we find the tangency points *H* and *K*, respectively, by causing the parallels through *N* to *BD* and *AC* to intersect with *DA* and *BC*, respectively. The fourth tangency point *M* is the point of intersection of *CD* with the parallel through *H* to *AC*.

Let the centers of the circle and ellipse be *o* and *O*, respectively.

We will now assume an arbitrary point *z* on the arc *nh* of the circle, connect this point with *m* and *n*, and designate the points of intersection of these connecting lines with *hk* and *da* as *x* and *y*. The two triangles *omx* and *any* are then similar, since the angles at *o* and *a*, as well as the angles at *m* and *n*, are equal because they are enclosed between pairs of orthogonal legs. From this similarity we obtain the proportion

$$ox/om = ay/an.$$

If we substitute *oh* for *om* and *ah* for *an* in this proportion, we obtain

$$ox/oh = ay/ah.$$

Let the normal projections of the points *x, y, z* be *X, Y, Z*. Since the ratio of parallel segments is not altered in normal projection, we have

$$OX/OH = AY/AH.$$

The points X *and* Y *accordingly divide the radius of the ellipse* OH *and the ellipse tangent* AH *in the same proportions.*

Quite similar proportions are naturally found to obtain for the other ellipse arcs *MH, MK, NK*.

We assign the tangents *AH, BK, DH, CK* to the arcs *NH, NK, MH, MK*, respectively.

In summary we can then say:

If we connect a point of one of the four arcs with *M* and *N*, the points of intersection of these connecting lines with the radius (*OH* or *OK*) and the corresponding tangents divide the radius and tangents in the same proportions.

This gives rise to the following elegant construction.

We divide the radii *OH* and *OK* and the tangents *AH, BK, DH, CK* each into *ν* equal segments (eight segments are shown in Figure 38) and number the segments from 1 to *ν*, beginning from the center of

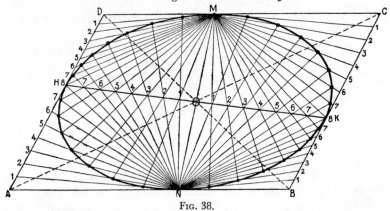

Fig. 38.

the ellipse with the radii and at the corners of the parallelogram with the tangents. We then connect M (N) with an arbitrary segment point of a radius and N (M) with the segment point with the same number of the tangent corresponding to the arc bounded by N (M) and the end point of the radius. The point of intersection of the two connecting lines is in each case a point on the ellipse.

44   A Parabola from Four Tangents

To draw a parabola four tangents to which are given.

The simplest solution of this beautiful problem is based upon

LAMBERT'S THEOREM: *The path of rotation of a parabola tangent triangle passes through the focus.*

(J. H. Lambert (1728–1777) was a German mathematician.)

In order to prove Lambert's theorem we need the

THEOREM OF SIMILAR TRIANGLES: *Two tangents* SA *and* SB *to a parabola, together with the lines from the focus to the contact points* A *and* B *and the point of intersection* S *of the tangents, form two similar triangles* FSA *and* FSB *such that the angle of the one triangle, situated at the point of tangency, is always equal to the angle of the other triangle that is situated at the point of intersection.*

PROOF. In accordance with the classical construction of the parabola, the mirror images H and K of the focus F on the tangents SA and SB, respectively, fall on the base points of the altitudes dropped from A and B, respectively, on the directrix L.

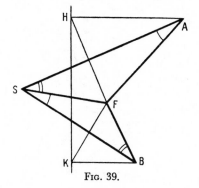

Fɪɢ. 39.

Since the angles *FAS* and *HAS* are symmetrical, and the angles *HAS* and *FHK*, as angles between pairs of orthogonal legs, are equal, it follows that

$$\angle FAS = \angle FHK$$

and likewise that

$$\angle FBS = \angle FKH.$$

The angles *FHK* and *FKH*, as the boundary angles opposite the chords *FK* and *FH*, respectively, on the circumference of rotation of the triangle *FHK* (whose center is the intersection *S* of the median perpendiculars *SA* and *SB* of the triangle) are half as great as the corresponding central angle and consequently equal to angles *FSB* and *FSA*, respectively. Consequently,

$$\angle FAS = \angle FSB \quad \text{and} \quad \angle FBS = \angle FSA. \qquad \text{Q.E.D.}$$

Lambert's theorem follows directly from the theorem we have just proved.

In fact: If *P* and *Q* are the points of intersection of a third tangent with the tangents *SA* and *SB* that touches the parabola at *O*, then, according to the theorem of similar triangles,

$$\angle FAS = \angle FSB \quad \text{and} \quad \angle FAP = \angle FPO$$

and consequently

$$\angle FSQ = \angle FPQ.$$

According to this equation, however, the quadrilateral *FPSQ* is a circle quadrilateral.

Lambert's theorem gives us directly the requisite *construction:* From the four tangent triangles that can be formed from the four given

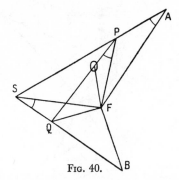

FIG. 40.

tangents, we choose two and draw the circumference for each. The point of intersection of the two circumferences is the focus. We then find the mirror image of the focus on two tangents and in this way obtain two points of the directrix, which gives us the directrix. The rest is extremely simple.

Note. The theorem of the circumference of the tangent triangle leads directly to the solution of the interesting problem:

Determine the locus of the foci of all parabolas that are tangent to three straight lines.

The sought-for locus is the circumference of the triangle formed from the lines.

45. A Parabola from Four Points

To draw a parabola that passes through four given points.

This lovely problem was first solved by Newton in his celebrated *Philosophiae naturalis principia mathematica*, 1687, and then once again in 1707 in his *Arithmetica universalis*.

It is commonly based upon the auxiliary problem:

To draw a parabola for which three points and direction of the axis are known.

The following solution of the auxiliary problem is based on the two theorems:

I. *The centers of parallel chords of a parabola lie on a parallel to an axis.*

II. *The perpendicular bisector of a parabola chord and the perpendicular to the axis through the center of the chord mark off the half parameter on the axis.*

PROOF. The equation for the amplitude of a parabola is commonly expressed in the form $y^2 = 2px$. If $x|y$ and $X|Y$ are the end points of a parabola chord, the slope of the chord with respect to the x-axis $\mathfrak{S} = (Y - y)/(X - x)$. From

$$y^2 = 2px \quad \text{and} \quad Y^2 = 2pX$$

it follows, however, by subtraction that

$$Y^2 - y^2 = 2p(X - x), \quad \text{i.e.,} \quad \mathfrak{S} = \frac{Y - y}{X - x} = \frac{2p}{Y + y}.$$

If we call the ordinate of the midpoint of the chord η, the last equation can be written (because $2\eta = Y + y$) in the form

$$\eta = \frac{p}{\mathfrak{S}}.$$

According to this equation, the midpoints of all chords with the same slope $\mathfrak{S}$ have the *same* ordinate, with the result that these midpoints lie on a line parallel to the axis of the parabola, and thus I. is proved.

To prove II., we take note of the fact that the segment marked off on the axis by the perpendicular bisector of our chords and the perpendicular to the axis through the chord midpoint is equal to $\eta\mathfrak{s}$, where $\mathfrak{s}$ is the slope of the perpendicular bisector of the chord with respect to the perpendicular to the axis. However, since $\mathfrak{s} = \mathfrak{S}$, the length of the segment is $\eta\mathfrak{S} = p$, which was to be proved. From II. it also follows that: *If the midpoints of two parabola chords lie on a perpendicular to the axis, the perpendicular bisectors of the chords intersect on the axis.*

Let A, B, C be the given parabola points, $\mathfrak{R}$ the direction of the axis. Let us draw through the center M of AB a parallel to the axis, through the center N of CA the perpendicular to the axis, and call their point of intersection M_0. Then according to I., M_0 is the midpoint of the parabola chord A_0B_0 that passes through M_0 and is parallel to AB. We draw the perpendicular bisectors of CA and A_0B_0 (the latter as a perpendicular dropped from M_0 to AB). According to II., their point of intersection is a point on the axis, its distance from the base point of the perpendicular dropped from M_0 or N is the half parameter p. The rest is simple. For example, making use of the subnormal (p) from A, we draw the normal AU and the tangent AV (both being drawn to the axis). The midpoint of UV is then the focus and the mirror image of the focus on the tangent is a point on the directrix.

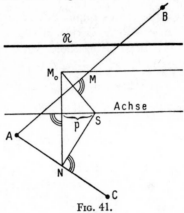

FIG. 41.

The solution of Newton's parabola problem is based upon the follow-ing auxiliary theorem: *In all parabola quadrilaterals the products of the diagonal segments are proportional to the squares of the segments on the diagonals that are bounded by their point of intersection and the axis of the parabola.*

PROOF.   Let AB be an arbitrary parabola chord, let M be its midpoint, U the point of intersection of the parallel to the parabola axis through M.   If we select UM as the x-axis and the parabola tangent through U as the y-axis, we obtain the usual parabola equation in the form

$$y^2 = 4kx,$$

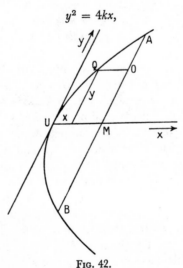

FIG. 42.

where k is the focal radius of the coordinate origin U. The coefficient $4k$ possesses the value $2p/\sin^2 \kappa$, where $2p$ is the parameter and κ the angle enclosed between the coordinate axes or the angle formed by the chord AB with the axis of the parabola.

We select an arbitrary point O on AB and designate the point of intersection of the parallel to the x-axis through O with the parabola as Q, the coordinates of Q as x and y, and the coordinates of A as X and Y, so that

$$QO = q = X - x, \qquad OA = Y - y, \qquad OB = Y + y.$$

From

$$Y^2 = 4kX \quad \text{and} \quad y^2 = 4kx$$

it follows by subtraction that

$$Y^2 - y^2 = 4k(X - x)$$

or

$$(Y + y)(Y - y) = 4k(X - x),$$

so that

(1) $$OA \cdot OB = 4kq.$$

If $A'B'$ is a second parabola chord through O, then accordingly

(2) $$OA' \cdot OB' = 4k'q,$$

with $4k' = 2p/\sin^2 \kappa'$, where κ' is the angle of the chord $A'B'$ with the parabola axis.

Division of (1) and (2) gives

$$OA \cdot OB/OA' \cdot OB' = k/k' = \sin^2 \kappa'/\sin^2 \kappa.$$

If H and H' are the points of intersection of the chords AB and $A'B'$ with the parabola axis, it follows from the sine theorem that

$$OH/OH' = \sin \kappa'/\sin \kappa.$$

From the last two equations we finally obtain

$$OA \cdot OB/OA' \cdot OB' = OH^2/OH'^2. \qquad \text{Q.E.D.}$$

With this theorem we can now obtain the following *solution to Newton's problem:* Let A, B, C, D be the given points. We draw the diagonals AC and BD of the quadrilateral $ABCD$ and call their point of intersection O. On the diagonals we mark off from O the mean proportionals $OP = \sqrt{OA \cdot OC}$ and $OQ = \sqrt{OB \cdot OD}$. The connecting line QP, according to the theorem we have just proved, is then parallel to the parabola axis, and the problem now reduces to the auxiliary problem treated above.

The following *projective solution of Newton's problem* also consists of the reduction of the problem to the preceding auxiliary problem. This transformation of the problem is accomplished by means of Desargues' involution theorem (No. 63). According to this theorem, every tangent to a parabola cuts the opposite sides of an inscribed quadrilateral in point pairs of an involution in which the point of tangency of the tangent is a double point.

As tangent T let us choose a *very* distant one. Let it be tangent to the parabola at O and let it be cut at P, Q, P', and Q' by the lines AB, BC, CD, DA connecting the four given parabola points. O is then the double point of the involution determined by the pairs (P, P') and (Q, Q'). Similarly, the rays drawn from an arbitrary point Z of the picture plane to P, Q, P', Q', O form an involution with the ray pairs (ZP, ZP') and (ZQ, ZQ') and the double ray ZO. Because of the very great distances of the points P, Q, P', Q', O the rays ZP, ZQ, ZP', ZQ' *on the drawing paper* run parallel to the quadrilateral sides AB, BC, CD, DA, and the ray ZO here runs parallel to the axis of the parabola. (The slope $(y - b)/(x - a) = (\sqrt{2px} - b)/(x - a)$ of the line connecting points $Z(a|b)$ and $O(x|y)$, because of the great value of x, is essentially equal to zero, so that the ray ZO appears parallel to the axis on the drawing paper.)

Accordingly we obtain the following *construction*. We draw through an arbitrary point Z of the paper the parallels p, q, p', q' to the lines AB, BC, CD, and DA and construct a double ray of the involution determined by the ray pairs (p, p') and (q, q'); this ray has the direction of the parabola axis. Thus, the problem is reduced to the auxiliary problem solved above.

Since in ray involution there are in general *two* double rays, there are in general *two* parabolas that can be drawn through four given points.

46   A Hyperbola from Four Points

To draw a right-angle (equilateral) hyperbola for which four points are given.

The construction is based upon the *auxiliary theorem: The Feuerbach circle of a triangle inscribed in an equilateral hyperbola passes through the center of the hyperbola.*

PROOF. Let ABC be a triangle inscribed in an equilateral hyperbola with the center at Z and the asymptotes I and II; let A', B', C' be the midpoints of the sides BC, CA, AB, and let A_1 and A_2 be the points of intersection of BC with I and II, and B_1 and B_2 the points of intersection of CA with I and II.

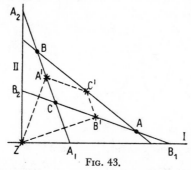

FIG. 43.

Since the asymptotes mark off equal segments on the extensions of a hyperbola chord, $BA_2 = CA_1$ and $CB_2 = AB_1$, and A' is the midpoint of A_1A_2 and B' the midpoint of B_1B_2. These midpoints are also the midpoints of the circumferences of rotation of the right triangles A_1ZA_2 and B_1ZB_2, so that

$$\angle A'ZA_1 = \angle A'A_1Z \quad \text{and} \quad \angle B'ZB_1 = \angle B'B_1Z.$$

Since the difference of the left sides of these equations represents angle $A'ZB'$ and the difference of the right sides angle A_1CB_1 (according to the theorem of external angles), both of these angles are equal or angles $A'ZB'$ and $A'CB'$ are supplementary. However, since the angles of the parallelogram $CA'C'B'$ at C and C' are equal, angles $A'ZB'$ and $A'C'B'$ are also supplementary. The quadrilateral $ZA'C'B'$ is therefore a circle quadrilateral. In other words: the circumference of rotation of the triangle $A'B'C'$, i.e., the Feuerbach circle of the triangle ABC (see No. 28), passes through the center of the hyperbola. Q.E.D.

CONSTRUCTION. Let the four given points be A, B, C, D. We draw the Feuerbach circle of the triangles ABC and ABD; the point of their intersection Z is the center of the hyperbola. We connect Z to the midpoint A' of BC, draw the circle $A'|A'Z$ and at its points of intersection A_1 and A_2 with the line BC we have two points of the asymptotes I and II, which gives us the asymptotes. The rest is easy. (To draw the hyperbola from points, for example, we pass an arbitrary line through one of the given points, for example A, and mark off on this line the segment between A and I from II to A; the point at the end of the marked-off segment is a new point of the hyperbola. Repetition of the construction with new lines through A gives us as many points of the hyperbola as desired.)

NOTE. The proved auxiliary theorem immediately gives, as well, the solution to the interesting

LOCUS PROBLEM: *Find the locus of the centers of all equilateral hyperbolas that can be circumscribed about a given triangle.*

The locus is the Feuerbach circle of the given triangle.

47   Van Schooten's Locus Problem

Two vertexes of a rigid triangle in a plane slide along the arms of an angle of the plane; what locus does the third vertex describe?

Franciscus van Schooten (the younger) (1615–1660), a Dutch mathematician, treated this beautiful problem in his *Exercitationes mathematicae*, which appeared in 1657.

SOLUTION. We will first consider a special case of van Schooten's problem, the solution to which had already been taught by the Byzantine Proclus (410–485).

On a rigid line three points are marked; two of these slide along the arms of a right angle; what locus does the third describe?

We select the arms I and II of the right angle as the x- and y-axes of a coordinate system. Let the three marked points of the rigid line be A, B, C, their mutual distances $BC = a, CA = b$, and $AB = c$. Then $c = a \pm b$, accordingly as C does or does not lie between A and B. Let the point A slide on I and B on II. Let the marked point C possess the coordinates x and y. Let the angle of the line with respect to the x-axis be v; thus x, as the projection from a on I, is equal to $a \cos v$; y, as the projection of b on II, is equal to $b \sin v$; and consequently, $x^2 = a^2 \cos^2 v, y^2 = b^2 \sin^2 v$, and

$$\frac{x^2}{a^2} + \frac{y^2}{b^2} = 1.$$

The locus of the marked point C is thus an ellipse with the half axes a *and* b.

This locus property is the basis of the so-called paper strip construction of the ellipse and trammel.

PAPER STRIP CONSTRUCTION OF THE ELLIPSE

On the sharp edge of a paper strip we mark off the three points in the sequence B, A, C in such manner that $BC = a$ and $AC = b$ ($< a$) are equal to the given half axes of an ellipse. We move the strips in such manner that A always remains on the x-axis and B on the y-axis and we constantly mark the place at which C is situated. The locus described by the point C is an ellipse with the prescribed half axes a and b.

THE TRAMMEL

A trammel consists of a cross with two grooves at right angles to each other in which two sliding pins A and B move. The pins are fixed to a beam to which at some point a movable pencil M can be attached. When the pins slide in the grooves the pencil describes an ellipse with the half axes AM and BM.

Now for the general van Schooten problem!

Let S be the apex of the fixed angle σ along the arms of which the vertexes A and B of the rigid triangle ABC slide. We draw the circle $\Re$ with AB as chord and σ as peripheral angle, join its midpoint M with C and determine the points of intersection P and Q of this connecting line with $\Re$. Let us consider this circle along with points P and Q as being firmly connected to the rigid triangle, so that it also participates in the motion of the triangle. Consequently, since σ is the peripheral angle opposite AB, it passes continuously through S. The arcs AP and AQ continuously change their position *but not their*

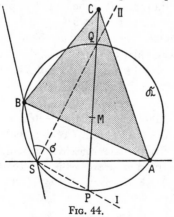

FIG. 44.

magnitude! This entails the invariance of the peripheral angles ASP and ASQ, which implies the invariance of the directions I and II that are determined by SP and SQ. Since PQ is a diameter of $\mathfrak{K}$, I and II are perpendicular to each other. We can therefore consider the motion of the vertex C as the motion of the marked point C of a rigid line PQC the other marked points of which P and Q slide along the arms I and II of a right angle. According to the above special case, C describes an ellipse.

RESULT: VAN SCHOOTEN'S THEOREM: *The locus of one corner of a three-cornered plate the other two corners of which slide along the arms of a fixed angle is an ellipse.*

The above derivation also gives the magnitudes and position of the ellipse. The axes of the ellipse have the positions I and II and the magnitudes $2 \cdot CP$ and $2 \cdot CQ$.

48   Cardan's Spur Wheel Problem

What is the locus described by a marked point on a circular disc that rolls along the inner edge of a disc of double its radius?

Jerome Cardan, an Italian mathematician (1501–1576), is known for the Cardan formula for solution of cubic equations.

SOLUTION. Let the boundary of the large disc be $\mathfrak{K}$ and that of the smaller disc $\mathfrak{k}$, and let their radii be equal to $R = 2r$ and r, respectively. First we will observe the motion of the marked disc diameter AB, which we give the mark M. At the beginning of the motion let A lie at the midpoint O and B at the boundary point H on $\mathfrak{K}$. When the circle $\mathfrak{k}$ is rolled forward within $\mathfrak{K}$ by the arc HT, let it cut the radius OH at X, and let Y be the point at which it cuts the radius OK of $\mathfrak{K}$, which is perpendicular to OH. Since the angle XOY is 90°, XY is a diameter of $\mathfrak{k}$, and the intersection S of XY with OT is the center of $\mathfrak{k}$. If w is a peripheral angle XOT of $\mathfrak{k}$ in radian measure, then the corresponding central angle XST is $2w$ and the arc XT is $2rw$. However, since w also represents the central angle HOT of $\mathfrak{K}$, the arc $HT = Rw = 2rw$. The arc XT of the smaller circle is exactly as long as the arc HT of the larger circle upon which the small circle is rolled forward. X must therefore be the end B of the marked diameter AB, consequently Y is the other end A of this diameter. *The rotation of a disc along the inner margin of a disc of double its width consequently means that the end points of a marked diameter of the smaller circle slide along two fixed orthogonal diameters of the larger circle.* The locus of our marked point *M* is therefore also the locus of the mark *M* of the diameter *AB* whose end points *A* and *B* slide along the arms *OK* and *OH* of the right angle *HOK*. In view of the paper strip construction of the ellipse (No. 47), *the locus we are seeking is thus an ellipse.*

The half axes of this ellipse are *MA* and *MB*.

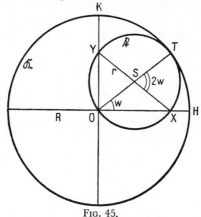

FIG. 45.

NOTE. Since a marked point on the boundary of the smaller disc describes a diameter of the larger disc, a gear consisting of two spur wheels the ratio of whose diameters is as 2:1 effects the *conversion of a circular motion into a reciprocal rectilinear motion.*

49 Newton's Ellipse Problem

To determine the locus of the centers of all ellipses that can be inscribed in a given (convex) quadrilateral.

Newton's very elegant solution to this problem is based upon the theorem, also stemming from Newton:

The line connecting the centers of the diagonals of a quadrilateral circumscribed about a circle passes through the center of the circle.

The proof of this property of a tangent quadrilateral is based upon the following auxiliary theorem: *The locus of the common vertex of two triangles with prescribed base lines and a prescribed area sum is a straight line.*

[PROOF: Let *f* and *g* be the two prescribed base lines, *x* and *y* the distances of the common vertex *S* of the two triangles from the prescribed base lines and, at the same time, the "coordinates" of the point *S*.   The prescribed sum of the areas of the two triangles we will call *K*.   Since the triangles have the area $\frac{1}{2}fx$ and $\frac{1}{2}gy$, we obtain the equation $fx + gy = 2K$, and this is the equation of a straight line.]

Let there be circumscribed about a circle of center *O* and radius *r* the tangent quadrilateral *ABCD* with the sides $AB = a$, $BC = b$, $CD = c$, $DA = d$, so that $a + c = b + d$.   Let *M* be the midpoint of the diagonal *AC* and *N* the midpoint of *BD*, 2*J* the area of the quadrilateral.   Since $\triangle MAB$ and $\triangle MCD$ have areas equal to one half $\triangle CAB$ and $\triangle ACD$, respectively, the sum of the areas of the two

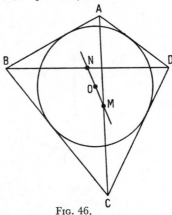

FIG. 46.

triangles *MAB* and *MCD* is equal to *J*, or half the area of the quadrilateral.   Consequently, the line *MN* is the locus of the common vertex *S* of all the pairs of triangles (*SAB*, *SCD*) having the area *J*. However, since the two triangles *OAB* and *OCD* also have the area sum *J* (specifically,

$$\text{I} = OAB + OCD = r\frac{a + c}{2} \quad \text{and} \quad \text{II} = OBC + ODA = r\frac{b + d}{2}$$

and I = II.   From I + II = 2*J* it then follows that I = II = *J*), thus *O* belongs to the locus.   Q.E.D.

Now for the solution to Newton's problem!

Let us consider *any* ellipse inscribed in the given quadrilateral as the normal projection of a circle.   In this reflection the quadrilateral appears as the image (the normal projection) of an object quadrilateral circumscribed about the circle.   Now, since: 1. in the object the center of the circle lies upon the line connecting the midpoints of the diagonals; 2. halving is preserved in the normal projection; 3. the center of the ellipse is the image of the center of the circle, then in the image also the ellipse center lies on the line joining the midpoints of the diagonals of the prescribed quadrilateral.

CONCLUSION: *The locus of the centers of all the ellipses that can be inscribed in a given quadrilateral is a straight line, specifically, the line connecting the midpoints of the diagonals of the quadrilateral.*

50   The Poncelet-Brianchon Hyperbola Problem

To determine the locus of the intersection of the altitudes of all the triangles that can be inscribed in a right-angle (equilateral) hyperbola.

Brianchon (1785–1864) and Poncelet (1788–1867) were French mathematicians. The solution is in vol. XI of the *Annales de Gergonne* (1820–1821).

We relate the hyperbola to its asymptotes, which will serve as coordinate axes (the x-axis and ξ-axis), and take the abscissa (ordinate) of the apex of the hyperbola as the unit length. The equation for the hyperbola then reads

$$x\xi = 1.$$

Let PQR be an arbitrary triangle inscribed in the hyperbola, i.e., a triangle whose vertexes P, Q, R lie on the hyperbola. Let the abscissas of the points P, Q, R be a, b, c, the ordinates thus being $\alpha = 1/a$, $\beta = 1/b$, $\gamma = 1/c$.

The slope of the side QR is $(\beta - \gamma)/(b - c)$ or, if we substitute $1/b$ and $1/c$ for β and γ, $-1/bc$. The slope of the altitude to QR is thus bc.

The equation of this altitude is thus $\xi - \alpha = bc(x - a)$ or

(1) $$\xi + abc = bc(x + \alpha\beta\gamma).$$

For the altitude passing through Q we obtain similarly

(2) $$\xi + abc = ca(x + \alpha\beta\gamma).$$

Now, if the coordinates of the altitude intersection are understood to be $x|\xi$, (1) and (2) both apply, and by equalizing the right sides we find the abscissa x of the point of intersection of the altitudes:

(I) $$x = -\alpha\beta\gamma.$$

If we introduce this value into (1) or (2), we obtain as the ordinate of the altitude intersection

(II) $$\xi = -abc.$$

Multiplying (I) and (II) finally gives us

$$x\xi = 1.$$

The altitude intersection thus lies on the hyperbola.   Consequently:

The locus of the point of intersection of the altitudes of all the triangles that can be inscribed in an equilateral hyperbola is the hyperbola itself.

51   A Parabola as Envelope

On one arm of an angle the arbitrary segment *e* and, on the other, the segment *f* are marked off *n* times in succession from the vertex of the angle, and the segment end points are numbered, beginning from the vertex, 0, 1, 2, ..., *n* and *n*, *n* − 1, ..., 2, 1, 0, respectively.

Prove that the lines joining the points with the same number envelop a parabola.

The proof is based upon the

THEOREM OF APOLLONIUS: *Two tangents to a parabola are divided into segments of like proportion by a third and this third is divided in the same proportion by its point of tangency.*

More precisely: If the two parabola tangents *SA* and *SB*, with the points of tangency *A* and *B*, are intersected by a third parabola tangent at *P* and *Q*, and if *O* is the point of tangency of this third tangent (Figure 40), we obtain the equation

$$\frac{SP}{PA} = \frac{OQ}{OP} = \frac{BQ}{SQ}.$$

The proof of the Apollonian theorem is based upon the known parabola property: *The point of intersection of two parabola tangents lies on a parallel to the parabola axis, passing through the midpoint of the chord connecting the points of tangency.*   (It follows directly from the situation that the three median perpendiculars of the triangle *FA'B'* whose vertexes are the focus *F* and the projections *A'* and *B'* of the points of tangency *A* and *B* on the directrix pass through a single point.   Two median perpendiculars are the tangents and the third is the parallel to the axis.)

Because of this property

(1)   (2) $q' = b'$,   (3) $b' + \beta' = a' + \alpha'$,

if we call the projections of the segments $AP = a$, $PS = \alpha$, $BQ = b$, $QS = \beta$, $OP = p$, $OQ = q$ on the directrix a', α', b', ... Moreover, as a result of the equality of the projections of the segment PQ and the traverse PSQ,

$$(4) \qquad p' + q' = \alpha' + \beta'.$$

If, in accordance with (1) and (2), we substitute a' and b' for p' and q' in (4), we obtain

$$\alpha' + \beta' = a' + b',$$

and this equation when combined with (3) shows that

$$\alpha' = b' \quad \text{and} \quad \beta' = a'.$$

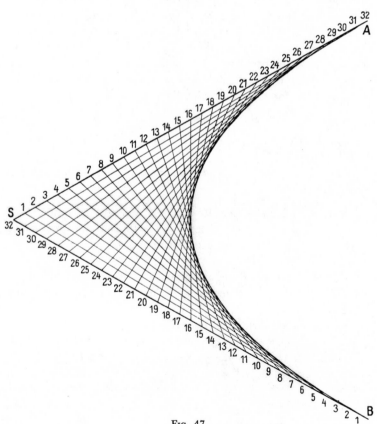

FIG. 47.

This now gives us

$$
\left.\begin{aligned}
\alpha/a &= \alpha'/a' = b'/a' \\
q/p &= q'/p' = b'/a' \\
b/\beta &= b'/\beta' = b'/a'
\end{aligned}\right\},
$$

which proves the theorem of Apollonius.

The execution of the envelope construction described above is now very simple. Let us call the apex angle S; we then select on the arms of the angle the points A and B in such manner that $SA = ne$ and $SB = nf$ (A and B are the same points that received the numbers n and 0 in the numbering process previously described), and consider the *parabola that is tangent to the arms of the angle at* A *and* B. According to Apollonius' theorem, *the line connecting the point* P *on* SA *to which the number v has been assigned with the point* Q *on* SB *is tangent to the parabola.* [The ratios $PS:PA$ and $QB:QS$ are both equal to $v:n - v$.] Consequently, the parabola is enveloped by the lines joining the points with the same numbers.

At the same time, Apollonius' theorem makes it possible to draw the tangency point for each connecting line.

52 The Astroid

To find the envelope of a straight line, two marked points on which slide along two fixed, mutually perpendicular axes.

Gottfried Wilhelm Leibniz (1646–1716), the inventor of infinitesimal calculus, founded the theory of envelopes in 1692 in his paper *De linea ex lineis numero infinitis ordinatim ductis inter se concurrentibus easque omnes tangente.*

SOLUTION. We seek the equation of the envelope in the coordinate system in which the two given axes are the x-axis and y-axis and their intersection O is the origin.

Let the constant distance between the designated points be represented by l. Let AB and $A'B'$ represent two positions of the marked-off distance l, M and N the midpoints of AA' and BB', $OM = a$, $ON = b$, $AA' = 2\alpha$, $BB' = 2\beta$, thus $OA = a + \alpha$, $OA' = a - \alpha$, $OB = b - \beta$, $OB' = b + \beta$. The conditions $AB = l$ and $A'B' = l$ can then be written

(1) $(a + \alpha)^2 + (b - \beta)^2 = l^2$ and $(a - \alpha)^2 + (b + \beta)^2 = l^2$,

from which we obtain by subtraction

(2)
$$a\alpha = b\beta.$$

The point of intersection $S(x, y)$ of the two straight lines AB and $A'B'$ is expressed by the two equations

$$\frac{x}{a + \alpha} + \frac{y}{b - \beta} = 1 \quad \text{and} \quad \frac{x}{a - \alpha} + \frac{y}{b + \beta} = 1,$$

and the following two equations:

(3)
$$\frac{ax}{a^2 - \alpha^2} + \frac{by}{b^2 - \beta^2} = 1$$

and

(4)
$$\frac{\alpha x}{a^2 - \alpha^2} = \frac{\beta y}{b^2 - \beta^2},$$

which are obtained from the first two by addition and subtraction.

If we then divide (4) by (2), we obtain

$$\frac{x}{a(a^2 - \alpha^2)} = \frac{y}{b(b^2 - \beta^2)}$$

and, with the use of (3),

(5)
$$x = a \frac{a^2 - \alpha^2}{a^2 + b^2}, \qquad y = b \frac{b^2 - \beta^2}{a^2 + b^2}.$$

If we then allow A and A' and B and B' to approach each other (naturally maintaining the conditions $AB = l$ and $A'B' = l$), then α and β become continuously smaller and the point of intersection S of the lines AB and $A'B'$ comes closer and closer to the envelope, finally reaching it when α and β are equal to zero. The point $x|y$ at which the envelope is reached is then represented, according to (5), by the equations

(5')
$$x = \frac{a^3}{a^2 + b^2}, \qquad y = \frac{b^3}{a^2 + b^2},$$

in which, in view of (1),

(1')
$$a^2 + b^2 = l^2$$

is true.

From (5′) it then follows that

$$a^3 = l^2x, \quad b^3 = l^2y \qquad \text{or} \qquad a^2 = l^{4/3}x^{2/3}, \quad b^2 = l^{4/3}y^{2/3},$$

from which

$$l^2 = l^{4/3}x^{2/3} + l^{4/3}y^{2/3}$$

is obtained by addition.

The equation of the envelope thus reads

$$x^{2/3} + y^{2/3} = l^{2/3},$$

or, in rational form,

$$(l^2 - x^2 - y^2)^3 = 27l^2x^2y^2.$$

(The second form is obtained from the first by cubing twice. The first cubing results in

$$x^2 + y^2 + 3x^{2/3}y^{2/3}(x^{2/3} + y^{2/3}) = l^2$$

or

$$3x^{2/3}y^{2/3}l^{2/3} = l^2 - x^2 - y^2,$$

and on the second cubing we obtain the indicated form.)

Because of its shape the curve $x^{2/3} + y^{2/3} = l^{2/3}$ is called an *astrois* or *astroid* in accordance with a proposal made by J. J. Littrow in 1838 or a *star line* after M. Simon's proposal.

The astroid is a hypocycloid* in which the radius of the fixed circle is four times that of the rolling circle.

PROOF. In Figure 49, let C be the center, l the radius, the arc JT a section of the fixed circle $\mathfrak{F}$, $\mathfrak{R}$ the rolling circle at the moment in which it touches $\mathfrak{F}$ at the point T, so that the center Z of the rolling circle cuts the radius CT into the two segments $ZT = r = \frac{1}{4}l$ and $CZ = 3r$. Also, let M be the point on the circumference of $\mathfrak{R}$ whose path we are to follow, x its abscissa and y its ordinate. We then select C as the origin of the coordinates and draw the (horizontal) x-axis through point J, at which the marked point was at the beginning of its motion. The arcs JT of $\mathfrak{F}$ and TM of $\mathfrak{R}$ are then of equal length; the sector angle $W = \angle TZM$ is therefore four times the sector angle $w = \angle JCT$. The slope of the radius ZM from the horizontal is $4w - w = 3w$, and the horizontal and vertical projections of ZM are $r\cos 3w$ and $r\sin 3w$, respectively. The

---

* If a circular disc rolls along the circumference of a fixed circle (without sliding), a marked point on the circumference of the rolling disc (the "rolling circle") describes an epicycloid when the disc rolls along the outside of the fixed circle and a hypocycloid when the disc rolls along the inside.

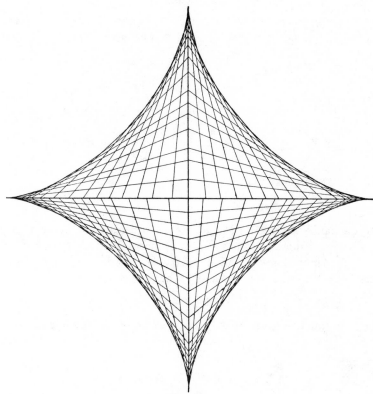

FIG. 48.

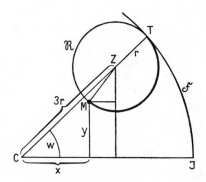

FIG. 49.

corresponding projections of CZ are $3r \cos w$ and $3r \sin w$.   Thus we obtain the equations (which can be read off the figure)

$$x = 3r \cos w + r \cos 3w,$$

$$y = 3r \sin w - r \sin 3w,$$

which, as a result of the relationships

$$\cos 3w = 4 \cos^3 w - 3 \cos w,$$

$$\sin 3w = 3 \sin w - 4 \sin^3 w,$$

can be transformed into

$$x = l \cos^3 w, \qquad y = l \sin^3 w.$$

In the pair of equations obtained the coordinates of the hypocycloid point $x | y$ are represented as functions of the so-called *rolling angle w*.

To obtain the curve equation in Cartesian coordinates, we solve for $\cos w$ and $\sin w$, square, and add.   Thus, we obtain

$$x^{2/3} + y^{2/3} = l^{2/3},$$

i.e., the equation of an astroid, which was to be demonstrated.

53   **Steiner's Three-pointed Hypocycloid**

To determine the envelope of the Wallace line of a triangle.

SOLUTION.   Let ABC be the given triangle, M the midpoint, and r the radius of the circle $\mathfrak{U}$ circumscribed about it.

A Wallace line of a triangle is the line connecting the three base points of the perpendiculars dropped from any point P on the circumference of the circle of circumscription to the sides of the triangle.

We will make M the origin of an X-Y coordinate system and preliminarily select the X-axis arbitrarily.   If we designate the angles formed by the radii MA, MB, MC, MP with the positive side of the X-axis as 2α, 2β, 2γ, 2φ, the coordinates of the three corners A, B, C are

$$(r \cos 2\alpha | r \sin 2\alpha), \quad (r \cos 2\beta | r \sin 2\beta), \quad (r \cos 2\gamma | r \sin 2\gamma),$$

and the coordinates of the point P are $(r \cos 2\varphi, r \sin 2\varphi)$.

In order to find the coordinates $X_1 | Y_1$ of the base point F_1 of the perpendicular dropped from P to BC, we form the equations of the line BC (in the two-point form) and the line PF_1 (in the slope form) and find from these equations that

$$X_1 = f(\cos 2\beta + \cos 2\gamma + \cos 2\varphi - \cos \overline{2\beta + 2\gamma - 2\varphi}),$$

$$Y_1 = f(\sin 2\beta + \sin 2\gamma + \sin 2\varphi - \sin \overline{2\beta + 2\gamma - 2\varphi}),$$

where f represents half of r.

Accordingly, the coordinates $X_2 | Y_2$ of the base point F_2 of the perpendicular dropped from P to CA will naturally be

$$X_2 = f(\cos 2\gamma + \cos 2\alpha + \cos 2\varphi - \cos \overline{2\gamma + 2\alpha - 2\varphi}),$$

$$Y_2 = f(\sin 2\gamma + \sin 2\alpha + \sin 2\varphi - \sin \overline{2\gamma + 2\alpha - 2\varphi}).$$

An appropriate parallel displacement of the coordinate system allows us to put the coordinates into a simpler form. This displacement of the coordinate system is based upon *Sylvester's theorem* (No. 27).

In accordance with this, the altitude intersection H of the triangle ABC has the coordinates

$$r(\cos 2\alpha + \cos 2\beta + \cos 2\gamma) \quad \text{and} \quad r(\sin 2\alpha + \sin 2\beta + \sin 2\gamma).$$

Since the center F of the Feuerbach circle lies halfway between M and H (No. 28), the coordinates of F are

$$X_0 = f(\cos 2\alpha + \cos 2\beta + \cos 2\gamma),$$

$$Y_0 = f(\sin 2\alpha + \sin 2\beta + \sin 2\gamma).$$

It is therefore convenient to select the center of the Feuerbach circle as the origin of the new coordinate system x, y. Between the coordinates $X | Y$ of a point in the old system and $x | y$ in the new system there exist the relations

$$X = X_0 + x, \qquad Y = Y_0 + y.$$

From these relations we obtain for the coordinates $(x_1 | y_1)$ and $(x_2 | y_2)$ of the points F_1 and F_2 in the new system the simpler values

$$x_1 = f(\cos 2\varphi - \cos 2\alpha - \cos \overline{2\beta + 2\gamma - 2\varphi}),$$

$$y_1 = f(\sin 2\varphi - \sin 2\alpha - \sin \overline{2\beta + 2\gamma - 2\varphi})$$

and

$$x_2 = f(\cos 2\varphi - \cos 2\beta - \cos \overline{2\gamma + 2\alpha - 2\varphi}),$$

$$y_2 = f(\sin 2\varphi - \sin 2\beta - \sin \overline{2\gamma + 2\alpha - 2\varphi}).$$

Now the equation for the Wallace line F_1F_2 reads

$$(y - y_1)/(x - x_1) = (y_2 - y_1)/(x_2 - x_1).$$

For the differences $x_2 - x_1$ and $y_2 - y_1$ appearing here, we obtain, in accordance with the coordinate values just given, the expressions

$$\begin{aligned}
x_2 - x_1 &= f(\cos 2\alpha - \cos 2\beta) \\
&\quad + f(\cos \overline{2\beta + 2\gamma - 2\varphi} - \cos \overline{2\gamma + 2\alpha - 2\varphi}) \\
&= -2f \sin \overline{\alpha + \beta} \sin \overline{\alpha - \beta} \\
&\quad + 2f \sin \overline{\alpha + \beta + 2\gamma - 2\varphi} \sin \overline{\alpha - \beta} \\
&= 4f \sin \overline{\alpha - \beta} \sin \overline{\gamma - \varphi} \cos \overline{\alpha + \beta + \gamma - \varphi}
\end{aligned}$$

and similarly

$$y_2 - y_1 = 4f \sin \overline{\alpha - \beta} \sin \overline{\gamma - \varphi} \sin \overline{\alpha + \beta + \gamma - \varphi}.$$

The quotient $(y_2 - y_1)/(x_2 - x_1)$ thus has the value $\sin \Phi/\cos \Phi$ with $\Phi = \alpha + \beta + \gamma - \varphi$, and the equation of the Wallace line assumes the form

$$x \sin \Phi - y \cos \Phi = x_1 \sin \Phi - y_1 \cos \Phi.$$

Using the above values for the coordinates x_1 and y_1, we are able to write the right side of this equation as

$$f(\sin \Phi \cos 2\varphi - \cos \Phi \sin 2\varphi) - f(\sin \Phi \cos 2\alpha - \cos \Phi \sin 2\alpha)$$

$$- f(\sin \Phi \cos \overline{2\beta + 2\gamma - 2\varphi} - \cos \Phi \sin \overline{2\beta + 2\gamma - 2\varphi}),$$

which expression becomes, according to the addition theorem of circular functions,

$$\begin{aligned}
f \sin (\alpha + \beta + \gamma - 3\varphi) &- f \sin (\beta + \gamma - \alpha - \varphi) \\
&- f \sin (\alpha - \beta - \gamma + \varphi) \\
&= f \sin (\alpha + \beta + \gamma - 3\varphi).
\end{aligned}$$

Now the equation of the Wallace line reads

$$x \sin \overline{\alpha + \beta + \gamma - \varphi} - y \cos \overline{\alpha + \beta + \gamma - \varphi}$$
$$= f \sin \overline{\alpha + \beta + \gamma - 3\varphi}.$$

For the sake of a final simplification we now choose the position of the hitherto arbitrary x-axis in such manner that the sum of the three angles α, β, γ is equal to an integral multiple of 2π. It is easily seen that with F as the point of origin there are only three rays, separated from each other by angles of $2\pi/3$, that satisfy this condition. We choose one of these three rays as the *x*-axis. In the coordinate system thus determined, the Wallace line has the simple equation

(1) $$x \sin \varphi + y \cos \varphi = f \sin 3\varphi.$$

To interpret this equation geometrically we draw a triangle *FQR* with the side $FQ = f$, with the angles 2φ at *F* and φ at *R*, thus, with the external angle 3φ at *Q*, whose side *FR* lies on the positive *x*-axis. The side *QR* of this triangle is then the Wallace line $\mathfrak{s}$ represented by (1). In fact: If $x = FU$ is the abscissa, $y = UV$ the ordinate of any point *V* of the line $\mathfrak{s}$, then the perpendicular *FW* dropped from *F* to $\mathfrak{s}$ is $f \sin 3\varphi$ as the projection of *FQ*; on the other hand, as the projection of the traverse $FU + UV$, it is $x \sin \varphi + y \cos \varphi$, so that equation (1) applies to the coordinates of *V*.

In particular, if *V* is the base point of the perpendicular *TV* dropped to $\mathfrak{s}$ from the end point *T* of the extension $QT = 2f$ of *FQ*, *V* lies on

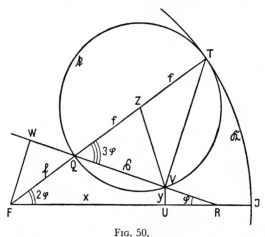

Fig. 50.

the circle $\mathfrak{k}$ whose center *Z* is the midpoint of the hypotenuse *QT* of the right triangle *QTV*, which has the radius *f*, and which is tangent to the Feuerbach circle at *Q* and to the circle $\mathfrak{K}$ of center *F* and radius $3T$ at *T*. Since $\measuredangle VZT$, as an external angle of the isosceles triangle *VZQ*, is equal to 6φ, the arc *VT* of the circle $\mathfrak{k}$ is equal to $f \cdot 6\varphi$. And since the arc *JT* stretching from the point of intersection *J* of circle $\mathfrak{K}$ with the *x*-axis to *T* is equal to $3f \cdot 2\varphi$, and is therefore also equal to $6f\varphi$, it follows that

arc *VT* of $\mathfrak{k}$ = arc *JT* of $\mathfrak{K}$.

If we then think of circle ſ as rolling along circle ℜ (along the inside) so that a point $\mathscr{M}$ marked off on ſ initially lies at J, the marked point arrives precisely at point V at the moment when the rolling circle ſ assumes the drawn position.

The locus of point V is consequently, as the path of the marked point $\mathscr{M}$, a hypocycloid (cf. No. 52), in which the radius of the fixed circle is three times as large as the radius of the rolling circle. And since *at the moment* depicted in the drawing the rolling circle is rotating precisely about the instantaneous point of rotation T, *at this moment* the marked point $\mathscr{M}$ at V is moving in a direction QV that is precisely perpendicular to TV, i.e., the Wallace line ſ is the tangent drawn to the hypocycloid at V! Thus the totality of Wallace lines represents the totality of all the hypocycloid tangents.

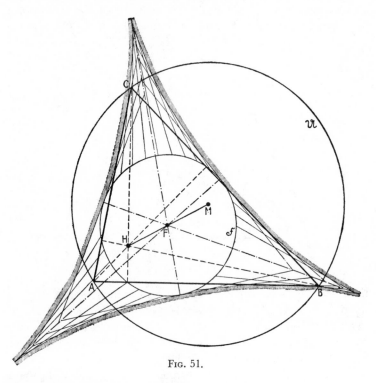

Fig. 51.

CONCLUSION: STEINER'S THEOREM: *The envelope of the Wallace lines of a triangle is a hypocycloid whose fixed circle possesses a radius that is three times as great as the radius of the rolling circle. The center of the fixed circle* is the center of the Feuerbach circle of the triangle, and the radius of the rolling circle is equal to the radius of the Feuerbach circle.

The three *points of the hypocycloid*—the three places at which the marked point on the rolling circle touches the fixed circle—are the end points of the three radii of the fixed circle, separated from each other by 120°, of which one lies on the positive *x*-axis.

The three *apexes of the hypocycloid*—the three places at which the marked point on the rolling circle touches the Feuerbach circle—divide the arcs of the Feuerbach circle lying outside the triangle, from the midpoints of the sides, into segments whose ratio to each other is as 1:2.

[This ratio follows easily from the position of the *x*-axis and from the fact that the peripheral angle opposite the arc of a Feuerbach circle cut off by a triangle side is equal to the difference between the two triangle angles at the end points of the side.]

54 The Most Nearly Circular Ellipse Circumscribing a Quadrilateral

Of all the ellipses circumscribing a given quadrilateral, which deviates least from a circle?

This problem, which was posed in the seventeenth volume of Gergonne's *Annales de Mathématiques*, was solved by J. Steiner (*Crelle's Journal*, vol. II; also: Steiner, *Gesammelte Werke*, vol. I).

SOLUTION (according to Steiner). To begin with, it is clear that the quadrilateral must be convex inasmuch as no ellipse can be circumscribed about a concave quadrilateral.

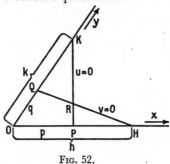

FIG. 52.

Let *OPRQ* be the given quadrilateral, let *QR* cut the extension of *OP* at *H* and *PR* cut the extension of *OQ* at *K*, and let $OP = p$,

$OQ = q$, $OH = h$, $OK = k$. We will take OP as the x-axis, OQ as the y-axis of an oblique-angle coordinate system. The equations for the sides OP and OQ of the quadrilateral are then $y = 0$ and $x = 0$, while the equations for the sides PR and QR are

$$\frac{x}{p} + \frac{y}{k} = 1 \quad \text{and} \quad \frac{x}{h} + \frac{y}{q} = 1$$

or, if we designate the expressions

$$kx + py - kp \quad \text{and} \quad qx + hy - hq$$

as u and v, $u = 0$ and $v = 0$.

The equation for every ellipse that can be circumscribed about the quadrilateral has the form

(1) $$\lambda xu + \mu yv = 0,$$

where λ and μ are two arbitrary constants or so-called parameters. [Since at O $x = 0$ and $y = 0$, at P $y = 0$ and $u = 0$, at Q $x = 0$ and $v = 0$, and, finally, at R $u = 0$ and $v = 0$, the second degree curve $\mathfrak{C}$ represented by (1) passes through all four corners. Thus, $\mathfrak{C}$ is an ellipse of circumscription, which, moreover, also passes through the fifth point $x_0|y_0$, and if we choose λ and μ in such manner that

$$\lambda x_0(kx_0 + py_0 - kp) + \mu y_0(qx_0 + hy_0 - hk) = 0,$$

then $x_0|y_0$ also lies on $\mathfrak{C}$. Since, however, only one second degree curve can pass through five points, $\mathfrak{C}$ is the ellipse $\mathfrak{E}$. Thus, every ellipse of circumscription can be represented by (1).]

We introduce the values of u and v into (1) and obtain the equation of an arbitrary ellipse of circumscription:

(1') $$Ax^2 + 2Bxy + Cy^2 + 2Dx + 2Ey = 0,$$

where

$$A = k\lambda, \quad 2B = p\lambda + q\mu, \quad C = h\mu, \quad D = -kp\lambda, \quad E = -hq\mu.$$

We begin by looking for the locus of the centers of all the parallel chords of the ellipse (1')

(2) $$y = \mathcal{M}x + n,$$

in which $\mathcal{M}$ is the common directional constant of the chords, n the segment cut off on the y-axis by one of these chords, chosen arbitrarily.

If we introduce y from (2) into (1'), we obtain the quadratic equation

$$(A + 2B\mathcal{M} + C\mathcal{M}^2)x^2 + 2[(Cn + E)\mathcal{M} + Bn + D]x + Cn^2 + 2En = 0$$

for the abscissas x_1 and x_2 of the points of intersection of the chord (3) with the ellipse (1). According to a well-known theorem from quadratic equation theory, the sum of the two roots x_1 and x_2 of this equation is

$$x_1 + x_2 = -2\frac{(Cn + E)\mathcal{M} + Bn + D}{A + 2B\mathcal{M} + C\mathcal{M}^2},$$

i.e., the abscissa of the chord midpoint is

$$X = -\frac{(C\mathcal{M} + B)n + E\mathcal{M} + D}{C\mathcal{M}^2 + 2B\mathcal{M} + A}.$$

Since the chord midpoint $X|Y$ satisfies the equation (2) of the chord, $Y = \mathcal{M}X + n$, so that we can substitute $Y - \mathcal{M}X$ for n in the equation found for X. If we do this, we obtain for the coordinates X and Y of the chord midpoint the equation

(3) $$Y = \mathcal{M}'X + n',$$

with

(3a) $$\mathcal{M}' = -\frac{A + B\mathcal{M}}{B + C\mathcal{M}}, \qquad n' = \frac{D + E\mathcal{M}}{B + C\mathcal{M}}.$$

Since (3) is the equation of a straight line, the following theorem applies:

The midpoints of all the parallel chords of an ellipse possessing the directional constant $\mathcal{M}$ lie on a straight line (a diameter of the ellipse) *with the directional constant $\mathcal{M}'$.* The two directional constants $\mathcal{M}$ and $\mathcal{M}'$, as well as their corresponding directions and the diameters of the ellipse possessing this direction are said to be *conjugate* to each other.

We will now prove two auxiliary theorems.

AUXILIARY THEOREM I: *There is only one pair of conjugate directions* (diameters) *that belong to all the ellipses circumscribing a quadrilateral.*

PROOF. We replace A, B, C in (3a) with their values and obtain

$$-\mathcal{M}' = \frac{(2k + p\mathcal{M}) \cdot \lambda + q\mathcal{M} \cdot \mu}{p \cdot \lambda + (2h\mathcal{M} + q) \cdot \mu}.$$

If $\mathcal{M}'$ (for a prescribed $\mathcal{M}$) is to maintain the same value no matter which ellipse of circumscription we are concerned with and consequently, no matter how great λ and μ are, then this value must be obtained when $\lambda = 1$ and $\mu = 0$ as well as when $\lambda = 0$ and $\mu = 1$. Consequently, it must be true that

$$\frac{2k + p\mathcal{M}}{p} = \frac{q\mathcal{M}}{2h\mathcal{M} + q}.$$

And if we are able to find a suitable $\mathcal{M}$ for this equation, then for *every* λ and *every* μ

$$-\mathcal{M}' = \frac{(2k + p\mathcal{M})\lambda + (2k + p\mathcal{M})\mu}{p\lambda + p\mu} = \frac{2k + p\mathcal{M}}{p}$$

or

(4) $$\mathcal{M}' = -\mathcal{M} - 2\frac{k}{p},$$

i.e., $\mathcal{M}'$ is *independent* of λ and μ. The equation giving the condition for $\mathcal{M}$ is written

$$hp\mathcal{M}^2 + 2hk\mathcal{M} + kq = 0$$

and gives the two $\mathcal{M}$-values

(5) $$\mathcal{M}_1 = -\frac{k}{p} + \frac{r}{hp}, \qquad \mathcal{M}_2 = -\frac{k}{p} - \frac{r}{hp}$$

with $r^2 = h^2k^2 - hp \cdot kq = hk(hk - pq)$.

Since, according to the drawing, $hk > pq$, r^2 is real, r is positive, and both $\mathcal{M}$-values are real. Moreover,

(5a) $$\mathcal{M}_1 + \mathcal{M}_2 = -2\frac{k}{p}.$$

Now, according to (4), the directional constant $\mathcal{M}_1'$ that is conjugate to $\mathcal{M}_1$ has the value $-\mathcal{M}_1 - 2(k/p)$, i.e., the value $\mathcal{M}_2$. In like manner,

$$\mathcal{M}_2' = \mathcal{M}_1.$$

Thus, there is only one pair of specific directions, determined by the directional constants $\mathcal{M}_1$ and $\mathcal{M}_2$, that will form a pair of conjugate directions for each ellipse of circumscription.

AUXILIARY THEOREM II: *The acute angle formed by two conjugate diameters of an ellipse attains a minimum when the two conjugate diameters are* equal, *and the tangent of the half angle-minimum is equal to the ratio* b:a *of the two half axes.*

Proof. If ψ and φ are the two acute angles that the two conjugate diameters of an ellipse with the half axes a and b form with the large axis, then obviously

(6)
$$\tan \psi \cdot \tan \varphi = \frac{b^2}{a^2}.$$

For the angle $\Omega = \psi + \varphi$ of the two conjugate diameters we therefore obtain

$$\tan \Omega = \tan (\psi + \varphi) = \frac{\tan \psi + \tan \varphi}{1 - \tan \psi \tan \varphi} = \frac{\tan \psi + \tan \varphi}{1 - \dfrac{b^2}{a^2}}.$$

But the left side of this equation, and therefore the angle Ω, attains a minimum when the numerator of the right side assumes its smallest value. This numerator is the sum of two numbers ($\tan \psi$ and $\tan \varphi$) of constant product and, according to No. 10, attains a minimum when the numbers are equal. From $\tan \psi = \tan \varphi$ it follows that $\psi = \varphi$ and from this that the two diameters are *equal*. At the same time from (6) we obtain the value b/a for the tangent of the half angle-minimum.

These preliminaries concluded, the solution of the problem is simple.

The circumscribed ellipse becomes more and more circular, the closer the ratio $b:a$ of the small to the large half axis comes to unity. Now, according to auxiliary theorem II., this ratio has the value $\tan (\omega/2)$, where ω is the smallest angle formed by conjugate diameters. The most nearly circular circumscribed ellipse is therefore the ellipse in which ω attains its maximum possible value. And this is the ellipse in which the directional constants of its *equal* conjugate diameters are determined by (5). Thus, if ω_0 is the angle between the equal conjugate diameters of this ellipse, then for every other ellipse of circumscription, ω_0, as the angle between two *unequal* conjugate diameters (with the directional constants $\mathscr{M}_1$ and $\mathscr{M}_2$), is greater than the angle ω of this ellipse enclosed between *equal* conjugate diameters, so that $\omega_{\max} = \omega_0$.

Consequently:

Of all the ellipses circumscribed about a quadrilateral the ellipse that deviates least from a circle is the one whose equal conjugate diameters possess the conjugate directions common to all the ellipses of circumscription.

The directional constants of these specific directions are determined by the quadratic equation

$$hp\mathscr{M}^2 + 2hk\mathscr{M} + kq = 0.$$

55   The Curvature of Conic Sections

To determine the curvature of a conic section.

By the curvature of a curve at a point is meant the reciprocal value of the radius of the circle of curvature, i.e., the radius of the circle that fits the curve most closely at the relevant point.

SOLUTION.   Let the conic section be called $\mathfrak{K}$, its parameter $2p$, its form number ε, its shortest focal radius k, so that $p = k(1 + \varepsilon)$, and finally, let the equation for its maximum be

$$qx^2 + y^2 - 2px = 0, \quad \text{with} \quad q = 1 - \varepsilon^2.$$

It is known that the coordinates of a point $\Pi(\xi|\eta)$ at a distance R from another point $P(x|y)$ and lying at a direction from P that forms the angle ϑ with the positive x-axis are

$$\xi = x + oR, \qquad \eta = y + iR,$$

where o is the cosine and i the sine of ϑ.

If Π lies on $\mathfrak{K}$, then from

$$q\xi^2 + \eta^2 - 2p\xi = 0$$

we obtain the quadratic equation for R

$$DR^2 - ER + F = 0$$

with the coefficients

$$D = i^2 + qo^2, \qquad E = 2(ou - iy), \qquad F = qx^2 + y^2 - 2px,$$

where $u = p - qx$.

In respect to the conic section, we will call the three expressions D, E, F the *directional number for the "direction"* ϑ, the *emanant at point* $x|y$ *for the direction* ϑ, and the *power at point* $x|y$.

If $P\Pi$ is a secant, the roots R_1 and R_2 of the quadratic equation are the segments generated on the secant by the conic section. The relations between the roots and the coefficients of a quadratic equation give us the following theorems:

I. *The emanant is the Dth sum of the secant segments.*

II. *The power is the Dth product of the secant segments.*

We now draw through an arbitrary point $P(x|y)$ of the conic section the tangent $\mathfrak{T}$ and the normal and designate the segment of the normal from P to the x-axis as n and the segment reaching from P to the conic section as N. If ϑ is the angle of $\mathfrak{T}$ with the x-axis, o the cosine, i the sine of ϑ, then the directional number for the tangent direction is

$$D = i^2 + qo^2 = \frac{u^2}{n^2} + q\frac{y^2}{n^2} = \frac{p^2}{n^2}$$

(since $u = p - qx$ represents the subnormal), while for the directional number of the inward-pointing normal we obtain the value

$$\Delta = o^2 + qi^2.$$

The emanant at P for the direction of the normals becomes

$$E = 2(oy + iu) = 2n.$$

Therefore, according to I.,

(1) $$2n = \Delta N.$$

On tangent $\mathfrak{T}$ we select a point O whose distance OP from P we set equal to t; and we draw through O perpendicular to $\mathfrak{T}$ through the conic section the secant $\mathfrak{S}$. Let the two segments of the secant created by $\mathfrak{K}$ and measured from O be s and let $S > s$. According to II., we can write for the power of $\mathfrak{K}$ at O both Dt^2 and ΔSs, so that

(2) $$Dt^2 = \Delta Ss.$$

We now draw a circle $\mathfrak{k}$ to which for the time being we will attribute the arbitrary radius ρ; the center of this circle lies on the internal normal and the circle is tangent to the conic section at P. If s_0 and $S_0 > s_0$ are the segments measured from O that the circle creates on the secant $\mathfrak{S}$, then, according to the tangent theorem,

(3) $$t^2 = S_0 s_0.$$

By division of (2) and (3) we obtain

$$DS_0 s_0 = \Delta Ss$$

and, using (1), we obtain

$$DNS_0 s_0 = 2nSs.$$

Now the closer the fraction s/s_0 is to unity, the closer the approximation of the circle to the conic section in the vicinity of point P. But this fraction, according to the last equation, has the value

$$\frac{s}{s_0} = \frac{N}{S} \cdot \frac{S_0}{2\rho} \cdot \frac{D\rho}{n}.$$

In the *immediate vicinity* of the point P, S becomes equal to N and $S_0 = 2\rho$, so that both the first and second factors on the right-hand side are equal to 1. Consequently, the fraction s/s_0 comes closest to unity when the third right-hand factor $D\rho/n$ is also equal to 1. Thus: *Of all circles † the one that most closely approximates the conic section is the one possessing the radius* $\rho = \mathrm{n}/\mathrm{D}$.

Since D was previously determined as equal to p^2/n^2, we obtain the *fundamental theorem*:

The radius of curvature of a conic section has the value

$$\rho = n^3/p^2.$$

To draw the circle of curvature we must consider that p/n is the cosine of the angle ψ formed by the normal n with the focal radius r of the point P,* and accordingly we write the obtained formula as

$$\rho = n/\cos^2 \psi.$$

From inspection of this equation we obtain the following

CONSTRUCTION OF THE RADIUS OF CURVATURE: At the point of intersection H of the normal with the x-axis we erect a perpendicular

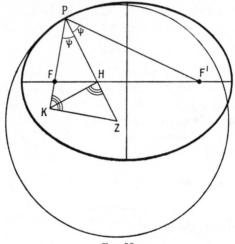

FIG. 53.

---

* From the triangle with sides n, r and the line w joining the end points of n and r lying on the x-axis, we obtain $\cos \psi = (n^2 + r^2 - w^2)/2nr$. If we express the numerator of this fraction entirely in terms of x, thus expressing n^2 by $y^2 + u^2 = 2px - qx^2 + (p - qx)^2$, r by $\varepsilon x + k$, and w by $(x - k) + u = \varepsilon^2 x + k\varepsilon$, and combine, the numerator then becomes equal to $2p(\varepsilon x + k) = 2pr$ and $\cos \psi$ becomes $2pr/2nr = p/n$.

to the normal. At its point of intersection K with the (extended) focal radius we then erect the perpendicular to the focal radius. The point of intersection Z of this second perpendicular with the normal is the center of curvature, its distance from P the desired radius of curvature.

56   Archimedes' Squaring of a Parabola

To determine the area enclosed in a parabola section.

The squaring of a parabola is one of Archimedes' most remarkable achievements. It was accomplished about 240 B.C. and is based upon the properties of Archimedes triangles.

An Archimedes triangle is a triangle whose sides consist of two tangents to a parabola and the chord connecting the points of tangency. The last-mentioned side is taken as the base line or the base

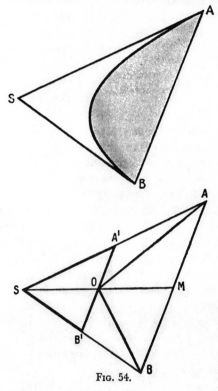

Fig. 54.

of the triangle.   In order to construct such a triangle we draw the parallels to the parabola axis through the two points H and K of the directrix and erect the perpendicular bisectors upon the lines connecting H and K with the focus F.   If we designate the point of intersection of the two perpendicular bisectors as S, the point of intersection of the first perpendicular bisector with the first parallel to the axis as A, and the point of intersection of the second perpendicular bisector with the second parallel to the axis as B, then A and B are points of the parabola and SA and SB are tangents of the parabola (classical construction of the parabola), and ASB is an Archimedes triangle (cf. Figure 39).

Since SA and SB are two perpendicular bisectors of the triangle FHK, the parallel to the axis through S is the third perpendicular bisector; it consequently passes through the center of HK, and, as the midline of the trapezoid $AHKB$, it also passes through the center M of AB.   This gives us the theorem: *The median to the base of an Archimedes triangle is parallel to the axis.*

Let the parabola tangents through the point of intersection O of the median SM to the base with the parabola cut SA at A', SB at B'. Then $AA'O$ and $BB'O$ are also Archimedes triangles.   Consequently, according to the above theorem, the medians to their bases are also parallel to the axis and are therefore also parallel to SO.   These medians are therefore midlines in the triangles SAO and SBO, so that A' and B' are the centers of SA and SB. $A'B'$ is consequently the midline of the triangle SAB and is therefore parallel to AB; also the point O on $A'B'$ must be the center of SM.

The result of our investigations is the

THEOREM OF ARCHIMEDES: *The median to the base of an Archimedes triangle is parallel to the axis, the midline parallel to the base is a tangent, and its point of intersection with the median to the base is a point of the parabola.*

Now we can determine the area J of the parabola section enclosed in our Archimedes triangle ASB with the base line AB.

The tangents $A'B'$ and the chords OA and OB divide the triangle ASB into four sections: 1. the "internal triangle" AOB enclosed within the parabola; 2. the "external triangle" $A'SB'$ lying outside the parabola; 3. and 4. two "residual triangles" AOA' and BOB', which are also *Archimedes triangles* and are penetrated by the parabola.

Since O lies at the center of SM, *the internal triangle is twice the size of the external triangle.*

In the same fashion, each of the two residual triangles in turn gives rise to an internal triangle, an external triangle and two new residual Archimedes triangles that are penetrated by the parabola, and once again each internal triangle is twice the size of the corresponding external triangle.

Thus, we can continue without end and cover the entire surface of the initial Archimedes triangle *ASB* with internal and external triangles. The sum of all the internal triangles must also be twice as great as the sum of all the external triangles. In other words:

THEOREM OF ARCHIMEDES: *The parabola divides the Archimedes triangle into sections whose ratio is* 2:1.

Or also:

The area enclosed by a parabola section is two thirds the area of the corresponding Archimedes triangle.

Archimedes arrived at this conclusion by a somewhat different method. He found the area of the section by adding together the areas of all the successive internal triangles.

If Δ represents the area of the initial Archimedes triangle *ASB*, then the area of the corresponding internal triangle is one half Δ, the area of the corresponding external triangle is one quarter of Δ, and the area of each of the two residual triangles is one eighth of Δ. The successive Archimedes triangles therefore have the areas

$$\Delta, \frac{\Delta}{8}, \frac{\Delta}{8^2}, \dots;$$

the corresponding internal triangles possess half this area; and since each internal triangle gives rise to *two* new internal triangles, we thus obtain for the sum of all the successive internal triangle areas the value

$$\frac{1}{2}\left[\Delta + 2\cdot\frac{\Delta}{8} + 4\cdot\frac{\Delta}{8^2} + 8\cdot\frac{\Delta}{8^3} + \cdots\right].$$

The bracket encloses a geometrical series with the quotient $\frac{1}{4}$, the sum of which is equal to $\Delta/(1 - \frac{1}{4}) = \frac{4}{3}\Delta$. Thus, we again obtain for the area of the section the value $J = \frac{2}{3}\Delta$.

Since $A'B'$ is tangent to the parabola at O, the perpendicular h dropped from O to the base line AB of the section is the altitude of the section. Since h is also half the altitude of the triangle ASB, $\Delta = AB\cdot h$ and $J = \frac{2}{3}\cdot AB\cdot h$, i.e.:

The area enclosed by a parabola section is equal to two thirds the product of the base and the altitude of the section.

Finally, we will express the area of the section in terms of the transverse q of the section, i.e., by the projection normal to the axis of the chord bounding the section.

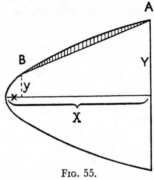

Fig. 55.

We use the equation for the amplitude of the parabola, calling the coordinates of the corners of the section $x|y$ and $X|Y$, and we have

$$y^2 = 2px \quad \text{and} \quad Y^2 = 2pX$$

with $2p$ representing the parameter. From Figure 55 it follows directly that

$$J = \tfrac{2}{3}XY - \tfrac{2}{3}xy - (X - x)\cdot\frac{Y+y}{2}.$$

If we replace X and x here with $Y^2/2p$ and $y^2/2p$, we obtain $12pJ = Y^3 - y^3 - 3Y^2y + 3Yy^2 = (Y - y)^3$. Since $Y - y$ is the section transverse q, we finally obtain

$$12pJ = q^3.$$

This important formula can be expressed verbally as follows:

Six times the product of the parameter and the area of the section is equal to the cube of the section transverse.

57   Squaring a Hyperbola

To determine the surface area enclosed by a section of a hyperbola.

We select the major axis of the hyperbola as the x-axis, the minor axis as the y-axis; the hyperbola equation then reads

(1)
$$\frac{x^2}{a^2} - \frac{y^2}{b^2} = 1,$$

where a and b are half the major and minor axes, respectively.

We must find the area A of the hyperbola section cut off at a distance of x from the apex of the hyperbola by the hyperbola chord $2y$ that is normal to the x-axis (Figure 56). The coordinates for the corners of the section H and K are thus $x|y$ and $x|-y$.

First we determine the area T of a so-called hyperbola trapezoid, i.e., the trapezoidal surface that is bounded by a hyperbola arc, the parallels to one of the asymptotes through the end points of the arc, and the segment cut off on the other asymptote by these parallels.

Let the asymptote angle be 2α, its sine J, the sine and cosine of its halves i and o, so that $i = b/e$ and $o = a/e$ (with $e = \sqrt{a^2 + b^2}$) and $J = 2io = 2ab/e^2$ (Figure 56).

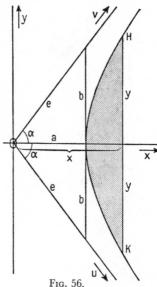

Fig. 56.

We choose as the asymptotes the u- and v-axis of a second (oblique-angle) coordinate system. Between the coordinates $x|y$ and $u|v$ of a hyperbola point in the two systems there then exist the transformation equations

$$(2) \qquad x = ou + ov, \qquad y = iv - iu,$$

as may be seen from Figure 57, so that for the left side of (1) we obtain the value $4uv/e^2$ and we have the equation of the hyperbola in the second system, the so-called *asymptote equation of the hyperbola*

(3) $$uv = P \quad \text{with} \quad P = \tfrac{1}{4}e^2,$$

in which P is the so-called *power of the hyperbola.*

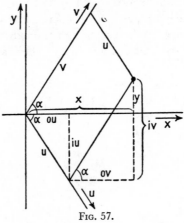

FIG. 57.

Let the trapezoid T to be calculated be bounded by the hyperbola arc with end point coordinates $u|v$ and $U|V$ (where we let $U > u$, $V < v$), by the two ordinates v and V and by the base line $U - u$ of the trapezoid (Figure 58).

We divide the trapezoid into n equal sections t by means of parallels to the v-axis, so that $T = nt$, and we designate the coordinates of the points marking off the segments on the trapezoid arc as $u_1|v_1$, $u_2|v_2$, ..., $u_{n-1}|v_{n-1}$.

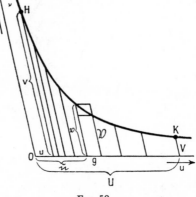

FIG. 58.

The asymptote parallels through the end points $\mathfrak{u}|\mathfrak{v}$ and $\mathfrak{U}|\mathfrak{V}$ of the hyperbola arc corresponding to an arbitrary trapezoidal section t determine two parallelograms with a common base line $g = \mathfrak{U} - \mathfrak{u}$ lying on the u-axis, one of which is larger and the other smaller than t. Since these parallelograms possess the areas $Jg\mathfrak{v}$ and $Jg\mathfrak{V}$, we obtain the inequality

$$Jg\mathfrak{v} > t > Jg\mathfrak{V}.$$

We introduce the so-called quotient of the trapezoid t, $q = \mathfrak{U}/\mathfrak{u}$, replace g on the left by $(q - 1)\mathfrak{u}$ and on the right by $[1 - (1/q)]\mathfrak{U}$, and obtain

$$J(q - 1)\mathfrak{u}\mathfrak{v} > t > J\left(1 - \frac{1}{q}\right)\mathfrak{U}\mathfrak{V}$$

or, as a result of (3),

$$PJ(q - 1) > t > PJ\left(1 - \frac{1}{q}\right).$$

If we replace t here with T/n, divide by PJ and abbreviate T/PJ as c, we obtain

$$q - 1 > \frac{c}{n} > 1 - \frac{1}{q}$$

or, solving for q,

$$1 + \frac{c}{n} < q < \frac{1}{1 - \dfrac{c}{n}}.$$

Using this inequality for all n trapezoidal sections, we obtain the n inequalities

$$1 + \frac{c}{n} < \frac{u_1}{u} < \frac{1}{1 - \dfrac{c}{n}},$$

$$1 + \frac{c}{n} < \frac{u_2}{u_1} < \frac{1}{1 - \dfrac{c}{n}},$$

$$\vdots$$

$$1 + \frac{c}{n} < \frac{U}{u_{n-1}} < \frac{1}{1 - \dfrac{c}{n}}.$$

Multiplication of these gives

$$\left(1 + \frac{c}{n}\right)^n < \frac{U}{u} < \frac{1}{\left(1 - \dfrac{c}{n}\right)^n}.$$

The mean of this inequality is the so-called quotient $Q = U/u$ of the hyperbola trapezoid T. The left and right side tend (according to No. 12) toward the value e^c for *infinitely increasing* n, e representing the Euler number $(2.71828\ldots)$. This gives us the *equality*

$$Q = e^c.$$

With logarithms we obtain

(I) $T = PJlQ,$

or verbally:

The area of the hyperbola trapezoid is proportional to the natural logarithm of the trapezoid quotient.

The proportionality constant is the product of the hyperbola power and the sine of the asymptote angle.

Since $4P = e^2$, $J = 2ab/e^2$, we also have

(Ia) $T = \dfrac{ab}{2}\, lQ.$

If we join the end points $u|v$ and $U|V$ of our hyperbola arc with the hyperbola center O, we obtain a hyperbola sector to which we can similarly assign the "quotient" Q. Since the two triangles that are formed by the connecting lines mentioned and the coordinates of the end points of the arc have the areas $\frac{1}{2}uvJ$ and $\frac{1}{2}UVJ$, which areas are equal in view of (3), the sector has the same area S as the trapezoid:

(II) $S = PJlQ = \dfrac{ab}{2}\, lQ.$

Now the determination of the area of the section A is simple. First, in accordance with (2), the abscissas u and U of the section corners H and K are found to be

$$u = \frac{e}{2}\left(\frac{x}{a} - \frac{y}{b}\right) \quad \text{and} \quad U = \frac{e}{2}\left(\frac{x}{a} + \frac{y}{b}\right).$$

From this it follows that the quotient of the sector OHK is

$$Q = \frac{U}{u} = \frac{\dfrac{x}{a} + \dfrac{y}{b}}{\dfrac{x}{a} - \dfrac{y}{b}} = \left(\frac{x}{a} + \frac{y}{b}\right)^2 \qquad \text{[cf. (1)]}$$

and, consequently, the area of the sector, according to (II), is

$$S = abl\left(\frac{x}{a} + \frac{y}{b}\right).$$

Finally, A is found to be the amount by which the triangle OHK is greater than the sector OHK, or

(III) $$A = xy - abl\left(\frac{x}{a} + \frac{y}{b}\right).$$

58   Rectification of a Parabola

To determine the length of a parabola arc.

SOLUTION. The following ingenious solution to this problem stems from the famous book *Lectiones Geometricae* of the English mathematician Isaac Barrow (1630–1677), which was published in 1670 in London. We refer the parabola to a coordinate system in which the x-axis is the axis of the parabola and the y-axis is tangent to the apex. The parabola equation then reads $y^2 = 2px$. We need only determine the length of an "*apex arc*," i.e., an arc of the parabola that takes its origin from the apex S, since *any* arc can be represented as the sum or difference of apex arcs. Let the end point P of the apex arc SP possess the coordinates X and Y, and let the sought-for length of the arc be L.

Since the subnormal of a parabola is equal to the half parameter p, there exists between the ordinate y of a point of the parabola and the normal n corresponding to this point the relation

$$n^2 - y^2 = p^2.$$

If we then assign to each parabola point $x|y$ of our coordinate system a point $n|y$ in a new $n|y$-coordinate system, we obtain in the new system an equilateral hyperbola with the half axis p.

We show that p times the length (pL) of the parabola arc SP is numerically equal to the surface area F of the hyperbola trapezoid that is bounded by the hyperbola, its axes, and the perpendicular N that is dropped from the hyperbola point P' corresponding to the point P onto the minor axis of the hyperbola. (N is at the same time the abscissa of the hyperbola point P' and the parabola normal at the parabola point P.)

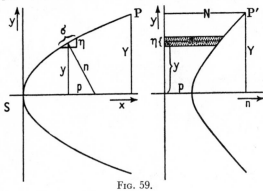

FIG. 59.

Let us consider a portion $\sigma = AB$ of the parabola arc SP that is short enough to be considered a rectilinear distance (a so-called arc element) and let us draw through its end points the parallel AC to the parabola axis and $BC = \eta$ to the apex tangent. At the same time we draw the ordinate y and the normal n of the midpoint of AB, which gives us a right triangle with the sides y, n, and p that is similar to the triangle ABC. As a result of this similarity we obtain the proportion $\eta:\sigma = p:n$, and this gives us the equation

(1) $$p\sigma = n\eta.$$

We then draw from the hyperbola points A' and B' corresponding to the points A and B the perpendiculars to the minor axis of the hyperbola, and we obtain a narrow hyperbola trapezoid that corresponds to the arc $A'B'$. The area φ of this trapezoid is the product of its altitude η and its midline n (the latter is n because it passes through the center of the altitude and thus through the end point of the hyperbola ordinate y):

(2) $$\varphi = n\eta$$

From (1) and (2) we get

$$p\sigma = \varphi.$$

If we form this equation for each element of the parabola arc SP and its corresponding minute hyperbola trapezoid, and if we add the resulting equations, we obtain on the left p times the arc length L and on the right the area F of the hyperbola trapezoid above described, i.e., the equation

$$pL = F.$$

Now from the concluding formula of No. 57 it follows that

$$F = \frac{NY}{2} + \frac{p^2}{2} l \frac{N + Y}{p}.$$

The sought-for arc length is thus

$$L = \frac{NY}{2p} + \frac{p}{2} l \frac{N + Y}{p}$$

where Y represents the ordinate, N the normal of the end point of the arc.

We now slightly transform the equation we have found.

Let T be the portion of the parabola tangent passing through P, bounded by P and the y-axis, let τ be the slope angle of the parabola at point P, i.e., the angle formed by the tangent with the x-axis (and, at the same time, by the normal N with the y-axis). Then

$$\frac{NY}{2p} = \frac{YY}{2p \cos \tau} = \frac{X}{\cos \tau} = T$$

and

$$\frac{N + Y}{P} = \frac{N + N \cos \tau}{N \sin \tau} = \frac{1 + \cos \tau}{\sin \tau} = \frac{2 \cos^2 \frac{\tau}{2}}{2 \sin \frac{\tau}{2} \cos \frac{\tau}{2}} = \cot \frac{\tau}{2},$$

consequently

$$L = T + kl \cot \frac{\tau}{2},$$

where we have replaced $\frac{1}{2}p$ by the shortest focal radius k.

CONCLUSION: *An apex arc of a parabola exceeds the length of the parabola tangent reaching from the end of the arc to the apex tangent by a quantity that is proportional to the natural logarithm of the cotangent of half the slope angle.*

The proportionality constant is the shortest focal radius.

59   Desargues' Homology Theorem (Theorem of Homologous Triangles)

If the lines connecting the homologous vertexes of two triangles pass through a point, the points of intersection of the homologous sides lie on a straight line.

And conversely:

If the points of intersection of the homologous sides of two triangles lie on a straight line, the lines connecting the homologous vertexes pass through a point.

One frequently has occasion to correlate to each other the vertexes and sides of two triangles (e.g., similar triangles), and in these cases for the sake of convenience the mutually correlated, so-called "homologous" vertexes and sides are usually designated by the same letter. Thus, one may have, for example, the homologous vertexes A and A', B and B', and finally C and C', as well as the homologous sides $BC = a$ and $B'C' = a'$, $CA = b$ and $C'A' = b'$, and finally $AB = c$ and $A'B' = c'$.

Two such triangles, for which we will assume that no pair of homologous vertexes or sides coincides, are called *copolar* [perspective from a point] when the lines AA', BB', CC' connecting the homologous vertexes pass through one point, the so-called *homology pole*. They are called *coaxial* [perspective from a line] when the points of intersection aa', bb', cc' of the homologous sides lie on a straight line, the so-called *homology axis*.

Using these terms, the above theorem can be expressed in the abbreviated form of:

DESARGUES' HOMOLOGY THEOREM: *Copolar triangles are coaxial, coaxial triangles are copolar.*

Triangles that are both copolar and coaxial are called *homologous triangles*.

The theorem of homologous triangles was discovered by the French mathematician and engineer Gérard Desargues (1593–1662) in about 1636 and is therefore known as Desargues' theorem. However, according to the Greek mathematician Pappus, this theorem was already contained in the lost treatise on porisms of Euclid.

Desargues' theorem plays a very important role in projective geometry. Consequently, we will prove it in a projective manner though other, shorter proofs are possible.

For the reader unfamiliar with projective geometry it may be appropriate to provide a short exposition of its most important concepts and its simplest theorems, especially as they will be encountered in the next few sections as well.

The totality of the points (considered as rigidly connected to each other) in a line is called a *range of points;* the line is called the *base* of the range. The totality of the lines (considered as rigidly connected to each other) that pass through one point is called a *ray pencil;* the point is called the *center* of the pencil. Similarly, the totality of the points of a circle or, more generally, of a conic section is called a circular or conic range of points or *field of points;* the totality of the tangents of a conic section is called a *field of tangents* of a conic section. Ranges of points, pencils, and tangent families are the *basic structures* of plane projective geometry, and the points, rays, and tangents are the *elements* of the corresponding structures.

Two basic figures are called *projective* (symbol: $\overline{\wedge}$) when their elements are unequivocally related to each other in such manner that every four elements of the one figure and the four corresponding or "homologous" elements of the other have the same double ratio. The relation existing between the figures is called *projectivity.*

[The *cross ratio* $(ABCD)$ of four points A, B, C, D of a straight line is the ratio

$$\frac{AC}{BC} : \frac{AD}{BD},$$

the cross ratio $(abcd)$ of four rays a, b, c, d of a pencil is the ratio

$$\frac{\sin ac}{\sin bc} : \frac{\sin ad}{\sin bd}.$$

The cross ratio of four points of a circle is the cross ratio of the four rays that connect the four points with a fifth point of the circle, where (according to the boundary angle theorem) this fifth point can be chosen at pleasure. The cross ratio of four points of a conic section is similarly the cross ratio of the four rays that join the four points with an arbitrarily chosen fifth point of the conic section (cf. No. 61). Finally, the ratio of four conic section tangents is the cross ratio of their points of tangency.]

A projectivity is completely determined if three elements of one structure and the corresponding elements of the other are given.

Two projective structures are called *conjective* when their bases (or centers) coincide.

A particularly important case of projectivity is *perspectivity.* A range of points and a ray pencil are called *perspective* $(\overline{\wedge})$ when each element of the range lies on the corresponding element of the pencil. Each ray is called the *reflection* of the homologous point, the whole pencil is called the *reflection* of the range. Two nonconjective ranges are called perspective (symbol: $\overline{\wedge}$) when the lines connecting the homologous points pass through one point, the *center of perspectivity*. Two ray pencils are called perspective if every pair of corresponding rays intersect on one straight line, the *axis of perspectivity*.

The projectivity of two perspective figures follows from

PAPPUS' THEOREM: *The cross ratio of four rays of a pencil is equal to the cross ratio of the four points at which an arbitrary line cuts the rays.*

(Pappus of Alexandria, fourth century A.D., *Collectiones mathematicae*.)

PROOF. Let A, B, C, D be the four points of intersection of a line with the pencil of four rays $OA = a$, $OB = b$, $OC = c$, $OD = d$. We designate the sine of the angle formed by two rays, for example, a and c, with each other as sine ac. Since the perpendiculars from A and B to c have the lengths $a \sin ac$ and $b \sin bc$ and are in the same ratio as AC to BC, we obtain the proportion

$$a \sin ac : b \sin bc = AC : BC.$$

Similarly,

$$a \sin ad : b \sin bd = AD : BD.$$

By division of these two equations we obtain

$$\frac{\sin ac}{\sin bc} : \frac{\sin ad}{\sin bd} = \frac{AC}{BC} : \frac{AD}{BD}. \qquad \text{Q.E.D.}$$

Two projective ranges or pencils can always be brought into a perspective position.

Two projective ranges (pencils) become perspective when they are placed in such a way that an element of one range (pencil) falls on the homologous element of the other range (pencil), though the bases (centers) do not coincide. We have the following two important theorems:

I. *If in the projectivity between two ranges the point of intersection of the two bases corresponds to itself, the ranges are perspective.*

II. *If in the projectivity between two pencils the line connecting the two centers corresponds to itself, the pencils are perspective.*

PROOF OF I. Let the bases of the two ranges be $\mathfrak{T}$ and $\mathfrak{T}'$, their point of intersection that corresponds to itself $O \equiv O'$. On $\mathfrak{T}$ we choose two fixed elements A, B and an arbitrary point P and we designate the homologous elements on $\mathfrak{T}'$ as A', B', and P'. We find the point of intersection S of the lines AA' and BB' and assign to the

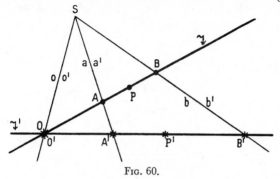

Fig. 60.

lines connecting the designated elements with S the same letters, but in lower case. Then, according to Pappus,

$$(oabp) = (OABP) \quad \text{and} \quad (o'a'b'p') = (O'A'B'P').$$

But since the right sides of these equations are equally great, according to our assumption, it follows that

$$(o'a'b'p') = (oabp).$$

But if two equal cross ratios agree in the first three elements $(o' \equiv o,\ a' \equiv a,\ b' \equiv b)$, then they also agree in the fourth. Consequently, p' falls on p, and thus PP' passes through S, and the ranges are perspective.

PROOF OF II. Let the centers of the two projective pencils $\mathfrak{Z}$ and $\mathfrak{Z}'$ be Z and Z', their self-corresponding connecting line $o \equiv o'$. We select on $\mathfrak{Z}$ two fixed elements a and b and an arbitrary element p and designate the homologous elements of $\mathfrak{Z}'$ as a', b', and p'. We find the connecting line g of the points aa' and bb' and assign to the points of intersection of the designated elements with g the same letters, but capitals. Then, according to Pappus,

$$(oabp) = (OABP) \quad \text{and} \quad (o'a'b'p') = (O'A'B'P').$$

But since the left sides of these equations are equal, in accordance with our initial assumption,

$$(O'A'B'P') = (OABP).$$

But if two equal cross ratios agree in the first three elements $(O' \equiv O,\ A' \equiv A,\ B' \equiv B)$, they also agree in the fourth. P' therefore falls on P, p and p' thus intersect on g, and the pencils $\mathfrak{Z}$ and $\mathfrak{Z}'$ are perspective.

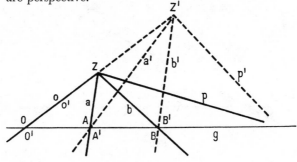

Fig. 61.

The *proof of Desargues' theorem* is now easily obtained (Figure 62). We call the vertexes of one triangle A, B, C, the sides opposite them a, b, c, the homologous vertexes of the other triangle A', B', C', the sides opposite them a', b', c'.

Let the points of intersection of the homologous sides a and a', b and b', c and c' be X, Y, and Z, respectively, and let the points of intersection of the line CC' with the two lines AB and $A'B'$ be H and H'.

The proof divides into two parts.

I. We assume that the connecting lines AA', BB', CC' pass through one point O. We project the range of points AB from O onto $A'B'$ and obtain two perspective ranges in which the elements A, B, H, Z of the first are homologous to the elements A', B', H', $Z' \equiv Z$ of the second. We then connect the points of these ranges with C and C', thereby obtaining two projective ray pencils in which the elements CA, CB, $CH \equiv CC'$, CZ correspond to the elements $C'A'$, $C'B'$, $C'H' \equiv C'C$, $C'Z'$. Since the line CC' connecting the pencil centers corresponds to itself in this projectivity, the projectivity of the pencil is perspective and the points of intersection of the homologous rays lie on a straight line. Thus, for example, the points of intersection Y (of CA and $C'A'$), X (of CB and $C'B'$), and Z (of CZ and $C'Z'$) lie on a straight line.

II. We assume that the points aa' (X), bb' (Y), cc' (Z) lie on a straight line g. We connect the points of the line g with C and C', thereby obtaining two perspective ray pencils in which the elements

a, b, CC', CZ of the first pencil correspond to the elements a', b', CC', $CZ' \equiv CZ$ of the second. We cut these pencils with the lines c and c' and obtain two projective ranges in which the elements B, A, H, Z of

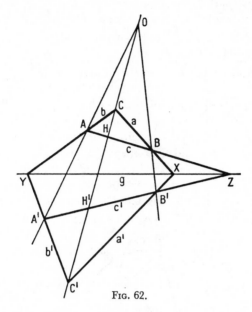

Fig. 62.

the first range correspond to the elements B', A', H', $Z' \equiv Z$ of the second. Since the point of intersection $Z \equiv Z'$ of the range bases corresponds to itself in this projectivity, the ranges are perspective and the connecting lines BB', AA', and $HH' \equiv CC'$ of the homologous elements thus pass through one point, which was to be proved.

60   Steiner's Double Element Construction

To draw the double elements of a conjective projection that are given by three pairs of homologous elements.

A double element of a conjective projectivity is an element that coincides with its homolog.

The following simple solution to this fundamental problem of projective geometry was discovered by the German mathematician Jakob Steiner (*Die geometrischen Konstruktionen*, etc. [cf. No. 34], Berlin, 1833).

Steiner's double element construction enriched the geometry of antiquity by providing it with a new and fruitful method for solving problems of geometric construction. This so-called *method of false position (regula falsi)* is based on the theorem:

If in the projectivity between two ray pencils the line connecting the pencil centers corresponds to itself, the pencils are perspective (No. 59).

We can distinguish three cases:

I. *Double elements of a projectivity on a circle.* Let the projectivity between the two ranges of points $\Re$ and $\Re'$ of the circle $\Re$ be given by the two corresponding point triplets (A, B, C) and (A', B', C'). We consider the ray pencils $\mathfrak{S}$ and $\mathfrak{S}'$, whose rays run from the points of ranges $\Re$ and $\Re'$, respectively, through the centers A' and A, respectively. Since $\Re \barwedge \mathfrak{S}$ and $\Re' \barwedge \mathfrak{S}'$, and, according to our assumption, $\Re \barwedge \Re'$, it is also true that $\mathfrak{S} \barwedge \mathfrak{S}'$. But since in the line AA' connecting the centers of the two pencils $\mathfrak{S}$ and $\mathfrak{S}'$ corresponding pencil elements coincide, the latter projectivity is a perspectivity. The axis of perspectivity is the line $\mathfrak{g}$ connecting the point of inter-

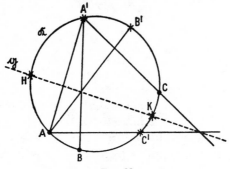

FIG. 63.

section of the rays $A'B$ and AB' with the point of intersection of the rays $A'C$ and AC'. Two corresponding rays of $\mathfrak{S}$ and $\mathfrak{S}'$ thus always intersect at $\mathfrak{g}$. Thus, in order to obtain a point P' of $\Re'$ corresponding to the arbitrary point P of $\Re$, we need only connect the point of intersection of $A'P$ and $\mathfrak{g}$ with A. The connecting line touches $\Re$ at P'. If we carry out this construction for the points of intersection H and K of the perspectivity axis with the circle, H' falls on H, K' on K. The double points of the projectivity on a circle are therefore the points of intersection of the circle with the above perspectivity axis.

II. *Double elements of two ray pencils.* We draw a circle $\mathfrak{K}$ through the common center of the two projective pencils and, in accordance with I., we draw the double points of the two ranges at which the rays of the two pencils cut $\mathfrak{K}$. The pencil rays passing to these double points are the double rays we are looking for.

III. *Double elements of two ranges of points.* We draw, in accordance with II., the double rays of the two pencils that are obtained from the lines connecting the points of the two conjective projective ranges with an arbitrary center Z outside the base of the range. The points of intersection of the two double rays with the base of the range are the double points we are looking for.

61 Pascal's Hexagon Theorem

To demonstrate that the three points of intersection of the opposite sides of a hexagon inscribed in a conic section lie on a straight line.

A hexagon inscribed in a conic section essentially consists of six points anywhere on the conic section 1, 2, 3, 4, 5, 6, the "vertexes" of the hexagon, and the six connecting lines 12, 23, 34, 45, 56, 61, the "sides" of the hexagon. The sides 12 and 45, the sides 23 and 56, and finally 34 and 61 are called the "opposite sides." The straight line on which the three points of intersection of the opposite sides lie is called the *Pascal line*, and the hexagon is called the *Pascal hexagon*. In a somewhat more abbreviated form the theorem to be proved can be stated as:

The three points of intersection of a Pascal hexagon lie on a straight line.

This fundamental theorem in conic section theory was published in 1640 by Blaise Pascal (1623–1662) at the age of 16 in his six-page *Essai sur les Coniques*.

There are a number of proofs of the Pascal theorem. The following projective proof is based upon the two *theorems of Steiner:*

I. *The points of a conic section are projected from pairs of themselves by projective pencils.*

II. *If in the projectivity between two ranges of points the point of intersection of their bases corresponds to itself, the ranges are perspective.*

PROOF OF I. The theorem applies most directly to the circle. (In circles the designated pencils are even congruent.) Now, since a conic section is the central projection of a circle, and since in this projection the pencils we are concerned with appear as projections of projective ray pencils in a circle, we need only show that the central projection of a pencil on a plane is projective with respect to the pencil.   Now this is the case according to Pappus' theorem. Specifically, if a, b, c, d are four rays lying in plane E, a', b', c', d' their central projections on plane E', and A, B, C, D the points of intersection of the ray pairs (a, a'), (b, b'), (c, c'), and (d, d') lying on the line of intersection of the two planes, then, according to Pappus,

$$(a'b'c'd') = (ABCD) \quad \text{and} \quad (abcd) = (ABCD),$$

thus, also

$$(a'b'c'd') = (abcd),$$

i.e., the pencil and the pencil projection are projective.

The proof of II. is in No. 59.

Now to prove the Pascal theorem!

Let the vertexes of the hexagon be 1, 2, 3, 4, 5, 6.   According to I., the rays from the *centers* 1 *and* 3 to the conic section points 2, 4, 5, 6 form projective pencils; thus the points of intersection 2', 4', 5', 6' and 2″, 4″, 5″, 6″ of these rays with the *straight lines* 54 *and* 56 form projective ranges.   Since at the point of intersection 5 of their bases the corresponding range elements are coincident ($5' \equiv 5''$), the ranges are perspective according to II., and consequently the lines 2′2″, 4′4″, and

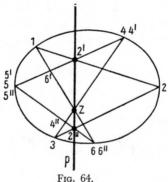

Fig. 64.

6′6″ pass through one point, the point of intersection Z of the lines 4′4″ and 6′6″, i.e., the lines 34 and 61.   In other words: The points of intersection of the opposite sides 2′ (intersection of 12 and 45), 2″ (intersection of 23 and 56), and Z (intersection of 34 and 61) lie on one straight line, the Pascal line $p \equiv 2'Z2''$.   Q.E.D.

THE CONVERSE OF PASCAL'S THEOREM: *If the opposite sides of a hexagon* (of which no three vertexes lie on a straight line) *intersect on a straight line, the six vertexes lie on a conic section.*

INDIRECT PROOF. Let the conic section that is unequivocally determined by the five vertexes 1, 2, 3, 4, 5 touch the fifth side of the hexagon 56 at 6*. According to Pascal's theorem, we obtain 6* by drawing the Pascal line (as the line connecting the points of intersection of the opposite sides 12 and 45, as well as 23 and 56 ≡ 56*), causing it to intersect with 34 at Z and determining the point of intersection (6*) of 1Z with 56* ≡ 56. But according to our assumption, this is 6, so that 6* ≡ 6.

If two vertexes of a Pascal hexagon coincide once or twice or three times, there follow the corollaries of the Pascal theorem, the most important of which we will now give.

I. *The vertexes 5 and 6 coincide:* this is to be considered as meaning that point 6 approaches point 5 ever more closely until it finally coincides with it. This transforms the chord 56 into the tangent at point 5 and the hexagon is transformed into the pentagon 1 2 3 4 5. Pascal's theorem then assumes the form:

COROLLARY 1 (Figure 65): *In every pentagon inscribed in a conic section the points of intersection of two pairs of nonadjacent sides and the point of intersection of the fifth side with the tangent passing through the opposite vertex lie on a straight line.*

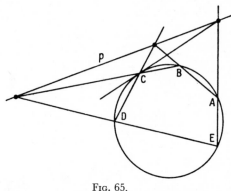

FIG. 65.

II. *The vertexes 5 and 6 coincide and the vertexes 2 and 3 coincide;* the hexagon thus becomes a tetragon 1 2 4 5. Now the opposite sides of the tetragon 12 and 45, and likewise 24 and 51, and the tangents at the opposite vertexes 2 and 5 intersect each other on a straight line.

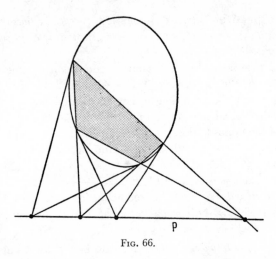

FIG. 66.

Since we could just as easily choose the two other opposite vertexes, the point of intersection of the tangents at these vertexes also lies on the Pascal line. We therefore obtain the following

COROLLARY 2 (Figure 66): *In every tetragon inscribed in a conic section all the pairs of opposite sides and tangents to the pairs of opposite vertexes intersect on a straight line.*

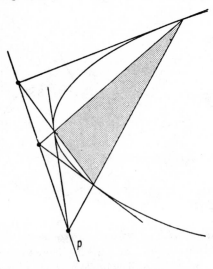

FIG. 67.

III. The vertexes 1 and 2 coincide, so do vertexes 3 and 4, and so do vertexes 5 and 6; the hexagon becomes a triangle, and we obtain

COROLLARY 3 (Figure 67): *In every triangle inscribed in a conic section the sides intersect with the tangents to the opposite vertexes on a straight line.*

62   Brianchon's Hexagram Theorem

To demonstrate that the three opposite vertex lines of a hexagram circumscribed about a conic section pass through a point.

A hexagram circumscribed about a conic section consists essentially of six tangents I, II, III, IV, V, VI to the conic section, which are the sides of the hexagram, and the six points of intersection I II, II III, III IV, IV V, V VI, VI I forming the vertexes of the hexagram. The vertexes I II and IV V, the vertexes II III and V VI, and the vertexes III IV and VI I are called opposite vertexes, and the lines connecting them are called opposite vertex lines.

The point through which the three opposite vertex lines pass is called the *Brianchon point* and the hexagram the *Brianchon hexagram*. The theorem to be proved can be stated in a somewhat shorter form as follows.

The three opposite vertex lines of a Brianchon hexagram pass through a point.

This theorem, which is as important in the theory of conic sections as the Pascal theorem, was published in 1810 by the French mathematician Brianchon (1785–1864) in the *Journal de l'École Polytechnique.*

The following projective proof of Brianchon's theorem is based on the two *theorems of Steiner*:

I. *The tangents of a conic section cut two of the tangents into projective ranges of points.*

II. *If in the projectivity between two ray pencils the line joining the pencil centers corresponds to itself, the pencils are perspective.*

PROOF OF I. We first prove I. for a circle. For this purpose let us consider the following structure: 1. the range of points $\Re$ through which a moving point P on the circle passes; 2. the pencil $\mathfrak{B}$ of the rays FP that run from the fixed circle point F to the moving point P; 3. the field $\mathfrak{S}$ of tangents t drawn to the different positions of P; 4. the range $\mathfrak{r}$ of the points of intersection S of these tangents with the fixed circle tangents *f* through *F*; 5. finally, the pencil 𝔟 of the rays *MS* that run from the center point *M* of the circle to *S*. Then ℜ, 𝔅, and 𝔖 are projective by definition, 𝔅 and 𝔟 are projective because they are congruent (every ray from 𝔅 is perpendicular to the corresponding ray from 𝔟), and finally 𝔯 and 𝔟 are projective because they are perspective. Consequently, 𝔖 and 𝔯 are projective. I.e.:

A field of tangents to a circle is projective with respect to the range of points that the tangents of the field generate on an arbitrary fixed tangent. From this it follows directly that:

The tangents of a circle cut two of them into projective ranges of points.

We will now prove theorem I. for a conic section. The conic section is the central projection of a circle in which its tangents are perspectives of circle tangents. In this projection the ranges of points mentioned appear as perspectives of the two ranges that the circle tangents generate on the two fixed circle tangents, which correspond to the chosen conic section tangents in the central projection. Now, since the latter ranges are projective, the former must also be.

Proof of II. is given in No. 59.

Now for the proof of Brianchon's theorem!

Let the sides of the hexagram be I, II, III, IV, V, VI. According to auxiliary theorem I., the points of intersection generated on tangents *I and III* by II, IV, V, VI form projective ranges of points, and consequently the junction lines II′, IV′, V′, VI′, and II″, IV″, V″, VI″ of these points with the points (*centers*) *V IV and V VI* form

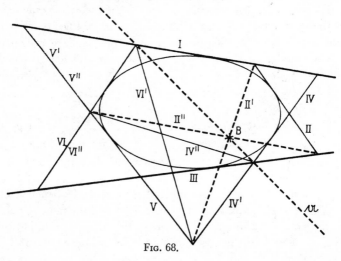

FIG. 68.

projective pencils. Since in the line V connecting the centers, corresponding rays $(V' \equiv v'')$ coincide, the pencils are perspective according to auxiliary theorem II., and the rays II' and II", IV' and IV", and VI' and VI" intersect on one straight line, the axis of perspectivity, the junction line $\mathfrak{a}$ of the points IV' IV" and VI' VI", i.e., of the points III IV and VI I. In other words: The opposite vertex lines II' (from I II to IV V), II" (from II III to V VI), and $\mathfrak{a}$ (from III IV to VI I) pass through one point, the Brianchon point. Q.E.D.

THE CONVERSE OF BRIANCHON'S THEOREM: *If the opposite vertex lines of a hexagram* (of which three sides do not pass through one point) *pass through a point, the sides of the hexagram form tangents of a conic section.*

Indirect proof, similar to the proof of the converse of Pascal's theorem (No. 61).

If two sides of the Brianchon hexagram coincide once or twice or three times, we obtain the corollaries of the Brianchon theorem, the most important of which we will here mention.

I. The sides V and VI coincide; this is to be considered as a situation in which side VI comes closer and closer to side V and finally coincides with it. The point of intersection V VI then becomes the point of tangency of the tangent V, and the hexagram becomes the pentagram I II III IV V. Brianchon's theorem then assumes the following form:

COROLLARY 1 (Figure 69): *In every pentagram circumscribed about a conic section the lines joining two pairs of nonadjacent vertexes and the junction line of the fifth vertex with the point of tangency of its opposite side pass through one point.*

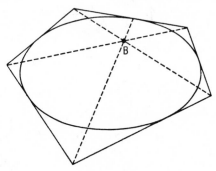

FIG. 69.

II. The sides V and VI coincide, and the sides II and III coincide;
here the hexagram becomes the tetragram I II IV V. Now the
junction lines of the opposite vertexes I II and IV V, as well as those
of II IV and V I, and also the junction lines of the tangency points
of II and V pass through one point.   Since we could as easily select

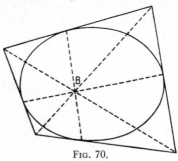

FIG. 70.

the tangency points of the opposite sides I and IV, their junction
line also passes through the Brianchon point.   Consequently, we
obtain

COROLLARY 2 (Figure 70): *In every tetragram circumscribed about a
conic section the two diagonals and the two tangency chords of the opposite sides
pass through one point.*

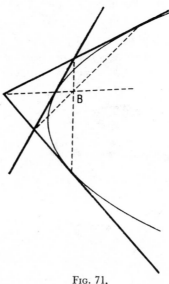

FIG. 71.

III. The sides I and II coincide, the sides III and IV coincide, and the sides V and VI also coincide; the hexagram becomes a trigram, and we obtain

COROLLARY 3 (Figure 71): *In every triangle circumscribed about a conic section the lines connecting the vertexes with the tangency points of the opposite sides pass through one point.*

63   Desargues' Involution Theorem

The points of intersection of a line with the three pairs of opposite sides of a complete tetragon* and a conic section circumscribed about this tetragon form four point pairs of an involution. The lines joining a point with the three pairs of opposite vertexes of a complete tetragram* and the tangents drawn from the point to a conic section inscribed in the tetragram form four ray pairs of an involution.

It is here assumed that the line does not pass through a corner of the tetragon and that the point does not lie on a side of the tetragram.

This double theorem was formulated and proved in 1639 by Desargues (No. 59) in his major work on conic sections. The work bears the strange title *Brouillon-Projet d'une atteinte aux événements des rencontres d'un cône avec un plan,* or approximately in English "First Draft of a Projected Essay on the Phenomena Arising from the Intersection of a Cone with a Plane."

Desargues was the source of the concept of involution and of an amazing series of involution theorems as well, so that it seems appropriate at this point to take up briefly for readers unfamiliar with it the most significant properties of involution.

In a conjective projectivity (No. 59) between two homologous structures I and II each element of a common base can be assigned to I as well as II. Now, if there are two elements A and B of the base such that to the element A of I there corresponds the element B of II and *simultaneously* to the element B of I there corresponds the element A of II, we say that the elements A and B are *conjugate* (to each other) or *correspond to each other in double fashion.*

---

* A complete tetragon (tetragram) consists essentially of four points (lines) 1, 2, 3, 4 and their six connecting lines (points of intersection) 23, 14, 31, 24, 12, 34, of which 23 and 14, 31 and 24, 12 and 34 are known as opposite sides (opposite vertexes).

Let us consider in addition to the conjugate point pair (A, B) another arbitrary pair of homologous elements: P from I and Q from II. From the equation

$$(ABPQ) = (BAQP)$$

it then follows that to the element Q from I there also corresponds the element P from II, i.e., P and Q are also conjugate. Thus, *if one pair of homologous elements in a conjective projectivity is composed of conjugate elements, then* every *pair is composed of conjugate elements.*

A conjective projectivity in which every two homologous elements are conjugate is called an *involution* or an *involutional projectivity.* Every pair of conjugate elements is called for short an *element pair of the involution.*

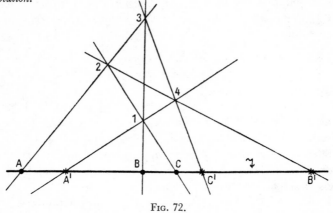

Fig. 72.

Since a projectivity is fixed by three elements of one structure and the homologous elements of the other, an involution is determined by two pairs A, A' and B, B' of conjugate elements insofar as the elements A, A', B of the one structure correspond to the elements A', A, B' of the other.

Construction of an involution, i.e., construction of an element P' corresponding to an arbitrary element P, is most effectively accomplished by means of Desargues' involution theorem (where conic sections do not enter into the picture). Let us say, for example, that we are concerned with the involution of two ranges of points. Let (A, A') and (B, B') be the given point pairs of the involution, C an additional given point of the base $\mathfrak{X}$, and C' the homolog of C we are looking for. We draw through A, B, C three lines that form a triangle 1 2 3 (*A* on 23, *B* on 31, *C* on 12), connect *A'* with 1, *B'* with 2, and the point of intersection 4 of these connecting lines with 3. Then 34 touches the base at *C'*. (The opposite side pairs 23 and 14, 31 and 24, 12 and 34 of the tetragon 1 2 3 4 cut 𝔛 at the point pairs (*A, A'*), (*B, B'*), and (*C, C'*) of the Desargues involution.) The construction of the involution between two ray pencils is carried out in a very similar fashion.

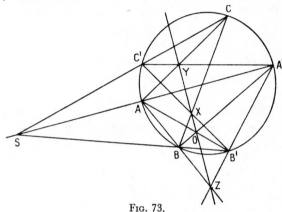

Fig. 73.

We will now consider the important case of the *involution on a circle*. Let (*A, A'*) and (*B, B'*) be two point pairs of an involution between two ranges of points of a circle (Figure 73).

We connect the points of both sets with the circle points *A* and *A'*. We thereby obtain two projective ray pencils in which the rays *AA'*, *AB*, *AB'* of the first pencil correspond to the rays *A'A*, *A'B'*, *A'B* of the second pencil. Since the junction line *AA'* of the pencil centers corresponds to itself, the pencils are perspective (No. 59). The axis of perspectivity is the junction line of the points of intersection *Z* of *AB* and *A'B'* and *O* of *AB'* and *BA'*.

In order to find the homolog *C'* in the involution of an arbitrary point *C*, we cause *AC* and *OZ* to intersect at *Y* and connect *Y* with *A'*; the connecting line touches the circle at *C'*.

Since we can just as well undertake the whole consideration with the pencil centers *B* and *B'* (instead of *A* and *A'*), we also obtain *C'* when we cause *BC* and *OZ* to intersect and connect the point of intersection *X* with *B'*.

Since the homologous sides (bearing the same letter designation) of triangles *ABC* and *A'B'C'* intersect on a straight line (*XYZ*), then, according to Desargues' homology theorem (No. 59), the junction lines AA', BB', and CC' of the homologous vertexes pass through one point S.   If we then draw through S any secant, this secant cuts the circle at two conjugate points of the involution.

The result of our consideration is the theorem:

The lines joining the conjugate points of an involution on a circle pass through a fixed point.

And conversely:

A secant rotated about a fixed point cuts a circle at the point pairs of an involution.

In quite similar fashion the following theorem is proved:

The points of intersection of conjugate tangents of an involution on a circle lie on a straight line.

And conversely:

If a point moves on a line, the tangents drawn from this point to a circle generate an involution on the circle (Figure 74).

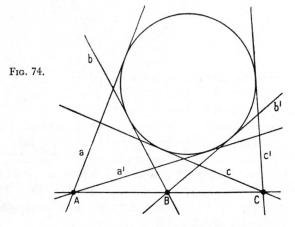

Fig. 74.

Moreover, since every conic section is the central projection of a circle, and projectivity, and thus also involution, between two structures is not annulled by projection of these structures (Pappus' theorem, No. 59), the two just stated theorems are valid for conic sections as well:

INVOLUTION ON A CONIC SECTION: *The lines connecting conjugate points of an involution on a conic section pass through a fixed point.*

The points of intersection of conjugate tangents of an involution on a conic section lie on a fixed straight line.

And conversely:

A secant rotated about a fixed point cuts a conic section at the point pairs of an involution. The tangents from a point moving along a fixed straight line to a conic section are tangent pairs of an involution.

The *proof of Desargues' involution theorem* is based on the theorems:

The points of a conic section are projected from pairs of themselves by projective pencils (No. 61).

The tangents of a conic section cut two of the tangents into projective ranges of points (No. 62).

Let 1 2 3 4 be an inscribed tetragon. Let the line *g* cut the sides 23, 31, 12 at *A*, *B*, *C*, the opposite sides 14, 24, 34 at *A'*, *B'*, *C'*, the conic section at *S* and *S'*.

We connect the conic section points 2, 3, *S*, *S'* with 1 and 4 and obtain two projective pencils with the centers 1 and 4, so that the projections 12 13 1*S* 1*S'* and 42 43 4*S* 4*S'* are projective.

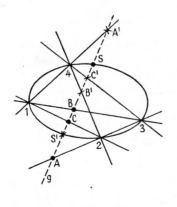

Fig. 75.

We cause these pencils to intersect with *g* and obtain two

Let I II III IV be a circumscribed tetragram. Let the lines connecting the point *P* with the vertexes II III, III I, I II be *a*, *b*, *c*, with the opposite angles I IV, II IV, III IV *a'*, *b'*, *c'*. Let the tangents from *P* to the conic section be *t* and *t'*. We cut the conic section tangents II, III, *t*, *t'* with I and IV and obtain two projective ranges of points on the bases I and IV, so that the projections I II I III I*t* I*t'* and IV II IV III IV*t* IV*t'* are projective.

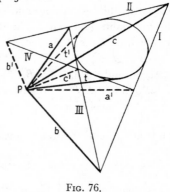

Fig. 76.

We project these ranges from *P* and obtain two conjective

conjective projective ranges of points with the base g in which

$$CBSS' \barwedge B'C'SS',$$

i.e.,

$$(CBSS') = (B'C'SS').$$

projective ray pencils with the center P in which

$$cbtt' \barwedge b'c'tt',$$

i.e.,

$$(cbtt') = (b'c'tt').$$

We now switch the first two terms with each other and the second two terms with each other on the right-hand side and obtain

$$(CBSS') = (C'B'S'S),$$

so that

$$CBSS' \barwedge C'B'S'S.$$

$$(cbtt') = (c'b't't),$$

so that

$$cbtt' \barwedge c'b't't.$$

In this projection there are two conjugate points S and S'. Consequently, the projectivity is an involution, and the points B and B', as well as the points C and C', are conjugate.

In this projection there are two conjugate rays t and t'. Consequently, the projectivity is an involution, and the rays b and b', as well as the rays c and c', are conjugate.

If we connect the conic section points 3, 1, S, S' with 2 and 4, and undertake the same considerations, we find that

$$(ACSS') = (A'C'S'S),$$

so that in the involution defined by the point pairs (S, S') and (C, C') the points A and A' are also conjugate.

If we cut the conic section tangents III, I, t, t' with II and IV, and undertake the same considerations, we find that

$$(actt') = (a'c't't),$$

so that in the involution defined by the ray pairs (t, t') and (c, c') the rays a and a' are also conjugate.

Accordingly, (A, A'), (B, B'), (C, C'), and (S, S') are point pairs of an involution.

Accordingly, (a, a'), (b, b'), (c, c'), and (t, t') are ray pairs of an involution.

Thus Desargues' theorem is proved.

Special Cases

We maintain fixed the conic section, the three vertexes 1, 2, 3, and the straight line g; we allow the vertex 4, on the other hand,

We maintain fixed the conic section, the three sides I, II, III, and the point P; we allow the side IV to roll along the conic

to travel on the conic section toward the point 3. The secant 34 then comes closer and closer to the tangent at 3, while at the same time point A' comes closer and closer to point B and point B' closer and closer to point A. When 4 reaches 3, 43 becomes a tangent through 3, and A' coincides with B and B' with A.

section into position III. The vertex III IV then comes closer and closer to the point of tangency of the tangent III, while at the same time the ray a' comes closer and closer to the ray b and the ray b' comes closer and closer to the ray a. When IV coincides with III, IV III becomes the tangency point of III, and a' coincides with b and b' with a.

Consequently, we obtain

COROLLARY 1

The points of intersection of a straight line: 1. *with a conic section,* 2. *with two sides of a triangle inscribed in a conic section,* 3. *with the third side of the triangle and the conic section tangent passing through its opposite vertex are three point pairs of an involution.*

1. *The tangents from a point to a conic section,* 2. *the lines joining the point with two vertexes of a trigram circumscribed about a conic section,* 3. *the lines joining the point with the third vertex of the trigram and the point of tangency on its opposite side are three ray pairs of an involution.*

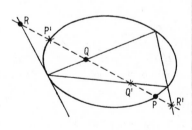

FIG. 77.

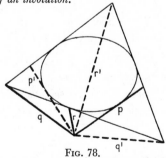

FIG. 78.

If we maintain fixed the conic section in the figure obtained, the line g, and the vertexes 1 and 3, and let 2 travel toward 1, then 12 approaches more and more closely the tangent through

If we maintain fixed the conic section in the figure obtained, the point P, and the sides I and III, and let II roll toward I, the point I II approaches more and more closely the tangency

1 and *A* the point *A'*. When 2 reaches 1, 12 becomes the tangent through 1, *A* coincides with *A'*, and *C* falls on the tangent through 1.

point of I and *a* the ray *a'*. When II reaches I, I II becomes the tangency point of I, *a* coincides with *a'*, and *c* passes through the tangency point of I.

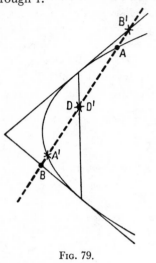

Fig. 79.

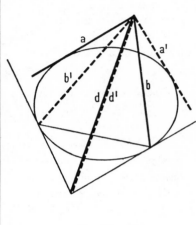

Fig. 80.

Thus, we have

COROLLARY 2

Given a conic section with two tangents and their corresponding tangency chord (Figures 79 and 80):

If the points of intersection of an arbitrary line with the conic section are chosen as the first pair, the points of intersection with the given tangents as the second pair of an involution, the point of intersection of the tangency chord with the line is a double point of the involution.

If the tangents drawn to a conic section from an arbitrary point are chosen as the first pair, and the rays from the point to the ends of the tangency chord as the second pair of an involution, the line joining the point with the point of intersection of the given tangents is a double ray of the involution.

NOTE.   Through the four corners of a tetragon there pass an infinite number of conic sections, which form a so-called conic section pencil. The (complete) tetragon is called a *fundamental tetragon* in this context.

Similarly, there are an infinite number of conic sections that are tangent to the four sides of a tetragram; they form a so-called field of conic sections. The (complete) tetragram in this context is called a *fundamental tetragram.*

Since Desargues' theorem applies to every one of these conic sections, we can state the theorem in the following manner, which is its most general and shortest form.

DESARGUES' INVOLUTION THEOREM: *The intersection point pairs of a line with the conic sections of a pencil are point pairs of an involution.*

The tangent pairs from a point to the conic sections of a field are ray pairs of an involution.

Here the opposite side pairs of the fundamental tetragon are to be considered as (degenerate) conic sections of the pencil, and the opposite vertex pairs of the fundamental tetragram as (degenerate) conic sections of the field.

64   A Conic Section from Five Elements

To draw a conic section of which five elements—points and tangents—are known.

In the solution of this fundamental problem we distinguish three cases:

I. the five elements are of the same type;

II. four elements are of the same type, but the fifth is of the other;

III. three elements are of one type, two are of the other.

In the following we will designate the conic section as $\Re$.

I. *To draw a conic section from five points.*

This problem is commonly solved by means of Pascal's theorem.

We number the points in an arbitrary sequence from 1 to 5 and designate as 6 the unknown point of intersection of an arbitrary line $\mathfrak{g} \equiv 56$, passing through 5, with $\Re$. We then draw the Pascal line p of the

I. *To draw a conic section from five tangents.*

This problem is commonly solved by means of Brianchon's theorem.

We number the tangents in an arbitrary sequence from I to V and designate as VI the unknown tangent drawn to $\Re$ from an arbitrary point $P \equiv V\,VI$ of tangent V. We then draw the Brianchon point B of the hexagram hexagon 1 2 3 4 5 6 as the line connecting the point of intersection of the opposite sides 12 and 45 with the point of intersection of the opposite sides 23 and 56 ≡ g.

The line joining the point of intersection of the two lines 34 and p with the vertex 1 cuts g (≡ 56) at the sought-for point 6.

By repeating the construction with another line g we can obtain as many points of K as we desire.

In order to draw the tangent to 𝔎 at one of the five known points 1, 2, 3, 4, 5 of a conic section, let us say at 5, we make use of the first corollary to Pascal's theorem.

We draw the point of intersection of the two sides 51 and 43, also the point of intersection of the sides 54 and 12, and allow the line p connecting these two points with the side 23 to intersect. The line connecting the resulting point of intersection with the vertex 5 is the sought-for tangent at 5.

II. *To draw a conic section of which four points* 1, 2, 3, 4 *and one tangent* t *are given.*

FIRST CASE: *The tangent* t *passes through one of the given points, for example, through* 4.

Let us consider the tangent *t* as the line connecting two infinitely close conic section points

I II III IV V VI as the point of intersection of the line connecting the opposite vertexes I II and IV V with the line connecting the opposite vertexes II III and V VI ≡ P.

The point of intersection of the line connecting the two points III IV and B with the side I is a second point of the sought-for tangent VI.

By repeating the construction with other points P we can obtain as many tangents of 𝔎 as we desire.

To draw on one of five known tangents I, II, III, IV, V to a conic section, let us say on V, the point of tangency with 𝔎, we make use of the first corollary to Brianchon's theorem.

We draw the line connecting the two vertexes V I and IV III and the line connecting the two vertexes V IV and I II, and connect the point of intersection B of the two lines with the vertex II III. This new junction line meets the tangent V at the sought-for point of tangency.

II. *To draw a conic section of which four tangents* I, II, III, IV *and one point* P *are given.*

FIRST CASE: *The point P lies on one of the given tangents, for example, on* IV.

Let us consider the point *P* as the point of intersection of two infinitely close conic section

4 and 5, so that $t \equiv 45$, and let us designate as 6 the point of intersection of $\Re$ with an arbitrary line x starting from 1, so that $x \equiv 16$. We then draw the Pascal line p of the hexagon 1 2 3 4 5 6 as the line connecting the point of intersection of opposite sides 12 and 45 $\equiv t$ with the point of intersection of the opposite sides 34 and 61 $\equiv x$. The line connecting the point of intersection of the lines p and 23 with the vertex 4 meets $\mathfrak{g}$ at the sought-for point 6.

We now have five known points of $\Re$, and the problem is reduced to I.

SECOND CASE: *The tangent* t *does not pass through any of the given points.*

To solve this problem we use the Desargues' involution theorem (No. 63), taking t as the involution base. We determine the points of intersection, let us say A, A', B, B', of the sides 12, 34, 23, 41 of the tetragon 1 2 3 4 with t and draw a double point of the involution determined on t by the two point pairs (A, A') and (B, B'); this is the point of tangency of the tangent t.

Now five points of $\Re$ are known and the problem is reduced to I.

tangents IV and V, so that $P \equiv$ IV V, and let us designate as VI a second tangent from an arbitrary point X of I to $\Re$, so that $X \equiv$ I VI. We then draw the Brianchon point B of the hexagram I II III IV V VI as the point of intersection of the line connecting the opposite vertexes I II and IV V $\equiv P$ and the line connecting the opposite vertexes III IV and VI I $\equiv X$. The point of intersection of the line connecting the points B and II III with the side IV is a second point of the sought-for tangent VI.

We now have five known tangents of $\Re$ and the problem is thereby reduced to I.

SECOND CASE: *The point* P *does not lie on any of the given tangents.*

To solve this problem we make use of Desargues' involution theorem (No. 63), taking P as the involution base. We determine the junction lines a, a', b, b' connecting the vertexes I II, III IV, II III, IV I of the tetragram I II III IV with P and construct the double ray of the involution determined on P by the two ray pairs (a, a') and (b, b'); this is the conic section tangent passing through P.

We now have five known tangents of $\Re$ and the problem thus reduces to I.

The second case of II. has two solutions if the involution has two double elements and no solution if the involution has no double elements.

III. *To draw a conic section of which three points* A, B, C *and two tangents* d *and* e *are given.*

FIRST CASE: d *passes through* A, *and* e *through* B.

We draw the point of intersection S of an arbitrary line g originating at *A* with $\mathfrak{K}$.

For our purpose we construct the Pascal line *p* of the hexagon 1 2 3 4 5 6 of which the vertexes 1 and 2 coincide with *A*, the vertexes 3 and 4 with *B*, the vertex 5 with *C*, and the vertex 6 with *S*, the sides 12 and 34 being represented by the tangents *d* and *e*, respectively. *p* is the line connecting the point of intersection of the sides 12 ≡ *d* and 45 ≡ *BC* with the point of intersection of the sides 34 ≡ *e* and 61 ≡ *g*. The line connecting the point of intersection of the lines *p* and 23 ≡ *AB* with the vertex 5 ≡ *C* meets *g* at the sought-for conic section point S.

In the same way we draw a fifth point of $\mathfrak{K}$ and thus reduce the problem to I.

SECOND CASE: d *passes through* A, *and* e *does not pass through any of the given points.*

III. *To draw a conic section of which three tangents* a, b, c *and two points* D *and* E *are given.*

FIRST CASE: D *lies on* a, *and* E *on* b.

We draw the (second) tangent *t* from an arbitrary point *P* of tangent *a* to $\mathfrak{K}$.

For our purpose we construct the Brianchon point *B* of the hexagram I II III IV V VI of which the sides I and II coincide with *a*, the sides III and IV with *b*, the side V with *c*, and the side VI with *t*, the vertexes I II and III IV being represented by the points *D* and *E*, respectively. *B* is the point of intersection of the line connecting the vertexes I II ≡ *D* and IV V ≡ *bc* and the line connecting the vertexes III IV ≡ *E* and VI I ≡ *P*. The point of intersection of the line connecting points *B* and II III ≡ *ab* with the side V = *c* is a second point of the sought-for tangent *t*.

In the same way we draw a fifth tangent of $\mathfrak{K}$ and thereby reduce the problem to I.

SECOND CASE: D *lies on* a, *and* E *does not lie on any of the given tangents.*

We solve this case with the second corollary to Desargues' involution theorem.

We determine the points of intersection D and E of the line BC with d and e and construct a double point of the involution defined by the point pairs (B, C) and (D, E). Its junction line with A passes through the point of tangency of e.

We determine the connecting lines d and e joining the point bc with D and E and draw a double ray of the involution determined by the ray pairs (b, c) and (d, e). Its point of intersection with a lies on the tangent passing through E; this tangent is thus determined.

The problem is now reduced to the preceding case.

THIRD CASE: *Neither of the two tangents passes through any of the given points.*

THIRD CASE: *Neither of the two points lies on any of the given tangents.*

In this case also the solution is based on the second corollary to Desargues' involution theorem.

We designate the points of intersection of BC with d and e as D and E and determine a double point P of the involution defined by the point pairs (B, C) and (D, E). It lies on the tangency chord of the tangents d and e.

We designate the lines joining bc with D and E as d and e and determine a double ray s of the involution determined by the ray pairs (b, c) and (d, e). It passes through the point of intersection of the tangents drawn through D and E.

We designate the points of intersection of CA with d and e as D' and E' and draw a double point P' of the involution determined by the point pairs (C, A) and (D', E'). This double point also lies on the tangency chord of the tangents d and e.

We designate the lines joining ca with D and E as d' and e' and draw a double ray s' of the involution determined by the ray pairs (c, a) and (d', e'). This double ray also passes through the point of intersection of the tangents through D and E.

The line joining the two double points P and P' is thus the tangency chord we have mentioned and meets the tangents d and e at their tangency points.

The point of intersection of the two double rays s and s' is thus the tangent intersection point that was mentioned before and the lines joining it to D and E are the tangents passing through D and E.

We now know *five* points of $\Re$ and thus return to I.

We now have *five* tangents of $\Re$ and thus return to I.

This last problem admits of a solution only when each of the two designated involutions has double elements. And since we can connect each of the two double elements of one of the involutions with each of the double elements of the other, we obtain *four* possible tangency chords and tangent intersection points, respectively, and thus *four* different conic sections.

65   A Conic Section and a Straight Line

To draw the points of intersection of a given straight line with a conic section of which five elements—points and tangents—are known.

In the solution of this problem we may assume, in view of No. 64, that five *points* of the conic section are known. The solution is then based on the theorem: *The points of a conic section are projected from pairs of themselves by projective pencils* (No. 61) and on *Steiner's double element construction* (No. 60).

Let the given line be called g, the given points of the conic section A, B, C, D, E. We can think of the points of the conic section as projected from D and E by the two projective pencils I and II. These pencils cut g into the two projective ranges of points 1 and 2. The points of intersection S and T of g with the conic section are the double elements of the projectivity $1 \overline{\wedge} 2$. This projectivity is, however, determined by the points of intersection A_1, B_1, C_1 of the rays DA, DB, DC with g and the homologous points of intersection A_2, B_2, C_2 of the rays EA, EB, EC with g.

We therefore draw according to Steiner the double elements of the projectivity defined on g by the homologous point triplets (A_1, B_1, C_1) and (A_2, B_2, C_2); they are the points of intersection we are looking for.

66°   A Conic Section and a Point

To draw the tangents from a given point to a conic section of which five elements—points and tangents—are known.

In view of the considerations of No. 64, we may assume the given conic section elements to be *tangents*.

The solution to this problem is based upon the theorem: *The tangents of a conic section mark off projective ranges of points on two of the tangents* (No. 62) and on *Steiner's double element construction* (No. 60).

Let the given point be P, the given tangents a, b, c, d, e. Let us consider the tangents of the conic section as intersecting with d and e, so that we obtain on d and e the projective ranges 1 and 2 in which the points of intersection A_1, B_1, C_1 of the tangents a, b, c with d and the points of intersection A_2, B_2, C_2 of the tangents a, b, c with e are homologous elements. The reflections of these ranges of points on P thus form two projective ray pencils I and II. The (conjective) projectivity is determined by the lines a_I, b_I, c_I connecting the points of intersection A_1, B_1, C_1 to P and the homologous connecting lines a_{II}, b_{II}, c_{II} joining the points of intersection A_2, B_2, C_2 to P. Since each of the two tangents s and t from P to the conic section cuts 1 and 2 into homologous elements, s and t are therefore the double elements of the projectivity I $\barwedge$ II.

We thus draw according to Steiner the double elements of the conjective projectivity determined by the homologous ray triplets (a_I, b_I, c_I) and (a_{II}, b_{II}, c_{II}); they are the sought-for tangents.

Stereometric Problems

67   Steiner's Division of Space by Planes

What is the maximum number of parts into which a space can be divided by n *planes?*

This very interesting problem appears in Steiner's paper "Several laws governing the division of planes and space" (*Crelle's Journal*, vol. I and Steiner's *Complete Works*, vol. I).

We first solve the PRELIMINARY PROBLEM: *What is the maximum number of parts into which a plane can be divided by* n *straight lines?*

The number of parts will evidently be maximal when no two lines are parallel and no more than two lines pass through one point. In the following we will assume these two conditions to be satisfied and we will designate the corresponding number of surface sections generated by the n lines as $\bar{n}$.

Thus, let the plane be divided by n lines into $\bar{n}$ surface sections. We now draw one additional line. This line is divided by the first n lines into n points, and thus traverses $n + 1$ of the available $\bar{n}$ surface sections, dividing each of them into *two* parts, so that the $(n + 1)$th line increases the number of surface sections by $n + 1$. Consequently, we obtain the equation

$$\overline{n + 1} = \bar{n} + (n + 1).$$

We then apply this equation to the cases in which $n = 0, 1, 2, \ldots$ and we form the n equations

$$
\begin{aligned}
\bar{1} &= 1 &&+ 1, \\
\bar{2} &= \bar{1} &&+ 2, \\
\bar{3} &= \bar{2} &&+ 3, \\
&\vdots \\
\bar{n} &= \overline{n - 1} + n.
\end{aligned}
$$

Addition of these equations results in

$$\bar{n} = 1 + (1 + 2 + 3 + \cdots + n)$$

or, since the sum of the first n natural numbers is $n(n + 1)/2$,

$$(1) \qquad \bar{n} = 1 + n\,\frac{n + 1}{2}.$$

Thus, the maximum number of parts into which a plane can be divided by n *lines is* (n² + n + 2)/2.

The obtained result is easily confirmed for the cases $n = 1, 2, 3, \ldots$.

Now for the space problem! It is apparent that the number of partial spaces attains a maximum when no more than three planes ever intersect at one point and when the lines of intersection of no more than two planes are ever parallel. We will therefore assume that these conditions are satisfied in the following and we designate the number of partial spaces formed by n planes as $\tilde{n}$.

Then, let the space be divided by n planes into $\tilde{n}$ partial spaces. To these planes we now add one additional plane. This plane is cut by the original n planes into n lines of which no more than two pass through a single point and no two or more are parallel. The new $(n + 1)$th plane is therefore divided by the n lines into $\bar{n}$ surface sections.

Each of these $\bar{n}$ surface sections cuts the partial space that it traverses into two smaller spaces, so that the addition of the $(n + 1)$th plane increases the number of the partial spaces originally present by $\bar{n}$. This gives us the equation

$$\widetilde{n + 1} = \tilde{n} + \bar{n}.$$

We form this equation for the cases $n = 1, 2, 3$, etc., and obtain the n equations

$$\tilde{1} = 1 \quad + 1,$$
$$\tilde{2} = \tilde{1} \quad + \bar{1},$$
$$\tilde{3} = \tilde{2} \quad + \bar{2},$$
$$\vdots$$
$$\tilde{n} = \widetilde{n - 1} + \overline{n - 1}.$$

Addition of these equations results in

$$\tilde{n} = 2 + \bar{1} + \bar{2} + \bar{3} + \cdots + \overline{n - 1}$$

or, according to (1),

$$\tilde{n} = n + 1 + \tfrac{1}{2}(1\cdot 2 + 2\cdot 3 + \cdots + (n - 1)n).$$

If we then divide each product $\nu(\nu + 1)$ into $\nu^2 + \nu$, we obtain

$$\tilde{n} = n + 1 + \tfrac{1}{2}\{[1^2 + 2^2 + \cdots + (n - 1)^2]$$
$$+ [1 + 2 + \cdots + (n - 1)]\}.$$

Now, according to No. 11, the sums in the first and second square brackets, respectively, are

$$\tfrac{1}{6}(n - 1)n(2n - 1) \quad \text{and} \quad \tfrac{1}{2}(n - 1)n, \quad \text{respectively;}$$

the brace thus equals $\frac{1}{3}(n-1)n(n+1)$, and

$$\tilde{n} = n + 1 + \frac{1}{6}(n-1)n(n+1)$$

or

$$\tilde{n} = \frac{n^3 + 5n + 6}{6}.$$

CONCLUSION: *The maximum number of parts into which a space can be divided by* n *planes is* $(n^3 + 5n + 6)/6$.

<hr>

68 Euler's Tetrahedron Problem

To express the area of a tetrahedron in terms of its six edges.

This fundamental problem was posed and solved by Leonhard Euler (*Novi Commentarii Academiae Petropolitanae ad annos 1752 et 1753*).

The following convenient and simple solution is based upon vector calculus.

We will designate the vertexes of the tetrahedron as A, B, C, O, the six edges BC, CA, AB, OA, OB, OC as $a, b, c, \mathfrak{p}, \mathfrak{q}, \mathfrak{r}$, the three vectors $\vec{OA}, \vec{OB}, \vec{OC}$ as $\mathfrak{p}, \mathfrak{q}, \mathfrak{r}$, and the area we are looking for as T. We will consider the edges $\mathfrak{p}, \mathfrak{q}, \mathfrak{r}$ originating from the vertex O as being so arranged that they form a right-handed system, i.e., that $\mathfrak{p}$ can be imagined as the thumb, $\mathfrak{q}$ as the index finger, and $\mathfrak{r}$ as the middle finger of the right hand.

If we take the triangle OAB as the base surface and the vertex C as the apex of the tetrahedron, then the double value of the base surface area S is given by the magnitude of the vector product $\mathfrak{S} = \mathfrak{p} \times \mathfrak{q}$, the altitude CF is the projection of the edge r on CF, i.e., ro, if we designate as o the cosine of the angle between CO and CF or also of the angle of the two vectors $\mathfrak{S}$ and $\mathfrak{r}$.

Consequently, six times the tetrahedron area is equal to $S \cdot ro$ or equal to the scalar product* $\mathfrak{S} \cdot \mathfrak{r}$ of the vector $\mathfrak{S}$ and $\mathfrak{r}$. Thus, we obtain the simple formula

$$6T = \mathfrak{p} \times \mathfrak{q} \cdot \mathfrak{r},$$

which can be stated verbally as follows:

Six times the area of a tetrahedron is equal to the mixed product of the three vectorial edges originating from one edge of the tetrahedron.

<hr>

* The scalar product of two vectors $\mathfrak{A}$ and $\mathfrak{B}$ is most conveniently written $\mathfrak{A} \cdot \mathfrak{B}$ or in the still simpler form $\mathfrak{A}\mathfrak{B}$.

Here the three factors of the mixed product must be written in such sequence as to form a right-handed system (for otherwise the mixed product would represent six times the *negative* tetrahedron area).

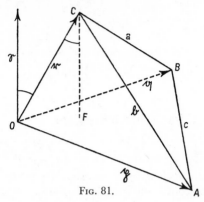

Fig. 81.

We now introduce a right-angle coordinate system with origin at O and designate the coordinates of the three vertexes A, B, C as $x|y|z$, $x'|y'|z'$, and $x''|y''|z''$. The three components of the vector $\mathfrak{S} = \mathfrak{p} \times \mathfrak{q}$ are then $yz' - zy'$, $zx' - xz'$, $xy' - yx'$, and the scalar product $\mathfrak{S} \cdot \mathfrak{r}$ is equal to $(yz' - zy')x'' + (zx' - xz')y'' + (xy' - yx')z''$, i.e., equal to the determinant whose columns are the components of the vectors $\mathfrak{p}$, $\mathfrak{q}$, $\mathfrak{r}$. Thus we obtain the elegant formula

$$6T = \begin{vmatrix} x & y & z \\ x' & y' & z' \\ x'' & y'' & z'' \end{vmatrix}.$$

On squaring this formula, multiplying the two (same) determinants row by row, we obtain $36T^2 = \triangle =$

$$\begin{vmatrix} xx + yy + zz & xx' + yy' + zz' & xx'' + yy'' + zz'' \\ x'x + y'y + z'z & x'x' + y'y' + z'z' & x'x'' + y'y'' + z'z'' \\ x''x + y''y + z''z & x''x' + y''y' + z''z' & x''x'' + y''y'' + z''z'' \end{vmatrix}$$

or, since the elements of this determinant are the scalar products of the vectors $\mathfrak{p}$, $\mathfrak{q}$, $\mathfrak{r}$ in pairs, or the squares of these vectors,

(I) $$36T^2 = \begin{vmatrix} \mathfrak{pp} & \mathfrak{pq} & \mathfrak{pr} \\ \mathfrak{qp} & \mathfrak{qq} & \mathfrak{qr} \\ \mathfrak{rp} & \mathfrak{rq} & \mathfrak{rr} \end{vmatrix}.$$

This is *Euler's tetrahedron formula.* (Euler, however, expressed the right-hand side as an algebraic sum rather than as a determinant.)

It contains the solution to the problem posed, since the elements of the determinant are simple expressions of the edges; specifically:

$$\mathfrak{pp} = p^2, \qquad \mathfrak{qq} = q^2, \qquad \mathfrak{rr} = r^2,$$

$$\mathfrak{qr} = \frac{q^2 + r^2 - a^2}{2}, \qquad \mathfrak{rp} = \frac{r^2 + p^2 - b^2}{2}, \qquad \mathfrak{pq} = \frac{p^2 + q^2 - c^2}{2}.$$

In the tetrahedron with the edges $a = 11$, $b = 10$, $c = 9$, $p = 8$, $q = 7$, $r = 6$, for example, we have

$$\mathfrak{pp} = 64, \quad \mathfrak{qq} = 49, \quad \mathfrak{rr} = 36, \quad \mathfrak{qr} = -18, \quad \mathfrak{rp} = 0, \quad \mathfrak{pq} = 16,$$

and

$$36T^2 = \begin{vmatrix} 64 & 16 & 0 \\ 16 & 49 & -18 \\ 0 & -18 & 36 \end{vmatrix} = 16 \cdot 36 \begin{vmatrix} 4 & 16 & 0 \\ 1 & 49 & -9 \\ 0 & -1 & 1 \end{vmatrix} = 16 \cdot 36 \cdot 9 \cdot 16$$

and $T = 48$.

We can put the obtained result into still another form.

If we multiply each element of $\triangle$ by 2 and express the doubled scalar product by the squares P, Q, R, A, B, C of the edge magnitudes p, q, r, a, b, c, we obtain

$$288T^2 = \begin{vmatrix} 2P & P + Q - C & P + R - B \\ Q + P - C & 2Q & Q + R - A \\ R + P - B & R + Q - A & 2R \end{vmatrix}.$$

Now we distribute zeros at the left and minus ones at the bottom and obtain

$$288T^2 = \begin{vmatrix} 0 & 2P & P + Q - C & P + R - B \\ 0 & Q + P - C & 2Q & Q + R - A \\ 0 & R + P - B & R + Q - A & 2R \\ -1 & -1 & -1 & -1 \end{vmatrix}.$$

If we add the P-, Q-, and R-multiples of the last row to the first, second, and third rows, respectively, we obtain the somewhat simpler

$$288\,T^2 = \begin{vmatrix} -P & P & Q-C & R-B \\ -Q & P-C & Q & R-A \\ -R & P-B & Q-A & R \\ -1 & -1 & -1 & -1 \end{vmatrix}.$$

We now distribute zeros and ones at the top and right:

$$288\,T^2 = \begin{vmatrix} 0 & 0 & 0 & 0 & 1 \\ -P & P & Q-C & R-B & 1 \\ -Q & P-C & Q & R-A & 1 \\ -R & P-B & Q-A & R & 1 \\ -1 & -1 & -1 & -1 & 0 \end{vmatrix}.$$

If we now subtract the P-, Q-, and R-multiples of the last column from the second, third, and fourth columns, respectively, we finally obtain

$$288\,T^2 = \begin{vmatrix} 0 & -P & -Q & -R & 1 \\ -P & 0 & -C & -B & 1 \\ -Q & -C & 0 & -A & 1 \\ -R & -B & -A & 0 & 1 \\ -1 & -1 & -1 & -1 & 0 \end{vmatrix}$$

or, if we reverse all the minus signs,

(II) $$288\,T^2 = \begin{vmatrix} 0 & P & Q & R & 1 \\ P & 0 & C & B & 1 \\ Q & C & 0 & A & 1 \\ R & B & A & 0 & 1 \\ 1 & 1 & 1 & 1 & 0 \end{vmatrix}.$$

In this remarkable formula P, Q, R, A, B, C are the squares of the edges p, q, r, a, b, c.

NOTE: THE FOUR-POINT RELATION: If A, B, C, O are four points of a plane, the area of the tetrahedron $ABCO$ is zero and (I) is transformed into the so-called *four-point relation*:

$$\begin{vmatrix} \mathfrak{p}\mathfrak{p} & \mathfrak{p}\mathfrak{q} & \mathfrak{p}\mathfrak{r} \\ \mathfrak{q}\mathfrak{p} & \mathfrak{q}\mathfrak{q} & \mathfrak{q}\mathfrak{r} \\ \mathfrak{r}\mathfrak{p} & \mathfrak{r}\mathfrak{q} & \mathfrak{r}\mathfrak{r} \end{vmatrix} = 0$$

for the six junction lines $BC = a$, $CA = b$, $AB = c$, $OA = p$, $OB = q$, $OC = r$ that are possible between the four points.

69   The Shortest Distance Between Skew Lines

To calculate the angle and distance between two given skew lines.

This important problem is usually encountered in one of the following two forms:

I. *To calculate the angle and distance between two skew lines when a point on each line and the direction of each line are given—the former by coordinates and the latter by the direction cosine of the lines.*

II. *To calculate the angle and distance between two opposite edges of a tetrahedron whose six edges are known.*

The distance between two skew lines is naturally the shortest distance between the lines, i.e., the length of the line perpendicular to *both* lines and joining a point on each.

SOLUTION OF I.   We designate the perpendicular coordinates of the two given points P and p as $A|B|C$ and $a|b|c$, the vector $\overrightarrow{pP}$ (with the components $A - a$, $B - b$, $C - c$) as $\mathfrak{d}$, the direction cosine of the two lines, together with the components of two unit vectors $\mathfrak{E}$ and e lying on the lines as L, M, N and l, m, n, the sought-for angle of the two lines as ω, and the sought-for minimum distance as k.

The solution to this problem, which is in itself not very simple, becomes astonishingly simple with the introduction of the scalar product $\mathfrak{E} \cdot e$ and the vector product $\mathfrak{E} \times e$ of the two vectors $\mathfrak{E}$ and e.

The *former* can be expressed on the one hand (since the vectors $\mathfrak{E}$ and e have a magnitude of 1) as $\cos \omega$, and, on the other, by the components of the factors as $Ll + Mm + Nn$.   We therefore obtain

(1) $$\cos \omega = Ll + Mm + Nn.$$

The *latter* is perpendicular to *both* lines, so that the projection of $\mathfrak{b}$ on the vector $\mathfrak{E} \times \mathfrak{e}$ represents the desired distance k (the shortest distance k between the two lines is specifically the projection of $\mathfrak{b}$ on k and at the same time the projection of $\mathfrak{b}$ on every parallel to k, for example, on $\mathfrak{E} \times \mathfrak{e}$). However, since the projection of a vector $\mathfrak{V}$ on a second vector $\mathfrak{v}$ of the magnitude v is $\mathfrak{V} \cdot \mathfrak{v}/v$, we obtain for k the value $\mathfrak{b} \cdot \mathfrak{E} \times \mathfrak{e}/\sin \omega$ ($\sin \omega$ is the magnitude of the vector $\mathfrak{E} \times \mathfrak{e}$).

Now the scalar product of the two vectors $\mathfrak{b}$ and $\mathfrak{E} \times \mathfrak{e}$ is nothing other than the so-called mixed product of the three vectors $\mathfrak{b}$, $\mathfrak{E}$, and $\mathfrak{e}$. And since the latter is equal to the determinant whose rows are the components of the three vectors (No. 68), we obtain the formula

$$(2) \qquad k = \begin{vmatrix} A - a & B - b & C - c \\ L & M & N \\ l & m & n \end{vmatrix} / \sin \omega.$$

NOTE. If we desire to calculate the coordinates $X/Y/Z$ and $x/y/z$ of the end points U and u of the shortest junction line k, we designate the segments PU and pu as R and r, the vector $\overrightarrow{uU}$ as $\mathfrak{k}$, and we then have

$$\overrightarrow{uU} = \overrightarrow{up} + \overrightarrow{pP} + \overrightarrow{PU},$$

or

$$\mathfrak{k} = -r\mathfrak{e} + \mathfrak{b} + R\mathfrak{E}.$$

If we multiply this equation in scalar fashion with $\mathfrak{E}$ and $\mathfrak{e}$, we obtain, as a result of $\mathfrak{E} \cdot \mathfrak{k} = 0$ and $\mathfrak{e} \cdot \mathfrak{k} = 0$, the two linear equations

$$\mathfrak{E}\mathfrak{E}R - \mathfrak{E}\mathfrak{e}r + \mathfrak{E}\mathfrak{b} = 0,$$

$$\mathfrak{E}\mathfrak{e}R - \mathfrak{e}\mathfrak{e}r + \mathfrak{e}\mathfrak{b} = 0,$$

from which the unknowns R and r are obtained.

SOLUTION OF II. Let the six edges of the tetrahedron be $BC = a$, $CA = b$, $AB = c$, $OA = p$, $OB = q$, $OC = r$, and let the vectors $\overrightarrow{BC}$, $\overrightarrow{CA}$, $\overrightarrow{AB}$, $\overrightarrow{OA}$, $\overrightarrow{OB}$, $\overrightarrow{OC}$ be $\mathfrak{a}$, $\mathfrak{b}$, $\mathfrak{c}$, $\mathfrak{p}$, $\mathfrak{q}$, $\mathfrak{r}$. Let the angle and distance between the two opposite edges $\mathfrak{c}$ and $\mathfrak{r}$ be called ω and k, respectively.

Determination of ω. We have

$$\mathfrak{c} + \mathfrak{r} = \overrightarrow{AB} + \overrightarrow{OC} = \overrightarrow{AO} + \overrightarrow{OB} + \overrightarrow{OA} + \overrightarrow{AC} = \overrightarrow{OB} + \overrightarrow{AC} = \mathfrak{q} - \mathfrak{b},$$

and thus

$$(\mathfrak{c} + \mathfrak{r})^2 = (\mathfrak{c} + \mathfrak{r}) \cdot (\mathfrak{q} - \mathfrak{b}) = \mathfrak{cq} + \mathfrak{qr} - \mathfrak{bc} - \mathfrak{br}.$$

However, since

$$(\mathfrak{c} + \mathfrak{r})^2 = \mathfrak{c}^2 + \mathfrak{r}^2 + 2\mathfrak{cr} = c^2 + r^2 + 2cr \cos \omega,$$

$$2\mathfrak{cq} = c^2 + q^2 - p^2, \qquad 2\mathfrak{qr} = q^2 + r^2 - a^2,$$

$$2\mathfrak{bc} = a^2 - b^2 - c^2, \qquad 2\mathfrak{br} = p^2 - b^2 - r^2,$$

the equation obtained is transformed into

(3) $$2cr \cos \omega = b^2 + q^2 - a^2 - p^2,$$

so that ω is determined.

CALCULATION OF k.   Let the area of the tetrahedron $ABCO$, which we can consider as known in accordance with Euler's formula (No. 68), be called T.   We displace the vector $\mathfrak{r}$ parallel to itself until it has a starting point A in common with $\mathfrak{c}$; its new end point we will call Q, and thus $AQ \,\#\!\!\!+ OC$.   Since the triangles CQA and COA are halves of the parallelogram $COAQ$, they are congruent, and thus the tetrahedrons $CQAB$ and $COAB$ have the same area (T).   If we now take QAB as the base surface of the tetrahedron $CQAB$ and C as the

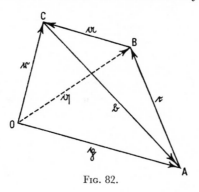

FIG. 82.

apex, the base surface has the area $\frac{1}{2}AQ \cdot AB \cdot \sin QAB = \frac{1}{2}rc \sin \omega$, and the altitude (as the distance of the point C from the plane QAB that contains the edge c and the line AQ that is parallel to the opposite edge OC) has a length of k.   The area of the tetrahedron is therefore $\frac{1}{3} \cdot \frac{1}{2}cr \sin \omega \cdot k$, and we obtain the formula

(4) $$6T = kcr \sin \omega.$$

Since all the magnitudes in this formula are known with the exception of k, it gives us the distance between the opposite edges k which we have been looking for.

NOTE. If we keep in mind that $cr \sin \omega$ is the magnitude of the vector $\mathfrak{c} \times \mathfrak{r}$ and that the shortest distance $\mathfrak{k}$ (conceived of as a vector) between the edges $\mathfrak{c}$ and $\mathfrak{r}$ is parallel to $\mathfrak{c} \times \mathfrak{r}$, we can write

$$6T = \mathfrak{k} \cdot \mathfrak{c} \times \mathfrak{r}$$

and we have the following

THEOREM: *The mixed product of two opposite sides of a tetrahedron and the distance between them is equal to six times the area of the tetrahedron.*

A direct consequence of this theorem is the famous

THEOREM OF STEINER: *All tetrahedrons having two opposite edges of prescribed length lying on two fixed lines have the same area.*

70   The Sphere Circumscribing a Tetrahedron

To determine the radius of the sphere circumscribing a tetrahedron of which all six edges are given.

One should compare the developments of Legendre in his *Éléments de Géométrie*, Note V.

We will first solve the

PRELIMINARY PROBLEM: *To find the relation between the six major arcs that connect the four points of a spherical surface.*

We will call the four points 0, 1, 2, 3, the *arcs* joining them 01, 02, 03, 23, 31, 12, the radii (considered as vectors) running to them $\mathfrak{r}_0$, $\mathfrak{r}_1$, $\mathfrak{r}_2$, $\mathfrak{r}_3$ and their common magnitude h. Since there is always a homogeneous linear relation between four vectors of a space, we have the equation

$$\alpha \mathfrak{r}_0 + \beta \mathfrak{r}_1 + \gamma \mathfrak{r}_2 + \delta \mathfrak{r}_3 = 0,$$

in which not all of the coefficients α, β, γ, δ vanish simultaneously. We multiply the relation sequentially in scalar fashion by $\mathfrak{r}_0$, $\mathfrak{r}_1$, $\mathfrak{r}_2$, $\mathfrak{r}_3$ and obtain the four equations

$$\mathfrak{r}_0\mathfrak{r}_0\alpha + \mathfrak{r}_0\mathfrak{r}_1\beta + \mathfrak{r}_0\mathfrak{r}_2\gamma + \mathfrak{r}_0\mathfrak{r}_3\delta = 0,$$
$$\mathfrak{r}_1\mathfrak{r}_0\alpha + \mathfrak{r}_1\mathfrak{r}_1\beta + \mathfrak{r}_1\mathfrak{r}_2\gamma + \mathfrak{r}_1\mathfrak{r}_3\delta = 0,$$
$$\mathfrak{r}_2\mathfrak{r}_0\alpha + \mathfrak{r}_2\mathfrak{r}_1\beta + \mathfrak{r}_2\mathfrak{r}_2\gamma + \mathfrak{r}_2\mathfrak{r}_3\delta = 0,$$
$$\mathfrak{r}_3\mathfrak{r}_0\alpha + \mathfrak{r}_3\mathfrak{r}_1\beta + \mathfrak{r}_3\mathfrak{r}_2\gamma + \mathfrak{r}_3\mathfrak{r}_3\delta = 0.$$

However, when four homogeneous linear equations with four unknowns $(\alpha, \beta, \gamma, \delta)$ possess an actual solution, the determinant of the coefficients of the equations must be equal to zero. Consequently

$$\begin{vmatrix} \mathfrak{r}_0\mathfrak{r}_0 & \mathfrak{r}_0\mathfrak{r}_1 & \mathfrak{r}_0\mathfrak{r}_2 & \mathfrak{r}_0\mathfrak{r}_3 \\ \mathfrak{r}_1\mathfrak{r}_0 & \mathfrak{r}_1\mathfrak{r}_1 & \mathfrak{r}_1\mathfrak{r}_2 & \mathfrak{r}_1\mathfrak{r}_3 \\ \mathfrak{r}_2\mathfrak{r}_0 & \mathfrak{r}_2\mathfrak{r}_1 & \mathfrak{r}_2\mathfrak{r}_2 & \mathfrak{r}_2\mathfrak{r}_3 \\ \mathfrak{r}_3\mathfrak{r}_0 & \mathfrak{r}_3\mathfrak{r}_1 & \mathfrak{r}_3\mathfrak{r}_2 & \mathfrak{r}_3\mathfrak{r}_3 \end{vmatrix} = 0.$$

Here we replace each product $\mathfrak{r}_n\mathfrak{r}_v$ by $h^2 \cos nv$, eliminate everywhere the factor h^2, and obtain the relation we are looking for

$$(1) \qquad \begin{vmatrix} \cos 00 & \cos 01 & \cos 02 & \cos 03 \\ \cos 10 & \cos 11 & \cos 12 & \cos 13 \\ \cos 20 & \cos 21 & \cos 22 & \cos 23 \\ \cos 30 & \cos 31 & \cos 32 & \cos 33 \end{vmatrix} = 0.$$

($\cos 00$, $\cos 11$, $\cos 22$, $\cos 33$ are naturally merely symmetrical ways of writing unity.)

The solution of the tetrahedron problem is now simple.

In order to maintain agreement with the designations of the preliminary problem we will call the vertexes of the tetrahedron 0, 1, 2, 3, the radius of the sphere of circumscription h. The edges 01, 02, 03, 23, 31, 12 we will call p, q, r, a, b, c, their squares P, Q, R, A, B, C, the area of the tetrahedron T.

We now introduce the four-point relation (1), assign to each cosine the factor $H = 2h^2$ and replace the new determinant elements in accordance with the cosine theorem, e.g., $H \cos 01$ by $H - P$, $H \cos 02$ by $H - Q$, $H \cos 23$ by $H - A$, etc. (naturally $H \cos 00$ and the other elements of the diagonals will be replaced by H). This gives us, after we reverse the sign of all the elements,

$$\begin{vmatrix} -H & P - H & Q - H & R - H \\ P - H & -H & C - H & B - H \\ Q - H & C - H & -H & A - H \\ R - H & B - H & A - H & -H \end{vmatrix} = 0.$$

We now line the bottom of this determinant with ones and the right-hand side with zeros and obtain

$$
\begin{vmatrix}
-H & P-H & Q-H & R-H & 0 \\
P-H & -H & C-H & B-H & 0 \\
Q-H & C-H & -H & A-H & 0 \\
R-H & B-H & A-H & -H & 0 \\
1 & 1 & 1 & 1 & 1
\end{vmatrix} = 0.
$$

We now add to the first, second, third, and fourth rows H times the last row; this gives us

$$
\begin{vmatrix}
0 & P & Q & R & H \\
P & 0 & C & B & H \\
Q & C & 0 & A & H \\
R & B & A & 0 & H \\
1 & 1 & 1 & 1 & 1
\end{vmatrix} = 0.
$$

If we call the minors of the last column M_1, M_2, M_3, M_4, M_5 and arrange them according to the elements of the last column, we obtain

$$
H(M_1 + M_2 + M_3 + M_4) + M_5 = 0.
$$

If we also arrange the determinant of equation (II) of No. 68 according to the elements of the last column, that equation assumes the form

$$
M_1 + M_2 + M_3 + M_4 = 288 T^2.
$$

From the last two equations we obtain

$$
288 H T^2 = -M_5,
$$

where

$$
M_5 = \begin{vmatrix}
0 & P & Q & R \\
P & 0 & C & B \\
Q & C & 0 & A \\
R & B & A & 0
\end{vmatrix}.
$$

Computation gives

$$
-M_5 = 2FG + 2GE + 2EF - E^2 - F^2 - G^2,
$$

where E, F, G are the three products AP, BQ, CR. If we replace A, B, C, P, Q, R once again by a^2, b^2, c^2, p^2, q^2, r^2 and designate the products ap, bq, cr of the opposite edges as e, f, g, the last formula can be written as

$$-M_5 = 2f^2g^2 + 2g^2e^2 + 2e^2f^2 - e^4 - f^4 - g^4.$$

If we consider e, f, g as sides of a triangle, the right side of this formula (according to Hero) represents 16 times the square of the area j of this triangle. Thus the equation found for $H = 2h^2$ is transformed into

$$576h^2T^2 = 16j^2,$$

and from this we can obtain the simple formula

$$6hT = j$$

for the radius of the sphere of circumscription. Verbally, this can be stated as follows:

Six times the product of a tetrahedron volume and the radius of its sphere of circumscription is equal to the area of a triangle whose sides are the products of the opposite edges of the tetrahedron.

NOTE. The question of the radius ρ of the sphere inscribed in a tetrahedron is much simpler. The lines joining the center Z of the inscribed sphere and the boundary points of the four triangles bounding the tetrahedron divide the tetrahedron into four pyramids with the common apex Z and the areas $\frac{1}{3}\rho I$, $\frac{1}{3}\rho II$, $\frac{1}{3}\rho III$, $\frac{1}{3}\rho IV$, where I, II, III, IV are the areas of the bounding triangles. We thus obtain the formula

$$T = \tfrac{1}{3}\rho(I + II + III + IV).$$

This equation represents ρ as a function of the tetrahedron edges, since I, II, III, IV, and T are known functions of the edges.

71　The Five Regular Solids

To divide the surface of a sphere into congruent regular spherical polygons.

SOLUTION. We will call the required division "regular" and we will first answer the question concerning the maximum possible number of regular divisions.

We will assume that the sphere is covered completely and without any gaps by z regular n-gons and that at every corner of such an n-gon ν sides come together. We divide each n-gon by means of the spherical radii running from the center to the vertexes into n isosceles triangles. Each of these triangles possesses the central angle $2\pi/n$ and the base angle π/ν (since at each vertex 2ν such base angles come together), and thus the spherical excess of each is

$$\varepsilon = \frac{2\pi}{n} + \frac{2\pi}{\nu} - \pi = \pi\left(\frac{2}{n} + \frac{2}{\nu} - 1\right).$$

Now, the area of such a triangle, when r is the spherical radius is $r^2\varepsilon$; the area of an n-gon is thus $nr^2\varepsilon$ and the area of the spherical surface consisting of z such n-gons is $znr^2\varepsilon$. Accordingly, we obtain the equation

$$znr^2\varepsilon = 4\pi r^2$$

or

$$zn\left(\frac{2}{n} + \frac{2}{\nu} - 1\right) = 4$$

or

$$\frac{2}{n} + \frac{2}{\nu} = 1 + \frac{4}{zn}.$$

Since the left side of this equation is >1 and at the same time n as well as ν must be >2, we obtain the following five possibilities for n, ν, and z:

n	ν	z
3	3	4
3	4	8
3	5	20
4	3	6
5	3	12

Thus, there are only five possible regular divisions of a spherical surface: by dividing the surface with

1. four regular triangles,
2. six regular tetragons,
3. eight regular triangles,
4. twenty regular triangles,
5. twelve regular pentagons.

If we connect every two adjacent corners of such a spherical n-gon by means of a line segment, we obtain a regular *plane n-gon* bounded by the n line segments that connect the corners. If we construct this plane n-gon for each of the z spherical n-gons, we obtain a regular polyhedron bounded by z regular n-gons, or a so-called *regular solid*.

There are accordingly only *five regular solids*, namely, the regular *tetrahedron, hexahedron* (the cube), *octahedron, icosahedron*, and *dodecahedron*.

In the following we will actually carry out the five regular divisions of the spherical surface, which we had initially only shown to be possible. For convenience in viewing the sphere we will imagine it as a globe with a north pole N and a south pole S and with meridians and latitudinal circles.

I. The *tetrahedron* ($n = 3$, $v = 3$, $z = 4$). On the three meridians $0°$, $120°$, $240°$ we lay off from N the three equal arcs NA, NB, NC such that the triangles NBC, NCA, NAB are equilateral. The three arcs BC, CA, AB enclosing the south pole then also form an equilateral triangle that is congruent to the designated triangles, and the spherical surface has been divided into the four regular triangles NBC, NCA, NAB, ABC.

II. The *hexahedron* ($n = 4$, $v = 3$, $z = 6$). On the four meridians $0°$, $90°$, $180°$, $270°$ we lay off from N and S the eight equal arcs NA, NB, NC, ND and SC', SD', SA', SB' (each one equal to h) such that each of the arcs AC', BD', CA', DB' is equal to AB ($= 2k$). k is obtained from the spherical triangle NAB by means of the equation

$$\cos 2k = \cos h \cos h.$$

Since on the one hand $2h + 2k = NA + SC' + AC' = NS = 180°$ or $h + k = 90°$, and thus $\cos h = \sin k$, and on the other hand $\cos 2k = 1 - 2 \sin^2 k$, we obtain

$$1 - 2 \sin^2 k = \sin^2 k$$

and consequently

$$\sin k = \sqrt{\tfrac{1}{3}}, \qquad \cos 2k = \tfrac{1}{3}, \qquad \cos h = \sqrt{\tfrac{1}{3}}.$$

The corners A, B, C, D, A', B', C', D' defined by these conditions are the eight corners of the cube.

III. The *octahedron* ($n = 3$, $v = 4$, $z = 8$). The corners of the octahedron are the points N, S and four equator points separated from each other by $90°$.

IV. The *icosahedron* ($n = 3$, $\nu = 5$, $z = 20$). We choose ten meridians 36° apart and call them 1, 2, 3, ..., 10. On the meridians 1, 3, 5, 7, 9 we lay off from N the equal arcs NA, NB, NC, ND, NE, and on the meridians 6, 8, 10, 2, 4 we lay off from S the equal arcs SA', SB', SC', SD', SE' such that the ten triangles NAB, NBC, NCD, NDE, NEA, $SA'B'$, $SB'C'$, $SC'D'$, $SD'E'$, $SE'A'$ are equilateral. The common length $2k$ of the marked-off arcs can be obtained, for example, from one of the right triangles NBO, NCO, into which the meridian 4 divides the equilateral triangle NBC. Since $\angle BNO = 36°$, $\angle OBN = 72°$, it follows from triangle NBO that

$$\cos BO = \cos k = \frac{\cos 36°}{\sin 72°} = \frac{1}{2 \sin 36°}$$

and from this that $2k = 63°26'$.

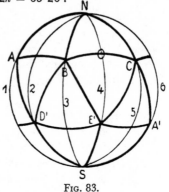

Fig. 83.

If we extend NO by its own length to H, we obtain the isosceles triangle NBH with the base $NH = 2h$ and the legs $BN = BH = 2k$, the base angle 36°, and the apex angle $HBN = 144°$. Since these angles have the same sine, the sines of their opposite sides NH and NB are equal according to the sine theorem. But since these opposite sides ($2h$ and $2k$) are not equal, $2h$ must be the supplement of $2k$. And since NE' is also the supplement of $2k$ ($= SE'$), then necessarily

$$NE' = 2h = NH.$$

Accordingly, point H coincides with E' and $E'B$ is equal to $2k$, i.e., equal to NB. In similar fashion each of the arcs AD', $D'B$, $E'C$, CA', $A'D$, DB', $B'E$, EC', $C'A$ is equal to $2k$, and the ten "encircling" triangles ABD', $D'E'B$, BCE', $E'A'C$, CDA', $A'B'D$, DEB', $B'C'E$, EAC', $C'D'A$ are likewise equilateral triangles and also congruent to the ten equilateral triangles above.

The 12 points N, S, A, B, C, D, E, A', B', C', D', E' are thus the vertexes of 20 equilateral triangles that completely cover the sphere; they are the 12 corners of the regular icosahedron.

V. The *dodecahedron* ($n = 5, \nu = 3, z = 12$). As in the icosahedron, we begin the construction of the dodecahedron by laying off a system of ten meridians $1, 2, 3, \ldots, 10$ that are $36°$ apart. About N as a common apex we group five congruent isosceles triangles NAB, NBC, NCD, NDE, NEA with the apex angle $72°$ and the base angle $60°$ ($= 180°/\nu$) whose base vertexes A, B, C, D, E lie on the meridians $1, 3, 5, 7, 9$. Thus we obtain the regular pentagon $ABCDE$. In the same way we draw about S as a common center point the regular pentagon $A'B'C'D'E'$ whose vertexes A', B', C', D', E' lie on the meridians $6, 8, 10, 2, 4$.

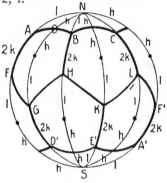

Fig. 84.

If O and O' represent the base midpoints of the isosceles triangles ABN and $D'E'S$, then NAO and $SD'O'$ are right triangles with the angles $60°$ and $36°$.

Our construction is now based on the theorem (proved below):
" *The perimeter of a spherical right triangle with angles of $60°$ and $36°$ is $90°$.*"
If we designate the hypotenuse, the long leg, and the short leg of such a triangle as l, h, and k, then

(1) $$l + h + k = 90°.$$

If we remember that

$$NA = SD' = l, \qquad NO = SO' = h, \qquad AO = D'O' = k,$$

we see that $2k$ is the side, l the radius of the circumscribed circle (on the sphere), h the radius of the inscribed circle, and $s = l + h$ the altitude of the pentagon $ABCDE$ or $A'B'C'D'E'$.

We now mark off on the meridians 1, 3, 5, 7, 9 from A, B, C, D, E southwards and on the meridians 6, 8, 10, 2, 4 from A', B', C', D', E' northwards the pentagon side $2k$, which gives us the points F, G, H, K, L, F', G', H', K', L'.

Now since, according to (1), each meridian consists of the four segments l, $2k$, s, and h, it follows that OG and $O'H$, for example, represent the pentagon altitude s; i.e., the pentagons $ABHGF$ and $D'E'KHG$ are congruent to the regular pentagon $ABCDE$. The same is naturally true of the pentagons $BCLKH$, $CDG'F'L$, $DEK'H'G'$, $EAFL'K'$, $E'A'F'LK$, $A'B'H'G'F'$, $B'C'L'K'H'$, $C'D'GFL'$.

With the 12 regular pentagons already designated the sphere is completely covered.

The points A, B, C, D, E, F, G, H, K, L, A', B', C', D', E', F', G', H', K', L' are accordingly the 20 corners of the regular dodecahedron.

SUPPLEMENT: PROOF OF THE THEOREM: " *The perimeter of a spherical right triangle with the angles* 60° *and* 36° *is* 90°."

Let the sides of the triangle be a, b, c, their opposite angles $\alpha = 60°$, $\beta = 36°$, $\gamma = 90°$. We express the tangents of the sides by the regular decagon side $z = 2 \sin 18°$ corresponding to the unit circle, for which it is known that $z^2 + z = 1$.

1. Firstly,

$$\cos \beta = 1 - 2 \sin^2 18° = 1 - \tfrac{1}{2}z^2 = \frac{1+z}{2} = \frac{1}{2z}$$

or

$$\sec \beta = 2z.$$

2. From $\sec c = \tan \alpha \tan \beta$ it follows that $\sec^2 c = 3 \tan^2 \beta$ or $(\tan^2 c + 1) = 3(\sec^2 \beta - 1)$ or $\tan^2 c = 4(3z^2 - 1)$. However, $3z^2 - 1 = z^2 + (2z^2 - 1) = z^2 + (1 - 2z) = [1 - z]^2 = z^4$, and thus

$$\tan c = 2z^2.$$

3. $\tan a = \tan c \cos \beta = 2z^2/2z = z$.
4. $\tan b = \tan c \cos \alpha = 2z^2 \cdot \tfrac{1}{2} = z^2$.
Now we have

$$\tan c \cdot \tan (a + b) = 2z^2 \cdot \frac{z + z^2}{1 - z^3} = \frac{2z^2}{1 - z^3}$$

$$= \frac{2z^2}{(1 - z)[1 + z + z^2]} = \frac{2z^2}{(z^2)[1 + 1]} = 1.$$

Consequently, $a + b$ is the complement of c. Q.E.D.

The regular solids were already known to the Pythagoreans and thus go back to the sixth century B.C. The proof that there are only five regular solids probably stems from Euclid (ca. 330–275 B.C.).

72   The Square as an Image of a Quadrilateral

To show that every quadrilateral can be considered as a perspective image of a square.

The perspective projection, perspectivity or central projection, the simplest and most important of all projections, can be explained as follows. Given are a fixed point Z, the *center of projection*, and a fixed plane E, the *plane of the image*. The *perspective image* or, more briefly, the *perspective* of an arbitrary point P_0 is understood to mean the point of intersection P of the "projection ray" ZP_0 with the plane of the image. P_0 is the "object," P the "image." The image of a figure is the totality of the images of the points of which the figure (the object) consists. Thus, the perspective of a straight line g_0 is a straight line g, namely the intersection of the plane Zg_0 with the plane of the image.

Of particular importance is the perspective projection in which only points of a plane E_0, the object plane, are projected onto the image plane. The line of intersection $\mathfrak{A}$ of the object plane and the image plane is called the *axis of perspectivity*. The axis of perspectivity is the locus of the object point that coincides with the point of its image. *An arbitrary object line and its image accordingly intersect at the axis.*

A noteworthy role in this perspectivity is played by the infinitely distant points of the object plane. Since the projection rays to the infinitely distant points of E_0 run parallel to E_0, they lie in a plane Δ passing through Z and parallel to E_0 and consequently meet the image plane at the line of intersection f of this plane with Δ. This line of intersection is called the *vanishing line* of the object plane E_0. *The vanishing line is parallel to the axis of perspectivity.*

In order to avoid limiting the general validity of the above theorem, "The perspective of a line is also a line," by a special case, we call the totality of infinitely distant points of E_0 the "infinitely distant line" of this plane and can then state briefly that:

The perspective of the infinitely distant line of a plane is the vanishing line of this plane.

The place at which the image g of an arbitrary line g_0 of E_0 intersects the vanishing line f and which is the image of the infinitely distant point of g_0 is called the *vanishing point* of g_0.

Now for the solution of our problem!

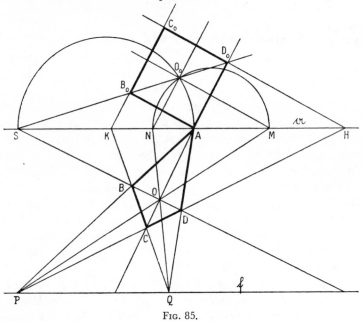

Fig. 85.

Let the quadrilateral $ABCD$ in the drawing plane E be the given quadrilateral, let O be the point of intersection of the diagonals AC and BD, P the point of intersection of the opposite sides AB and CD, Q the point of intersection of the opposite sides BC and DA. Let the square we are looking for be called accordingly $A_0B_0C_0D_0$, the point of intersection of its diagonals O_0, its plane E_0. Since the points of intersection P_0 and Q_0 of the two pairs of opposite sides lie on the infinitely distant line of E_0, their images P and Q must lie on the vanishing line f of the perspectivity passing from E_0 to E. We accordingly choose the line PQ as the vanishing line f. It makes no difference which parallel to f we choose as the axis of perspectivity a. We choose the parallel through A. The points of intersection of the axis with the lines CD, BC, OP, OQ, and BD we designate as H, K, M, N, and S. Since each object line meets the corresponding image line at the axis, these points may also be called H_0, K_0, M_0, N_0, S_0.

In the quadrilateral $ABCD$ the opposite sides PBA and PCD and the diagonals PO and PQ form a harmonic ray pencil. Since the ray PQ runs parallel to the line $\mathfrak{a}$, the segments MA and MH are of equal length.

In the quadrilateral $ABCD$ the opposite sides QCB and QDA and the diagonals QO and QP also form a harmonic ray pencil. Since $QP\|\mathfrak{a}$, the segments NA and NK are also equally long.

Since the diagonals of the sought-for square must meet the diagonals of the given quadrilateral at the axis, the diagonals of the square must pass through A and S. The point of intersection O_0 of the diagonals accordingly lies on the semicircle with the diameter AS belonging to the plane E_0.

Since the midlines M_0O_0 and N_0O_0 of the square pass through O_0, O_0 also lies on the semicircle with the diameter MN in the plane E_0.

The point of intersection of the two semicircles is the center point O_0 of the square.

The sides A_0B_0 and C_0D_0 of the square are the parallels through A and H to MO_0, the sides B_0C_0 and D_0A_0 of the square are the parallels through K and A to NO_0.

For convenience we execute the drawing (cf. Figure 85) in the drawing plane itself. Then, in order to obtain the spatial perspectivity we are looking for, we rotate the square about the axis $\mathfrak{a}$ as an axis of rotation into a new plane E_0, draw through f the plane Δ parallel to E_0, join the point of intersection of the diagonals, O_0, now lying in E_0, with O, and designate the point of intersection of this connecting line with Δ as Z.

If we now project the square $A_0B_0C_0D_0$ lying in E_0 from the center Z onto E, we thereby obtain as a perspective image the square $ABCD$.

73   The Pohlke-Schwarz Theorem

Four arbitrary points of a plane that do not all lie on the same line can be considered as an oblique image of the corners of a tetrahedron that is similar to a given tetrahedron.

This fundamental theorem of oblique parallel projection, proved by H. A. Schwarz (1843–1921) in 1864 (*Crelle's Journal*, vol. 63; also, Schwarz, *Gesammelte Abhandlungen*), includes as a special case the theorem formulated in 1853 by K. Pohlke (1810–1876):

THE FUNDAMENTAL THEOREM OF OBLIQUE AXONOMETRY: *Three arbitrary segments originating from a single point in a plane that do not all belong to the same line can be considered as the oblique image of a tripod.*

Before taking up the proof of this theorem we shall make several prefatory remarks about oblique projection, affinity, and axonometry.

An oblique projection is a projection of a plane or three-dimensional figure, an object figure, onto the drawing plane or image plane in which each object point is projected onto the image plane by a "projection" ray drawn in a fixed direction. If the projection rays are perpendicular to the image plane, the oblique projection is called a normal or orthogonal projection.

The oblique projection of points of a plane (the object plane) onto the image plane is a so-called affinity.

An *affinity* or *affine projection* is understood to mean a projection of an object plane onto the picture plane (which may also lie in the object plane) in which the points of the object plane are transformed into points of the image plane in such manner that they exhibit the following fundamental properties:

I. *The affine image of a line is also a line.*

II. *Parallelism is not annulled by affine projection.* (The image of a parallelogram is a parallelogram.)

III. *The ratio of parallel segments is not altered by affine projection.* In other words: *Parallel segments are projected in the same proportion.* (This third property is a consequence of I. and II.)

It is therefore immediately evident that the oblique projection of a plane onto a second plane possesses these three fundamental properties.

The most general affinity between two arbitrary planes E and E' is determined by the mutual correspondence between two arbitrary triangles ABC and $A'B'C'$ of these planes, where A', B', C' are determined as the affine images of A, B, C, respectively. The affine image P' of an arbitrary object point P (of E) is drawn by letting AP intersect with the side BC at H, then (according to III.) determining the affine image H' of H on the line $B'C'$ by means of the condition $B'H':C'H' = BH:CH$, and finally determining P' on $A'H'$ by means of the condition $A'P':H'P' = AP:HP$.

A frequently employed method of drawing the oblique projection of a three-dimensional figure is the axonometric method. In this method the points P of the three-dimensional figure are determined by their coordinates $x|y|z$ most commonly in a perpendicular coordinate system. Three equal segments OA, OB, and OC are laid off from the origin O on the axes; these segments form a so-called *tripod*. The oblique outline $O'A'B'C'$ of the tripod is drawn, and this also gives us the oblique images of the coordinate axes. We then construct, in accordance with III., the oblique image of the point P, which in this context is called the *axonometric image*.

It is now of fundamental importance to know whether three arbitrary segments $O'A'$, $O'B'$, $O'C'$ originating from a point O' of the drawing plane can be considered as the oblique projection of a tripod $OABC$. This question was answered by Pohlke and, in a somewhat more general fashion, by Schwarz, as mentioned above.

Of the numerous proofs of the Pohlke-Schwarz fundamental theorem the following (stemming from Schwarz) is quite elementary. It is based upon the *theorem of Lhuilier*, which is in itself very interesting: *The sections of an arbitrary three-edged prism include all the possible forms of triangles.* In other words: *Every triangle can be considered as the normal projection of a triangle of given form.* This theorem was stated in 1811 by the French-Swiss mathematician Simon Lhuilier (1750–1840).

PROOF. Since parallel sections of a prism are congruent, we can assume that the prescribed triangle $A_0B_0C_0$, which is also the cross section of the prism, and the sought-for prism section ABC, which possesses a prescribed form, have a common vertex, $C \equiv C_0$. If we now drop the perpendiculars A_0X and B_0Y from A_0 and B_0 to the intersection line (axis) g of the two planes E_0 of A_0B_0C and E of ABC and

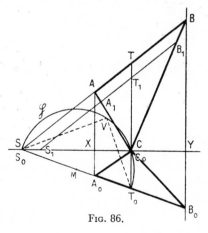

FIG. 86.

rotate the plane E about g as the rotation axis to the plane E_0, then A and B, as the figure shows, fall on the perpendiculars A_0X and B_0Y, respectively, and the point of intersection $S \equiv S_0$ of the lines A_0B_0 and AB falls on the axis.

We now draw the perpendicular to the axis through C and let it touch A_0B_0 at T_0 and AB at T. If we designate the cosine of the angle formed by the plane E in its original position with E_0 as μ, then $A_0X = \mu \cdot AX$, $B_0Y = \mu \cdot BY$, $T_0C = \mu \cdot TC$.

Now according to the ray theorem,

$$SA:AT:TB = S_0A_0:A_0T_0:T_0B_0.$$

We can therefore draw a parallel $S_1A_1T_1B_1$ to $SATB$ that cuts the lines g, CA, CT, CB at S_1, A_1, T_1, B_1, and is congruent to $S_0A_0T_0B_0$ (so that $S_1A_1 = S_0A_0$, $A_1T_1 = A_0T_0$, $T_1B_1 = T_0B_0$). We displace the triangle S_1B_1C in such a way that S_1 falls on S_0, A_1 on A_0, T_1 on T_0, B_1 on B_0. The vertex C then falls on a point V of the semicircle $\mathfrak{H}$ described about the diameter S_0T_0 (since $\triangle S_1CT_1$ is a right triangle), on which C lies, also.

From this fact we obtain the following simple method for constructing the described figure when the triangle $A_0B_0C_0$ and the form of the triangle ABC are given.

We draw over A_0B_0 the triangle A_0B_0V that is similar to the triangle ABC (with A_0, B_0, V being homologous to A, B, C, respectively). We let the median perpendicular of CV intersect with A_0B_0 at M and draw the semicircle $\mathfrak{H}$ with the center M and the radius $MC = MV$. The end points S_0 and T_0 of the semicircle, which lie on the line A_0B_0, we designate in such manner that S_0V and T_0C become sides (not diagonals) of the chord quadrilateral S_0T_0CV. We then choose CS_0 as the axis and CT_0 as the perpendicular to the axis. On the axis we make $CS_1 = VS_0$, on the perpendicular to the axis $CT_1 = VT_0$, and we draw the line $S_1A_1T_1B_1 \cong S_0A_0T_0B_0$. Finally, we draw parallel to $S_1A_1T_1B_1$ the line $SATB$ of which S, A, T, B lie on the perpendiculars through S_0, A_0, T_0, B_0, respectively, while at the same time A lies on CA_1 and B lies on CB_1.

If we rotate the triangle ABC about CS as the axis of rotation by the angle whose cosine $\mu = C_0T_0/CT$ as the angle of rotation, $A_0B_0C_0$ then appears as the normal projection of the rotated triangle ABC, which possesses the prescribed form.

That the ratio $\mu = C_0T_0/CT$ can be considered as a cosine, i.e., is a *proper* fraction, is shown as follows. According to the ray theorem, $CT = CT_1 \cdot (CS/CS_1)$, i.e., according to the construction, $= VT_0 \cdot CS/VS$. If we introduce this value into the equation for μ, we obtain

$$\mu = \frac{CT_0}{CT} = \frac{CT_0 \cdot VS}{CS \cdot VT_0}.$$

However, since, according to the theory of Ptolemy, in the chord quadrilateral ST_0CV the product $CT_0 \cdot VS$ of the opposite sides is smaller than the product $CS \cdot VT_0$ of the diagonals, μ represents a proper fraction.

This proves the auxiliary theorem concerning the prism.

The proof of the Pohlke-Schwarz theorem is now easy. We can state the theorem in the following manner:

The oblique image of a given tetrahedron can always be determined in such manner that it is similar to a given quadrilateral.

Let the tetrahedron be $ABCS$, the quadrilateral $A'B'C'D'$.

In the affinity between the planes ABC and $A'B'C'$, in which A', B', C' are correlated to the points A, B, and C, respectively, let the point D correspond to the point D'. We select SD as the direction of the affinity (projection ray).

We construct the triangular prism whose edges are parallel to SD through A, B, and C, and determine the section $A''B''C''$ that is parallel to $A'B'C'$.

In the affinity in which the points A'', B'', C'' are correlated to the points A', B', C', let the point D'' correspond to the point D'. Then $A''B''C''D''$ is similar to $A'B'C'D'$. Now, since $A''B''C''D''$ and also $ABCD$ are affine with respect to $A'B'C'D'$, then $A''B''C''D''$ is also affine to $ABCD$.

The latter affinity, however, arises from the projection rays parallel to SD. In this affinity the quadrilateral $A''B''C''D''$ that is similar to $A'B'C'D'$ is thus the oblique image of the given tetrahedron $ABCS$.

---

74 Gauss' Fundamental Theorem of Axonometry

Though three segments OA, OB, OC originating from a point O in the drawing plane (image plane) all three of which do not belong to the same straight line can always, according to Pohlke's fundamental theorem (No. 73), be considered as an *oblique* projection of a tripod, this is no longer the case for the *normal* projection of a tripod.

Moreover, there exists between the lengths and directions of the normal projections *OA*, *OB*, *OC* of the three legs a definite relationship. Thus we come to

GAUSS' PROBLEM: *What is the relation between the normal projections* OA, OB, OC *of the legs of a tripod?*

SOLUTION. We select the image plane *E* as the *xy*-plane, the perpendicular to this plane from the apex of the tripod as the *z*-axis of a triaxial orthogonal coordinate system; we take the common length of the three legs as the unit length and call the direction cosines of the legs $\lambda|\lambda'|\lambda''$, $\mu|\mu'|\mu''$, and $\nu|\nu'|\nu''$. At the same time we take the *xy*-plane as the Gauss plane (the plane of complex numbers) and designate the complex number represented by any point (*P*) of *E* by the corresponding small gothic letter ($\mathfrak{p}$).

Since the three points *A*, *B*, *C* in *E* have the coordinates $\lambda|\lambda'$, $\mu|\mu'$, $\nu|\nu'$,

$$\mathfrak{a} = \lambda + i\lambda', \qquad \mathfrak{b} = \mu + i\mu', \qquad \mathfrak{c} = \nu + i\nu'.$$

Squaring and adding, we obtain

$$\mathfrak{a}^2 + \mathfrak{b}^2 + \mathfrak{c}^2 = (\lambda^2 + \mu^2 + \nu^2) - [\lambda'^2 + \mu'^2 + \nu'^2] + 2i\{\lambda\lambda' + \mu\mu' + \nu\nu'\}.$$

According to the well-known relations between the direction cosines of three mutually perpendicular lines, the expression within parentheses and the expression within brackets both equal one, while the expression within the braces is equal to zero. This gives us the Gauss equation

$$\mathfrak{a}^2 + \mathfrak{b}^2 + \mathfrak{c}^2 = 0.$$

This formula forms

GAUSS' FUNDAMENTAL THEOREM OF NORMAL AXONOMETRY: *If in the normal projection of a tripod the image plane is considered as the plane of complex numbers, the projection of the apex of the tripod as the null point, and the projections of the leg ends as complex numbers of the plane, the quadratic sum of these numbers is equal to zero.*

The Gauss theorem immediately provides the solution of the

FUNDAMENTAL PROBLEM OF NORMAL AXONOMETRY: *To complete the normal projection* OABC *of a tripod of which the normal projections* OA *and* OB *of two of the legs are already drawn.*

SOLUTION. We select (as above) the point *O* as the null point of the complex number plane and the direction of *OA* as the direction of the positive real number axis. The magnitudes of the three numbers $\mathfrak{a}$, $\mathfrak{b}$, $\mathfrak{c}$ we will designate as a, b, c, and the three angles BOC, COA, AOB as α, β, γ.

We write the Gauss equation

$$\mathfrak{a} + \frac{\mathfrak{b}^2}{\mathfrak{a}} = -\frac{\mathfrak{c}^2}{\mathfrak{a}}.$$

In order to construct $\mathfrak{p} = \mathfrak{b}^2/\mathfrak{a}$, we lay off at O on OB the angle γ, at B on BO the angle OAB; the point of intersection P of the free legs of the angle drawn gives us $\mathfrak{p}$. We then draw through A the parallel to OP, through P the parallel to OA and obtain at the point of intersection Q of the two parallels the complex number $\mathfrak{q} = \mathfrak{a} + (\mathfrak{b}^2/\mathfrak{a})$. Consequently, the end point R of the extension of QO by itself is the

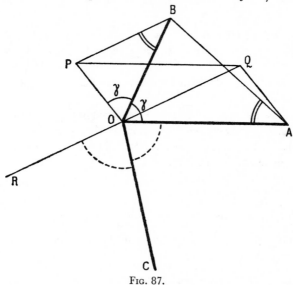

Fig. 87.

number $\mathfrak{r} = \mathfrak{c}^2/\mathfrak{a}$. From $\mathfrak{c} = \sqrt{\mathfrak{a}\mathfrak{r}}$ it follows that:

1. The magnitude of $\mathfrak{c}$ is the mean proportion of the magnitudes of $\mathfrak{a}$ and $\mathfrak{r}$;
2. the direction of $\mathfrak{c}$ is the direction of the bisector of the angle (2β) enclosed between OA and OR.

Accordingly, we bisect the angle AOR and mark off on the bisector from O the mean proportion of OA and OR; the end point of the marked-off segment is the sought-for point C. Since we can choose the bisector of the concave angle AOR as well as that of the convex angle (in accordance with the two values of $\sqrt{ar}$), there are two possible positions for C.

NOTE. WEISBACH'S THEOREM. Since the square of a complex number has an angle twice as great as the number itself, the vectors of the squares of two complex numbers form with each other an angle that is twice as great as the vectors of the numbers. Thus the vectors of the squares $a^2 \cdot b^2 \cdot c^2$ form the angles 2α, 2β, 2γ with each other. Thus, if we group these vectors (by magnitude and direction), we obtain (in accordance with the Gauss formula) a triangle with the external angles 2α, 2β, 2γ. Since the sides of this triangle are a^2, b^2, c^2, the sine theorem gives us the equation

$$a^2 : b^2 : c^2 = \sin 2\alpha : \sin 2\beta : \sin 2\gamma.$$

This formula is

WEISBACH'S THEOREM: *The squares of the normal projections of the legs of a tripod relate to each other as the sine of twice the angles enclosed by the projections.*

Thus, Weisbach's theorem appears as the direct consequence of the Gauss theorem.

The Gauss theorem can be found unproved in the second volume of Gauss' *Werke*, the Weisbach theorem in Weisbach's paper on axonometry, which was published in 1844 at Tübingen in the *Polytechnische Mitteilungen* of Volz and Karmarsch.

75   Hipparchus' Stereographic Projection

To present a conformal map projection that transforms the circles of the globe into circles of the map.

The projection we are looking for, which is called a *stereographic* or *polar projection*, is very important in cartography. In all probability the source of this problem is the astronomer Hipparchus (of Nicaea in Bithynia), one of the most amazing men of antiquity, who was making astronomical observations in the period from 160–125 B.C. in Rhodes, Alexandria, Syracuse, and Babylon.

The problem is solved by the following *projection directive*:

One selects as the projection plane or image plane (map plane) the plane E tangent to the globe at an appropriate point O—the so-called map center—of the area to be projected, and as the center of a central projection the end point Z of the globe diameter OZ originating at O.

The stereographic image P' of an arbitrary point P of the globe is the point of intersection of the projection ray ZP with the image plane E.

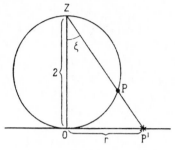

FIG. 88.

The distance $r = OP'$ from the map center is given by the equation

$$r = 2 \tan \zeta,$$

where ζ represents the angle formed by the projection ray ZP with the center ray ZO, and the radius of the globe is chosen as the unit length.

The stereographic projection thus defined has the following *two properties:*

I. *Every image circle of a globe circle is a circle.*

II. *The stereographic map is conformal.* (I.e., the map image of an angle located on the globe is an equally great angle.)

The proofs of these properties are both based on the following auxiliary theorem:

The image of a globe tangent bounded by globe and map is just as long as the tangent.

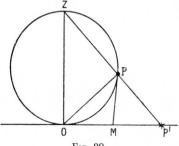

FIG. 89.

PROOF OF THE AUXILIARY THEOREM.   Let P be a point on the globe, P' its image, M the place at which the globe tangent passing through P and lying in the drawing plane ZOP meets the image plane, and at the same time (since the two tangents MO and MP are equal) the midpoint of the hypotenuse of the right triangle OPP'.   The intersection point D of any other globe tangent passing through P with the image plane will then lie perpendicularly above (below) M.   The image D' of D is D itself, and the image of the tangent DP is thus DP'. Now the two right triangles at M, DMP and DMP', are congruent ($MD = MD$ and $MP = MP'$).   Consequently, $D'P' = DP$, which was to be proved.

PROOF OF I.   We will now prove the somewhat more general *Chasles theorem:** *The stereographic image of a globe circle $\Re$ is a circle whose midpoint is the stereographic projection* S' *of the apex* S *of the cone that is tangent to the globe along the circle* $\Re$.

PROOF.   In Figure 90 let P be an arbitrary point of $\Re$, let P' be its image, D the point of intersection of the tangent to the sphere and cone-generator SP with the image plane E. According to the auxiliary theorem, DP then equals DP'.   Thus, if H is the point of

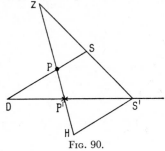

FIG. 90.

intersection of the parallel through S' to DP with the projection ray ZP, it follows from the similarity of the triangle $S'P'H$ to the isosceles triangle $DP'P$ that the two segments $S'P'$ and $S'H$ are equal. Consequently, in the relation

$$S'H:SP = ZS':ZS$$

derived from the ray theorem, we can replace $S'H$ with $S'P'$, obtaining

$$S'P' = SP \cdot \frac{ZS'}{ZS}.$$

---

* Michel Chasles (1793–1880), French mathematician, especially well-known for his brilliantly written *Aperçu historique sur l'origine et le développement des méthodes en géométrie.*

Now, if P describes the circle $\Re$, SP (as the distance of the apex S of the cone from $\Re$) remains constant, and consequently, in view of the last equation, $S'P'$ also remains constant and P' describes a circle in E.

If the object circle $\Re$ is a great circle of the globe, the apex S of the cone lies at infinity.

In this case let F be the place at which the perpendicular from Z on the plane of $\Re$ touches the map plane E, and let V be the place at which the globe tangent through P parallel to this perpendicular touches the map plane E. Since, according to the auxiliary theorem, $VP' = VP$, the triangle VPP' is isosceles; and since VP is parallel to FZ, the triangle FZP' is also isosceles; therefore,

$$FP' = ZF.$$

The locus of the image point P' is thus a circle with the midpoint F and the radius ZF.

In those great circles of the globe that pass through the projection center and the map center, the midpoint F of the image circle recedes to infinity. In fact, these circles, as direct inspection will show, are transformed into straight lines by projection.

PROOF OF II. Let ω be an arbitrary angle on the globe, its apex P, therefore, a point on the globe, and each of its legs a globe tangent. If X and Y are accordingly the points at which the two tangents intersect the image plane E, then $\omega = \angle XPY$.

The image ω' of this angle is the angle $XP'Y$.

Now, since the triangles XPY and $XP'Y$ are congruent ($XY = XY$; also, according to the auxiliary theorem, $XP = XP'$ and $YP = YP'$), we immediately obtain

$$\omega' = \omega,$$

which was to be proved.

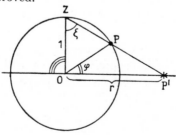

FIG. 91.

Note. If instead of the tangential plane E we choose a plane parallel to it as our map plane, we obtain a similar stereographic projection, which, naturally, also possesses the fundamental properties I. and II.   Of particular importance is a picture plane passing through the center of the globe, especially when the north pole is chosen as the projection center and the equatorial plane is accordingly chosen as the image plane.   In this case we obtain for the distance r of the image point P' from the map center O lying at the center of the globe the formula

$$r = \tan\left(45° + \frac{\varphi}{2}\right),$$

where φ is the geographic latitude of the point P.   (The above cited angle $\zeta = \measuredangle OZP$ is the base angle of the isosceles triangle OPZ in which the apex angle situated at O is the complement of the latitude φ.)

76   The Mercator Projection

To draw a conformal geographic map whose grid is composed of right-angle compartments.

The Mercator map, which is equally important for both geography and nautical science, was conceived by Gerhard Kremer, called Mercator (1512–1594).

On the Mercator map the equator is a segment AB, the length of which agrees with the length (2π) of the globe equator.   If we divide AB into 360 equal parts and erect at the dividing points perpendiculars to AB, we thereby obtain the map meridians.   The latitude parallel on the map that corresponds to the globe parallel of latitude φ is a line parallel to AB whose distance Φ from the map equator is called the *exaggerated latitude*.   The core of the problem consists of *representing the exaggerated latitude Φ as a function of the geographic latitude φ.*

In order to solve this problem we will compare the Mercator map with the—also conformal—Hipparchus map (No. 75), in which the north pole of the globe is the projection center and the plane E of the globe equator is the map plane, and in which, therefore, the globe equator is projected isometrically.   Here also the globe radius will serve as the unit length.

On the Mercator map we divide the distance Φ of the latitude parallel from the equator into n equal parts, where n is a very large

number; we draw through the dividing points the latitude parallels
1, 2, 3, ..., $n - 1$ and call their corresponding geographic latitudes
$\varphi_1, \varphi_2, \ldots, \varphi_{n-1}$, so that instead of φ we write φ_n also. We then draw
the two parallel map meridians λ' and Λ' corresponding to the globe
meridians λ and Λ, whose difference in longitude measured in radian
measure $\varepsilon = \Lambda - \lambda$ we will make very small. We thereby obtain on
the map a series of successive, very small, congruent rectangles with
the base line ε and the altitude Φ/n.

We now do the same on the Hipparchus map. Thus, we draw the
concentric map latitudes corresponding to the latitudes $\varphi_1, \varphi_2, \ldots,$
φ_{n-1} and call their radii $r_1, r_2, \ldots, r_n = r$. According to No. 75,

$$(1) \qquad r = \tan\left(45° + \frac{\varphi}{2}\right).$$

Similarly, we draw the map meridians λ'' and Λ'' corresponding to the
two longitudes λ and Λ; these meridians are at the same time the radii
of the circle of latitude of radius r. Thus, we obtain on the Hipparchus map a series of n successive, very small compartments, which

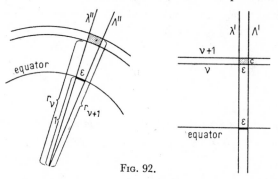

Fig. 92.

we can consider as rectangles if n is sufficiently great. We single out
the compartment situated between the latitude circles of radii r_ν and
$r_{\nu+1}$. Since its base line parallel to the map equator is r_ν times as
great as the base line ε of the first compartment, and thus also r_ν
times as great as the base line ε of the compartment of the Mercator
map, then as a result of the conformal nature of the two maps, the
altitude $r_{\nu+1} - r_\nu$ of the Hipparchus map compartment must also be
r_ν times as great as the altitude Φ/n of the corresponding compartment
of the Mercator map:

$$r_{\nu+1} - r_\nu = r_\nu \cdot \frac{\Phi}{n}.$$

From this it follows that

$$r_{v+1} = r_v\left(1 + \frac{\Phi}{n}\right).$$

If we construct this equation for all n compartments, r_0 being equal to 1, and multiply the resulting n equations together, we obtain

(2) $$r = \left(1 + \frac{\Phi}{n}\right)^n.$$

However, since for sufficiently great n the right side of this equation does not deviate noticeably from e^Φ (No. 12), we obtain the equation

(2a) $$r = e^\Phi.$$

From this we get $\Phi = lr$ or, because of (1),

(3) $$\Phi = l\tan\left(45° + \frac{\varphi}{2}\right),$$

and thus the exaggerated latitude Φ is represented as a function of the geographic latitude φ.

As a result of our investigation we obtain the following

DIRECTIVE FOR DRAWING A MERCATOR MAP: *The map image of a point on the earth of longitude λ and latitude φ has a distance λ from the zero meridian on the map and a distance of*

$$l\tan\left(\frac{\pi}{4} + \frac{\varphi}{2}\right)$$

from the map equator.

Here the angles λ and φ are taken as being in radian measure and the radius of the globe on which the map is based is taken as the unit length.

Nautical and Astronomical Problems

The Problem of the Loxodrome

To determine the longitude of the loxodromic line joining two points on the surface of the earth.

A *loxodrome* is understood to mean a line on the earth's surface that makes the same angle with all the meridians that it cuts.   As long as a ship does not alter its course it is sailing on a loxodrome.   The angle κ formed by the loxodrome with the meridians it cuts is therefore called the azimuth of course.   *On a Mercator map* (No. 76), *which is conformal and possesses rectilinear parallel meridians, the loxodrome appears as a* straight line *that cuts the map meridians at the* angle κ.

In our study of the Mercator map we chose the radius of the globe as the unit length.   Sailors use as the unit length the nautical mile (nm), which is the length of one minute latitude on a meridian of the earth's surface or, also, the length of a minute longitude on the equator (each being 1852 meters).   Since a meridian is π earth radians long and 180 degrees of latitude is equal to 10800 latitude minutes, the earth radius is $n = 10800/\pi$ nm long.   If we think of a Mercator map with $1:1$ scale (i.e., a map whose equator is as long as the real equator), the distance between the map circle corresponding to the latitude φ and the map equator, the so-called *exaggerated latitude* (according to No. 76), is

$$\Phi = nl \tan \left(45° + \frac{\varphi}{2}\right) \text{ nm.}$$

The two earth points O and O' whose loxodromic distance d is to be determined are given by their longitudes λ, λ' and latitudes φ, φ' $(>\varphi)$.

The exaggerated latitudes on the map are

$$\Phi = nl \tan \left(45° + \frac{\varphi}{2}\right) \quad \text{and} \quad \Phi' = nl \tan \left(45° + \frac{\varphi'}{2}\right) \text{ nm,}$$

the distances of the map meridians from the zero meridian Λ and Λ' nm, where Λ represents the number of longitude minutes comprising λ and Λ' the number of longitude minutes comprising λ'.

Let us say that the map meridian through O and the map parallel through O' intersect at S. Then $OS = B$ is the exaggerated latitude difference $\Phi' - \Phi$, $O'S = L = \Lambda' - \Lambda$ (nm), OO' is the map loxodrome and $\angle O'OS = \kappa$ is the azimuth of course.

From the right map triangle $OO'S$ we find the azimuth of course κ by means of the equation

$$(1) \qquad\qquad \tan \kappa = \frac{L}{B}.$$

In order to determine the loxodromic distance d of the two positions on the surface of the earth we divide d into N very small equal segments e considered as being rectilinear. If we draw the meridian through one of two adjacent division points and the circle of latitude through the other, we obtain thereby a very small right triangle with the hypotenuse e, whose meridional leg is the latitude difference β (measured in nm) of the two division points and forms the angle κ with the loxodrome, so that $\beta = e \cos \kappa$. Every two adjacent points thus possess the same latitude difference β. The total (measured in nm) latitude difference b of the two positions O and O' on the earth's surface is therefore $b = N\beta = Ne \cos \kappa = d \cos \kappa$. Consequently, the sought-for loxodromic distance is

$$(2) \qquad\qquad d = b \sec \kappa.$$

Formulas (1) and (2) contain the solution to the problem.

EXAMPLE. How great is the loxodromic distance from Valdivia ($\lambda = 286° 34.9'$ E, $\varphi = -39° 53.1'$) to Yokohama ($\lambda' = 139° 39.2'$ E, $\varphi' = +35° 26.6'$)? Here the longitudinal difference $L = 8815.7$ minutes; the latitudinal difference $b = 4519.7$ minutes or nautical miles; the exaggerated latitude difference $B = \Phi' - \Phi = 4890$ nm; κ, according to (1), is $60° 58' 50''$; and the loxodromic distance d, according to (2), is *9317 nm*.

NOTE. The *shortest* distance k between the two positions can be found by applying the cosine theorem to the spherical triangle NVY (North Pole–Valdivia–Yokohama). In this triangle $NV = 90° - \varphi = 129° 53.1'$, $NY = 90° - \varphi'$, $\angle VNY = \lambda - \lambda'$, and $VY = k$.

According to the cosine theorem

$$\cos k = \cos NV \cos NY + \sin NV \sin NY \cos (\lambda - \lambda')$$

or

$$\cos k = \sin \varphi \sin \varphi' + \cos \varphi \cos \varphi' \cos (\lambda - \lambda').$$

This yields

$$k = 153° \ 36.1' = 9216.1' = 9216.1 \ \text{nm}.$$

The shortest distance is consequently 101 nm shorter than the loxodromic distance.

The name loxodrome stems from the Dutchman Willebrord Snell (Snellius, 1581–1626). The Portuguese mathematician Pedro Nunes (1492–1577) was the first to recognize that the loxodromic line connecting two points of the earth's surface is not the shortest connecting line and that a loxodrome continuously approaches the pole without ever reaching it.

78 Determining the Position of a Ship at Sea

One of the most important problems in nautical science is that of *determining the position of a ship at sea.* The solution is usually obtained by the method of the so-called *astronomical meridian reckoning*, which will be analyzed in the following example.

PROBLEM: *On board a ship in the Pacific Ocean in the north latitude on October* 20, 1923 *at* 6:50 P.M. *mean Greenwich time by the chronometer the sun's altitude was taken in the morning as* h = 21° 40.5'; *the Nautical Almanac gave the declination of the sun for the time of observation as* δ = 10° 10.2' S, *the equation of time as* e = −15 min 3 sec. *The ship then sailed till noon* 15.2 nm WNW, *and the altitude of the sun at zenith was then measured as* H = 35° 2.7' *and the sun's declination determined at* Δ = 10° 13'.

Where was the ship?

The solution to this problem consists of four steps.

I. DETERMINATION OF THE MERIDIONAL LATITUDE Φ. At culmination the successive arcs—the altitude of the sun, the pole distance, the pole altitude—cover the meridional half circle above the horizon in such manner that $H + (90° + \Delta) + \Phi = 180°$. This gives us

$$\Phi = 90° - H - \Delta = 44° \ 44.3'.$$

II. DETERMINATION OF THE LATITUDE DIFFERENCE β AND THE LONGITUDE DIFFERENCE l OF THE TWO OBSERVATION POINTS, AS WELL AS THE A.M. LATITUDE φ.

If one imagines two sufficiently close points A and B on the earth's surface, the distance between which is d nm and the line connecting which forms the angle κ with the longitudinal circle passing through the center M of AB, then the latitudinal difference of the two points is $d \cos \kappa$ nm, the longitudinal difference $d \sin \kappa$ nm.   Since one nautical mile of latitudinal difference is equivalent to one minute latitude difference and one nautical mile longitudinal difference at the latitude φ corresponds to sec φ minutes longitudinal difference, then the latitudinal and longitudinal differences of A and B in minutes are:

$$\beta = d \cos \kappa, \qquad l = d \sin \kappa \sec \mu,$$

where μ is the latitude of M, the so-called *mean latitude* of A and B.

In our example ($d = 15.2$, $\kappa = 67.5°$) we find first that

$$\beta = 5.8'.$$

From this it follows that the A.M. latitude is

$$\varphi = \Phi - \beta = 44° \, 38.5',$$

and the mean latitude is

$$\mu = \frac{\Phi + \varphi}{2} = 44° \, 41.4'.$$

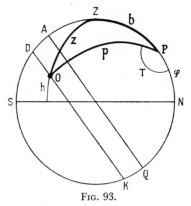

Fig. 93.

Accordingly we find the longitude difference to be

$$l = 19.75'.$$

III. Determination of the A.M. longitude λ.

In the formula (see Figure 93) corresponding to the nautical triangle PZO (pole–zenith–sun) of the A.M. observation

$$\cos z = \cos p \cos b + \sin p \sin b \cos ZPO,$$

we replace z, p, b, and $\measuredangle ZPO$ with $90° - h$, $90° + \delta$, $90° - \varphi$, and $180° - T$ (T being understood to represent the time angle of the sun), and we obtain

$$-\cos T = \tan \delta \tan \varphi + \frac{\sin h}{\cos \delta \cos \varphi}.$$

This yields the true local time T of the A.M. observation

$$\text{T.L.T.} = T = 134° \ 47.5' = 8 \text{ hr } 59 \text{ min } 10 \text{ sec.}$$

From this and the time equation e we obtain the mean local time of the observation

$$\text{M.L.T.} = \text{T.L.T.} + e = 8 \text{ hr } 44 \text{ min } 7 \text{ sec.}$$

If we reduce the mean Greenwich time of the observation by the mean local time, we obtain the western longitude λ of the observation point in time:

$$\lambda = \text{M.G.T.} - \text{M.L.T.} = 10 \text{ hr } 5 \text{ min } 53 \text{ sec.}$$

In angular measure (1 hr time longitude = 15 degrees longitude), this comes to

$$\lambda = 151° \ 28.25' \text{ W.}$$

IV. DETERMINATION OF THE MERIDIAN LONGITUDE Λ.

$$\Lambda = \lambda + l = 151° \ 48'.$$

RESULT: A.M. Position: 44° 38.5′ N, 151° 28.25′ W,
 Noon Position: 44° 44.3′ N, 151° 48′ W.

79　Gauss' Two-Altitude Problem

From the altitudes of two known stars determine the time and position.

This problem, which is very important for astronomers, geographers, and mariners, was solved by Gauss in 1812 in Bode's *Astronomisches Jahrbuch*.

Two stars are said to be known when their equatorial coordinates—the right ascension and declination—are known. Let these coordinates of the two stars S and S' be $\alpha|\delta$ and $\alpha'|\delta'$. In the present problem all we need in addition is the right ascension difference $\alpha' - \alpha$. In the figure let P be the world pole; thus $PS = p = 90° - \delta$

will be the pole distance from S; $PS' = p' = 90° - \delta'$ will be the pole distance from S'; and $\angle SPS' = \tau$ will be the angle between the hour circles of the two stars, as well as the magnitude of the right ascension difference; let Z be the zenith of the observation point, so that $PZ = b = 90° - \varphi$ is the complement of the latitude φ, $ZS = z$ the zenith distance from S, and $ZS' = z'$ the zenith distance from S', the last two being as well the complements of the altitudes h and h', respectively.

We still need the auxiliary magnitudes $\angle PSS' = \sigma$, $\angle PS'S = \sigma'$, $\angle PSZ = \psi$, $\angle ZSS' = \zeta$, $\angle ZPS = t$, and the side $SS' = s$.

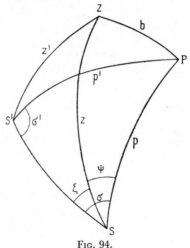

Fig. 94.

The computation, which is very simple, consists of three steps corresponding to the three triangles PSS', ZSS', PZS, which are taken up in that order.

I. TRIANGLE PSS'. The angles σ and σ' are determined according to Napier's formulas

$$\tan \frac{\sigma + \sigma'}{2} = \frac{\cos \dfrac{p' - p}{2}}{\cos \dfrac{p' + p}{2}} \cot \frac{\tau}{2} \qquad \tan \frac{\sigma - \sigma'}{2} = \frac{\sin \dfrac{p' - p}{2}}{\sin \dfrac{p' + p}{2}} \cot \frac{\tau}{2}$$

and the side s is determined according to the sine formula

$$\sin s : \sin p = \sin \tau : \sin \sigma'.$$

II. TRIANGLE ZSS'. The angle ζ is calculated according to the tangent theorem for the half angle:

$$\tan \frac{\zeta}{2} = \sqrt{\frac{\sin (\Sigma - z) \sin (\Sigma - s)}{\sin \Sigma \sin (\Sigma - z')}},$$

where Σ is half the sum of the triangle sides z, z', s. In connection with this we determine $\psi = \sigma - \zeta$.

III. TRIANGLE PZS, determination of the locale and the time.

The *sought-for latitude* can be obtained from

$$\cos b = \cos p \cos z + \sin p \sin z \cos \psi$$

or

$$\sin \varphi = \sin \delta \sin h + \cos \delta \cos h \cos \psi.$$

The sought-for time angle T, i.e., the angle at the pole that has been described by the hour circle of the star S since its lower culmination, follows from

$$\cos t = \frac{\cos z - \cos p \cos b}{\sin p \sin b} = \frac{\sin h - \sin \delta \sin \varphi}{\cos \delta \cos \varphi}$$

and

$$T = 12 \text{ hr} \pm t,$$

where the upper sign applies when the star S at the moment of observation is in the western celestial hemisphere and the lower when it is in the eastern celestial hemisphere. From this we obtain directly the *sought-for time*—sidereal time $\mathfrak{S}$ (the time angle of the Aries point)—of the observation when we add the right ascension α to the time angle T: $\mathfrak{S} = T + \alpha$.

In order to obtain the mean local time—M.L.T.—of the observation we first determine with an approximate value $\tilde{\alpha}_0$ of the right ascension of the mean sun for the moment of the observation the *approximate* mean local time $\mathfrak{S} - \tilde{\alpha}_0$ of the observation; then, using this already fairly exact mean local time we determine the *exact* right ascension α_0 of the mean sun for the moment of observation and finally the exact *mean local time*

$$\text{M.L.T.} = \mathfrak{S} - \alpha_0.$$

We can apply this solution of the Gauss two-altitude problem directly to the solution of the very important navigational problem,

Douwes'* problem: *From two altitudes of a star (the sun) with known declination and the interval between the two observations determine the latitude of the place of observation.*

We need only consider S and S', respectively, as the place, δ and δ', respectively, as the declination of the star at the first and second observations. For fixed stars $\delta = \delta'$, while for the sun and the planets δ' differs somewhat from δ. (τ is the angle determined by the known time interval between the hour circles of the star corresponding to the two moments of observation.)

Since the two measured altitudes are usually observed at *different* places A and B, while the above calculation is related to only *one* place, let us say B, the altitude measured at A must be "reduced to place B." For this purpose we solve the problem:

At a place A *the altitude of a star is observed at a given time* $\mathfrak{Z}$*; at the same moment in time what is the altitude of the star at place* B*?*

To begin with, it is clear that all places on the earth's surface at which the star has the same altitude or the same zenith distance at moment $\mathfrak{Z}$ lie on a *circle* of the geosphere the spherical midpoint of which is the end point S_0 of the earth radius from the geocenter to the star. This circle is called the *equal altitude circle* of the star, its midpoint S_0 the *star image*.

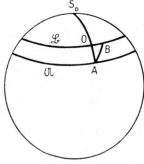

Fig. 95.

In Figure 95 let $\mathfrak{A}$ and $\mathfrak{B}$ be the two equal altitude circles of the star at moment $\mathfrak{Z}$ on which the observation points A and B lie; let S_0 be the star image, O the point of intersection of the great arc S_0A with $\mathfrak{B}$. We will assume that the distance AB is so small that the triangle AOB can be considered plane. This gives for the difference between

---

* Douwes was a Dutch admiralty mathematician.

the zenith distances and, consequently, also for the difference in the altitudes of the star at A and B

$$AO = AB \cos \omega,$$

where ω is the angle between the ship's course AB and the bearing AO of the star at A.

We accordingly obtain the sought-for star altitude h at B at the time β of the observation made at A if we increase or reduce the star altitude measured at A by the product of the traversed distance AB and the cosine of the angle between the course and the bearing of the star at A, accordingly as the ship draws nearer to or recedes from the star.

The "reduced" altitude thus obtained must then be substituted for h in the above Gauss equation, while the altitude measured at B must be used for h'.

The value for φ obtained by this calculation is naturally the latitude of the second observation point B.

80   Gauss' Three-Altitude Problem

From the time intervals between the moments at which three known stars attain the same altitude, determine the moments of the observations, the latitude of the observation point and the altitude of the stars.

The significance of this Gauss method for determining time and location resides in the fact that it eliminates all observational error resulting from atmospheric refraction.

SOLUTION. We designate the equatorial coordinates (right ascension and declination) of the three stars as $\alpha|\delta$, $\alpha'|\delta'$, $\alpha''|\delta''$, the latitude of the observation point as φ, the moments of the observations as t, t', t'', the time angles of the three stars at these moments as T, T', T'', so that the differences $T' - T = t' - t$ and $T'' - T = t'' - t$ are known.   This gives us the three equations

(1) $\sin h = \sin \delta \sin \varphi - \cos \delta \cos \varphi \cos T,$

(2) $\sin h = \sin \delta' \sin \varphi - \cos \delta' \cos \varphi \cos T',$

(3) $\sin h = \sin \delta'' \sin \varphi - \cos \delta'' \cos \varphi \cos T''.$

By subtracting the two first equations we obtain

(4) $\sin \varphi(\sin \delta - \sin \delta') = \cos \varphi(\cos \delta \cos T - \cos \delta' \cos T').$

We now introduce the half sum and half difference

$$s = \frac{\delta' + \delta}{2} \quad \text{and} \quad u = \frac{\delta' - \delta}{2}$$

and

$$S = \frac{T' + T}{2} \quad \text{and} \quad U = \frac{T' - T}{2}$$

of the declinations δ' and δ and the time angles T' and T, respectively, and accordingly replace δ' and δ in (4) by $s + u$ and $s - u$, and replace T' and T by $S + U$ and $S - U$. In the transformed equation (4) we then apply the addition theorem throughout and obtain

$$-\sin \varphi \cos s \sin u =$$
$$\cos \varphi (\sin S \sin U \cos s \cos u + \cos S \cos U \sin s \sin u).$$

Here we divide by $\cos \varphi \cos s \sin u$ and obtain

$$-\tan \varphi = \sin S \cdot \sin U \cot u + \cos S \cdot \cos U \tan s.$$

Since U, u, and s are known, we determine the auxiliary magnitudes r and w such that

$$r \cos w = \sin U \cot u \quad \text{and} \quad r \sin w = \cos U \tan s.$$

(First w is determined from $\tan w = \tan s \tan u \cot U$ and then r from one of the two auxiliary equations.) The equation obtained then assumes the simple form

(I) $$-\tan \varphi = r \sin [S + w].$$

In precisely the same way, by subtracting the two equations (1) and (3), introducing the half sums

$$\hat{s} = \frac{\delta'' + \delta}{2}, \qquad \mathfrak{S} = \frac{T'' + T}{2}$$

and half differences

$$\mathfrak{u} = \frac{\delta'' - \delta}{2}, \qquad \mathfrak{U} = \frac{T'' - T}{2},$$

and introducing the auxiliary magnitudes $\mathfrak{r}$ and $\mathfrak{w}$ determined by the conditions

$$\mathfrak{r} \cos \mathfrak{w} = \sin \mathfrak{U} \cot \mathfrak{u}, \qquad \mathfrak{r} \sin \mathfrak{w} = \cos \mathfrak{U} \tan \hat{s},$$

we find the equation

(II) $$-\tan \varphi = \mathfrak{r} \sin (\mathfrak{S} + \mathfrak{w}).$$

By division of II and I we obtain the *sine ratio of the two unknown angles* ($\mathfrak{S} + \mathfrak{w}$) and $[S + w]$,

(III) $$\frac{\sin (\mathfrak{S} + \mathfrak{w})}{\sin [S + w]} = \frac{r}{\mathfrak{r}}.$$

However, since the difference

$$(\mathfrak{S} + \mathfrak{w}) - [S + w] = \frac{T'' - T'}{2} + \mathfrak{w} - w$$

of these angles is known, it is easy to calculate the *sum* of the angles by applying the sine tangent theorem (No. 40) to (III). From the sum and the difference we obtain directly the angles $\mathfrak{S} + \mathfrak{w}$ and $S + w$ themselves and consequently also the unknown angles

$$\mathfrak{S} = \frac{T'' + T}{2} \quad \text{and} \quad S = \frac{T' + T}{2}.$$

From S and the known difference $T' - T$ we then obtain the *sought-for time angles* T and T'; from $\mathfrak{S}$ and the known difference $T'' - T$ we obtain in similar fashion the time angles T and T''. By adding the right ascension to the time angle we finally obtain the *moments of the observations in sidereal time*.

The *sought-for latitude* then follows from (I) or (II), the *sought-for altitude h* from (1), (2), or (3).

Note. If the latitude is to be determined from *two* observations of the same star altitude and the time interval between them, we have at our disposal only equations (1) and (2) and must assume that the time angle T for one of the observations is known. Equation (I), all the magnitudes on the right side of which are known, then gives φ.

A remarkable special case of this situation is the

Problem of Riccioli: *From the time between the culminations of two known stars that rise or set at the same time, find the latitude of the observation point.*

This problem posed by Riccioli in 1651 is especially noteworthy in that the method employed makes possible determinations of latitude without an angle-measuring instrument.

If T and T' are the time angles of star risings, their difference $2U = T' - T$ is also the time between their culminations. Our initial equations (1) and (2) are simplified here (because $h = 0$) to

$$\cos T = \tan \delta \tan \varphi \quad \text{and} \quad \cos T' = \tan \delta' \tan \varphi.$$

We introduce the complements τ and τ' of the time angles and obtain

$$\sin \tau = \tan \delta \tan \varphi, \qquad \sin \tau' = \tan \delta' \tan \varphi,$$

and from this by division we get the sine ratio of the angles τ and τ':

$$\sin \tau : \sin \tau' = \tan \delta : \tan \delta'.$$

Since $\tau - \tau' = T' - T$ is known, we obtain $\tau + \tau'$ from this equation, in accordance with the sine-tangent theorem. We then get $2\tau = (\tau + \tau') + (\tau - \tau')$ and finally φ from $\sin \tau = \tan \delta \tan \varphi$.

81 The Kepler Equation

From the mean anomaly of a planet calculate the eccentric and true anomaly.

Johannes Kepler (1571–1630) was one of the greatest astronomers of all time. The famous problem named after him is to be found in the 60th chapter of Kepler's major work *Astronomia nova*, published in Prague in 1609, a book that, according to Lalande, every astronomer must read at least once.

Before taking up the solution we will present a short explanation of the three anomalies.

Let S and P be the midpoints of the sun and a planet, respectively, let N be the point of the planet's orbit at which the planet is nearest to the sun, the so-called perihelion, let O be the midpoint of the elliptical orbit and of its circle of circumscription, P_0 the point of intersection of the circle of circumscription with the parallel drawn through P to the minor orbit axis, a and b the major and minor axes of the ellipse, respectively, $OS = e$ the linear eccentricity, $\varepsilon = e/a$ the

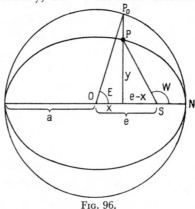

Fig. 96.

astronomic eccentricity or form number, T the period of revolution of the planet, and t the time elapsed at the planet's position P since its passage through the perihelion.

The *true anomaly* W is the angle NSP, i.e., the angle described by the focal radius of the planet in the time t, the *mean anomaly* M the angle that the focal radius would describe in the time t if it were to revolve uniformly (with the same period of revolution T), so that in angular measure

$$M = \frac{2\pi}{T}\, t.$$

Finally, the *eccentric anomaly* E is the angle NOP_0 formed by the radius of the circle of circumscription to P_0 with the radius of the circle of circumscription ON.

With E as a variable parameter we have

$x = a \cos E,$ $\qquad y = b \sin E$ the equation of the orbit

$x = a \cos E,$ $\qquad y_0 = a \sin E$ the equation of its circle of circumscription.

There exists between the eccentric and true anomaly the relation (obtainable from the right triangle with the legs $e - x$ and y)

$$\tan W = \frac{b \sin E}{a \cos E - e};$$

after squaring and use of the formulas $b^2 = a^2 - e^2$, $e = a\varepsilon$, and $\cos^2 E + \sin^2 E = 1$, $\sec^2 W - \tan^2 W = 1$, this relation is transformed into

$$\cos W = \frac{\cos E - \varepsilon}{1 - \varepsilon \cos E}.$$

In order to obtain, in addition, a formula that is convenient for logarithmic treatment, Gauss introduced the half angles $\frac{1}{2}W$ and $\frac{1}{2}E$ and made use of the formulas

$$1 + \cos \varphi = 2 \cos^2 \frac{\varphi}{2} \quad \text{and} \quad 1 - \cos \varphi = 2 \sin^2 \frac{\varphi}{2}.$$

We write the above equation

$$\frac{1 - \cos W}{1 + \cos W} = \frac{1 + \varepsilon}{1 - \varepsilon} \frac{1 - \cos E}{1 + \cos E}$$

and obtain the

GAUSS FORMULA:

$$\tan \frac{W}{2} = \sqrt{\frac{1 + \varepsilon}{1 - \varepsilon}} \tan \frac{E}{2}.$$

There exists between the eccentric and mean anomaly (in radian measure) the famous *Kepler equation*:

$$E - \varepsilon \sin E = M.$$

This equation is a consequence of the formula

$$J = \frac{ab}{2} (E - \varepsilon \sin E)^*$$

for the area J of the elliptical sector SNP and of the Kepler surface theorem: "The focal radius of a planet sweeps equal surfaces in equal times." [According to the area formula, the area of the half ellipse $(E = \pi)$ is $\frac{1}{2}\pi ab$; the area of the whole ellipse is thus πab. According to Kepler's surface theorem, there exists the proportion $J : \pi ab = t : T$. Consequently, $E - \varepsilon \sin E = 2\pi t : T = M$.]

The crux of the Kepler problem now consists of the solution of the Kepler equation

$$E - \varepsilon \sin E = M$$

for the unknown E (when M and ε are assumed to be known).

The following determination of E rests upon the assumption that the form number ε is a proper fraction and consists in the calculation of a series $E_1, E_2, E_3, \ldots$ of approximate values for the eccentric anomaly that deviate progressively less and less from the true value E as the index number increases and approximate the true value sufficiently closely at a relatively low index number.

For the first approximation value we choose

$$E_1 = M + \varepsilon \sin M.$$

Its deviation from the true value E is

$$E - E_1 = \varepsilon(\sin E - \sin M).$$

However, since

$$|\sin E - \sin M| < |E - M| = |\varepsilon \sin E| < \varepsilon,$$

it follows that

$$|E - E_1| < \varepsilon^2.$$

---

* This formula is obtained as follows: Since the circle sector ONP_0 has the area $J_0 = \frac{1}{2}a^2 E$ and *each* ordinate of the elliptical sector ONP is equal to b/a times the circle ordinate at that point, the area of the sector ONP is also equal to b/a times J_0, i.e., $\frac{1}{2}abE$. Consequently, the area J of the elliptical sector SNP that is smaller than ONP by the area $\frac{1}{2}ey = \frac{1}{2}abe \sin E$ of the triangle OSP, is $J = \frac{1}{2}abE - \frac{1}{2}ab \cdot \varepsilon \cdot \sin E$.

As the second approximation value we choose

$$E_2 = M + \varepsilon \sin E_1.$$

Its deviation from E is $E - E_2 = \varepsilon(\sin E - \sin E_1)$. However, since

$$|\sin E - \sin E_1| < |E - E_1|$$

and the latter magnitude, as was just shown, is $< \varepsilon^2$, it follows that

$$|E - E_2| < \varepsilon^3.$$

The third approximation value is

$$E_3 = M + \varepsilon \sin E_2.$$

Its deviation from E, absolutely considered, is $< \varepsilon^4$, etc.

The nth approximation value deviates from the true value by less than the $(n + 1)th$ power of the form number ε. The approximation values accordingly approach the true value progressively more rapidly as ε diminishes.

In the earth's orbit, for example, $\varepsilon = 0.01674$, $\varepsilon^3 = 0.00000469$, arc $1'' = 0.00000485$. Consequently:

For the earth's orbit the second approximation value is already exact to seconds!

In the orbit of Mars, which has the fairly high form number of 0.0933, $\varepsilon^5 = 0.0000071$, so that the fourth approximation value E results in an error of less than $2''$.

After E is determined the true anomaly is calculated by the Gauss formula.

NOTE. Kepler's problem is of the greatest importance for astronomy. It forms the basis, for example, for the *determination of the equation of time for a given moment of time.*

[The equation of time is conventionally understood to be the difference between mean and true local time or also the difference between the right ascensions α and α_0 of the true and mean sun:

$$e = \text{M.L.T.} - \text{T.L.T.} = \alpha - \alpha_0.]$$

The calculation is based on the following seven steps:

1. Determination of the right ascension α_0 of the mean sun for the given moment of time from its daily increase of 3 m 56.55536 s and its value for a *fixed* moment of time (on January 1, 1925, at midnight, M.G.T. was $\alpha_0 = 18$ hr 40 min 30 sec).

2. Calculation of the mean anomaly M according to the (definition) equation $\alpha_0 = M + \Pi$, where Π is the longitude of the true sun at perigee. (Π on January 1, 1925, was $281° 39' 2''$ and it increases annually by $1' 1.9''$.)

3. Determination of the eccentric anomaly E from Kepler's equation $E - \varepsilon \sin E = M$ with $e = 0.01674$.

4. Calculation of the true anomaly W from the Gauss formula

$$\tan \tfrac{1}{2}W = \sqrt{\frac{1 + \varepsilon}{1 - \varepsilon}} \tan \tfrac{1}{2}.$$

5. Determination of the longitude L of the true sun according to the equation $L = W + \Pi$.

6. Determination of the right ascension α of the true sun in accordance with the equation $\tan \alpha = \cos i \tan L$ obtained from the astronomical triangle having the hypotenuse L and the legs α and δ; in the equation, i represents the inclination of the ecliptic.

7. Calculation of the equation of time e from $e = \alpha - \alpha_0$.

EXAMPLE. The equation of time for the 2nd of December, 1925 at 4:00 P.M. Central European Time.

$$\alpha_0 = 16 \text{ hr } 43 \text{ min } 44 \text{ sec} = 250° 56', \quad M = 329° 16' 1'',$$
$$E_1 = 328° 46' 38'', \quad E_2 = E = 328° 46' 12'', \quad W = 328° 16' 10'',$$
$$L = 249° 56' 9'', \quad \alpha = 248° 17' 28'' = 16 \text{ hr } 33 \text{ min } 10 \text{ sec},$$
$$e = -10 \text{ min } 34 \text{ sec}.$$

82 Star Setting

Calculate the time and azimuth of setting of a known star for a given place and day.

SOLUTION. The method of calculation can best be illustrated by a numerical example. Thus, let us consider a more definite form of the problem:

On the 31st of December, 1932, when did Saturn set in Nördlingen, Bavaria ($\varphi = 48° 51.1'$, $\lambda = 10° 29.4'$)? The nautical almanac gives the following data for December 31, 1932 at midnight, mean Greenwich time: right ascension of Saturn $\alpha = 20$ hr 25 min 30 sec (hourly increase = 1.2 sec), declination of Saturn $\delta = 19° 47.4'$ S (hourly decrease 0.06'), right ascension of the mean sun $\alpha_0 = 18$ hr 36 min 50 sec (hourly increase = 9.86 sec).

At the moment of setting the star is already in reality a certain distance h below the horizon (SN) as a result of atmospheric refraction. The horizontal refraction h can be set at an average of 35', but in precise measurements special refraction tables must be consulted.

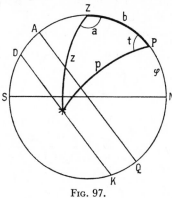

FIG. 97.

It follows from the nautical triangle $PZ*$ (in which $PZ = b = 90° - \varphi$ represents the complement of the latitude φ, $P* = p = 90° + \delta$ the pole distance, $Z* = z = 90° + h$ the zenith distance, $\angle ZP* = t$ the hour angle, and $\angle PZ* = a$ the azimuth of the star), according to the cosine theorem, that

$$\cos z = \cos b \cos p + \sin b \sin p \cos t.$$

If we introduce the magnitudes h, φ, δ here instead of z, b, p, we obtain

$$\cos t = \tan \varphi \tan \delta - \frac{\sin h}{\cos \varphi \cos \delta}.$$

First we calculate the *approximate* time t of setting, taking for the moment of setting $\delta = 19° 47.4'$. We then obtain from the formula we have found (assuming $h = 35'$), $t = 66° 42.8' = 4\,\text{hr}\,26\,\text{min}\,51\,\text{sec}$ and for the time angle T of the moment of setting

$$T = 16\,\text{hr}\,26\,\text{min}\,51\,\text{sec}.$$

From this we get for the sidereal time $\mathfrak{S}$ (i.e., the time angle at the vernal equinox) the approximate value

$$\mathfrak{S} = T + \alpha = 36\,\text{hr}\,52\,\text{min}\,21\,\text{sec},$$

and thus for the mean local time of setting

$$\text{M.L.T.} = \mathfrak{S} - \alpha_0 = 18\,\text{hr}\,15\,\text{min}\,31\,\text{sec}$$

and for the mean Greenwich time

  M.G.T. = M.L.T. − (λ = 41 min 58 sec) = 17 hr 33 min 33 sec.

At the moment of setting, then, approximately 17.55 hr have gone by since midnight mean Greenwich time. In these 17.55 hr the three magnitudes α, δ, and α_0 increase by 21 sec, −1.1′, 2 min 53 sec, so that at the moment of setting they have the values

$$\alpha = 20 \text{ hr } 25 \text{ min } 51 \text{ sec}, \quad \delta = 19° 46.3′,$$
$$\alpha_0 = 18 \text{ hr } 39 \text{ min } 43 \text{ sec}.$$

The calculation must now be repeated with these *exact values*. This gives

$$
\begin{array}{rl}
T = & 16 \text{ hr } 26 \text{ min } 57 \text{ sec} \\
\alpha = & 20 \text{ hr } 25 \text{ min } 51 \text{ sec} \\
\hline
\text{☉} = & 36 \text{ hr } 52 \text{ min } 48 \text{ sec} \\
\alpha_0 = & 18 \text{ hr } 39 \text{ min } 43 \text{ sec} \\
\hline
\text{M.L.T.} = & 18 \text{ hr } 13 \text{ min } 5 \text{ sec} \\
\text{M.G.T.} = & 17 \text{ hr } 31 \text{ min } 7 \text{ sec}.
\end{array}
$$

The sought-for azimuth a is computed from the sine formula

$$\sin \alpha : \sin t = \sin p : \sin z$$

and comes out to be

$$a = 120° 10′.$$

RESULT. *Saturn set at* 18 hr 31.1 min C.E.T. *at an azimuth of* S 59° 50′ W.

NOTE. The method described is naturally just as well suited to the determination of the rising time or the time at which a star attains a prescribed altitude. If it is specifically desired to determine the moment of culmination, the logarithmic calculation can be dispensed with, since the time angle of culmination, T = 12 hr, is known.

83 The Problem of the Sundial

To construct a sundial.

First we will consider the two simplest forms of sundial: the *horizontal dial* and the *vertical meridional dial*. In the first the plane of the dial E is horizontal, in the second vertical, specifically through the eastern and western points of the horizon. The earth's axis is represented by a pin, the *gnomon* or *style* that casts a shadow on E. At noon the shadow is situated at its center position, the *meridian line of the dial plane*, and at *t* hr before or after noon forms the "shadow angle" *s* or *σ*, respectively, with the meridian line.

The problem is to determine the relation between the time *t* and the shadow angle.

We will call the plane formed by the sun and the earth's axis (the gnomon) the shadow plane, since the shadow must lie in this plane. At noon the shadow plane at its central position passes through the north and south points of the horizon and at time *t* forms the angle *t* (t hr $= 15t°$) with its central position.

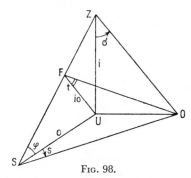

Fig. 98.

In the figure let *US*, *UO*, and *UZ* be segments running from *U* toward the southern point, the eastern point, and the zenith of the horizon, specifically in such manner that *SZ* represents the gnomon; thus $\angle USZ$ represents the latitude *φ* of the place and *SOZ* the shadow plane, so that *SO* is the shadow; $\angle USO$ is the shadow angle *s* of the horizontal dial, *ZO* the shadow, $\angle UZO$ the shadow angle *σ* of the vertical meridional dial. The angle *t* between the shadow plane *SOZ* and its meridional position *SUZ* is the angle *UFO* that is formed with *UF* by the perpendicular *OF* dropped from *O* to *SZ*. If we select *SZ* as the unit length and, for the sake of brevity, set $\cos φ = o$, $\sin φ = i$, it follows from the right triangle *SUZ* that $US = o$, $UZ = i$, $UF = oi$, from the right triangle *UOF* that $UO = oi \tan t$, and from the right triangles *USO* and *UZO* that $UO = o \tan s$ and $UO = i \tan σ$. If we set the three values for *UO* equal to each other, we get the equations

$$(1) \quad \tan s = i \tan t, \quad (2) \quad \tan σ = o \tan t,$$

which contain the sought-for relations between the time *t* and the shadow angles *s* and *σ*, respectively.

In order to construct the dial we compute, in accordance with (1) or (2), the shadow angle corresponding to different times t, draw them in, but write on their free leg not s or σ, but the corresponding times t.

It is also possible to use a purely graphic method. On an arbitrary segment AB we begin at B and mark off i or o times its length to C, draw the semicircle with the center C and the arc center B, and draw the tangent through B which is at the same time perpendicular to AC.

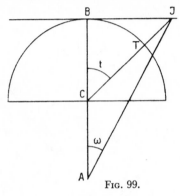

Fɪɢ. 99.

If we now make the arc BT equal to the time angle t (thus, for example, $45°$ for 3 hr), extend CT to the intersection J with the tangent, and connect J with A, then $\angle BAJ = \omega$ is the shadow angle s or σ for time t. [From $\triangle BJA$ it follows that $BJ = BA \tan \omega$, from $\triangle BJC$ that $BJ = BC \tan t$, so that $BA \tan \omega = BC \tan t$ or, since BC is i or o times BA,

$$\tan \omega = i \tan t \quad \text{or} \quad \tan \omega = o \tan t.$$

According to (1), ω is equal to s and according to (2), $\omega = \sigma$.]

We carry out the described construction for as many time angles t as possible and obtain the dial as the totality of lines AJ each of which bears written on it its corresponding time. In order to install it, we place the drawing plane horizontally, so that BA points from the northern point of the horizon to the southern point, or vertically, so that BA points perpendicularly upward and the tangent runs from west to east, and fix the style parallel to the earth's axis at A.

A Vᴇʀᴛɪᴄᴀʟ Sᴜɴᴅɪᴀʟ ᴀᴛ ᴀɴ Aʀʙɪᴛʀᴀʀʏ Aᴢɪᴍᴜᴛʜ

Let us now consider the case in which a sundial is to be fastened to a vertical house wall that does *not* run east and west.

In Figure 100, let UZ be a vertical line on the wall and UH a horizontal line on the wall, US a horizontal pointing south, ZS the gnomon, so that $\angle USZ = \varphi$ and $\angle UZS = b = 90° - \varphi$; UZS is the meridian plane and $\angle SUH = a$ the azimuth (calculated from the south point) of the wall; ZH is the shadow at time t, so that ZSH is the shadow plane, and the angle that it forms with the meridian plane ZSU is the time angle t; finally, the angle that ZH forms with ZU is

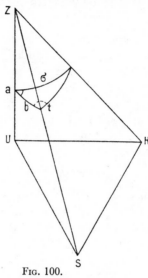

Fɪɢ. 100.

the shadow angle σ. The three-dimensional vertex Z with the edges ZU, ZH, ZS cuts out of the sphere with the center Z a spherical triangle (shown in the figure) in which the side σ, the angle a, the side b, and the angle t are four successive elements. According to the cotangent theorem, therefore,

$$\cos b \cos a = \sin b \cot \sigma - \sin a \cot t$$

or

$$\cos \varphi \cot \sigma - \sin a \cot t = \sin \varphi \cos a.$$

This is the relation between the time t and the shadow angle σ. This relation makes it possible to calculate a corresponding σ for every t.

The invention of the sundial is lost in antiquity. A statement by Vitruvius (which was also found engraved on an ancient sundial unearthed on the Via Flaminia), according to which the inventor is the Chaldaean Berosus, is not reliable in view of the fact that sundials were known in ancient Babylonia many centuries before Berosus.

84   The Shadow Curve

To determine the curve described by the shadow of a point of a rod in the course of a day, when the rod is erected at a place of latitude φ and the declination of the sun for the day has a value of δ.

SOLUTION.   We select the perpendicular from the point of the rod to the horizon of the place as the unit length and the base point O of the perpendicular as the origin of a right-angle coordinate system whose x-axis runs toward the north point and whose y-axis runs toward the west point of the horizon.   At the moment in which the sun ($\odot$) has the azimuth S $a°$ E and the zenith distance z, the distance of the shadow from O is $\tan z$, and the abscissa and ordinate, respectively, of the shadow are

$$x = \tan z \cos a, \qquad y = \tan z \sin a.$$

In the nautical triangle $PZ\odot$ the latitude complement $PZ = b$ and the pole distance $P\odot = p = 90° - \delta$ are constant.   The zenith distance $Z\odot = z$, the azimuth supplement $PZ\odot = 180° - a$ and the hour angle $ZP\odot = t$ are variable.   We find the equation of the shadow curve by expressing $\sin t$ and $\cos t$ in terms of x and y and introducing the resulting expressions into the equation

$$\cos^2 t + \sin^2 t = 1.$$

We abbreviate $\sin \varphi$, $\cos \varphi$, and $\tan \varphi$, as i, o, and q, respectively, and $\sin p$, $\cos p$, and $\tan p$, as I, O, and Q, respectively.   If we then apply to the nautical triangle the sine theorem, cosine theorem, and cotangent theorem in that order, we obtain the three equations

$$\sin a \sin z = \sin p \sin t,$$
$$\cos z = \cos p \cos b + \sin p \sin b \cos t,$$
$$-\cos b \cos a = \sin b \cot z - \sin a \cot t.$$

We divide the first by the second and obtain

$$\sin a \tan z = \frac{\tan p \sin t}{\sin \varphi + \cos \varphi \tan p \cos t}$$

or

(1) $$y = \frac{Q \sin t}{i + oQ \cos t}.$$

We multiply the third by $-\tan z$ and obtain

$$\sin \varphi \cdot \cos a \tan z = \sin a \tan z \cdot \cot t - \cos \varphi$$

or

(2) $$ix = y \frac{\cos t}{\sin t} - o.$$

From (1) and (2) we find

$$Q \cos t = \frac{o + ix}{i - ox}, \qquad Q \sin t = \frac{y}{i - ox}$$

and from this, in accordance with what was stated above, we obtain

$$(o + ix)^2 + y^2 = Q^2(i - ox)^2$$

as the equation of the shadow curve. We solve for y^2 and obtain

$$y^2 = (Q^2 i^2 - o^2) - 2io(Q^2 + 1)x + (Q^2 o^2 - i^2)x^2$$

or, if we go on to divide by o^2,

$$\frac{y^2}{o^2} = (Q^2 q^2 - 1) - 2q(Q^2 + 1)x + (Q^2 - q^2)x^2.$$

To put this equation into a simpler form, we introduce a new co-ordinate system X, Y whose origin U is situated at the apex of the curve, i.e., at the point where the shadow lies at noon; the X-axis runs toward the south and the Y-axis toward the west. When the sun is at meridian, its zenith distance is $p - b$, and thus

$$Uo = \alpha = \tan (p - b) = \frac{\tan p - \tan b}{1 + \tan p \tan b} = \frac{Qq - 1}{Q + q}.$$

We accordingly introduce

$$x = \alpha - X, \qquad y = Y$$

into the above curve equation and obtain

$$\frac{Y^2}{o^2} = 2Q(1 + q^2)X + (Q^2 - q^2)X^2$$

or, if we write the first parenthesis as $1/o^2$ and the second as

$$\frac{1}{O^2} - \frac{1}{o^2}$$

and multiply the equation by o^2,

$$Y^2 = 2QX - \left(1 - \frac{o^2}{O^2}\right)X^2.$$

The amplitude equation of the shadow curve thus reads

$$Y^2 = 2 \tan p X - \left(1 - \frac{\cos^2 \varphi}{\cos^2 p}\right) X^2.$$

The curve is consequently a conic section with the half parameter $\tan p$ *and the form number (eccentricity)* $\cos \varphi / \cos p$.

If the latitude is equal to the polar distance of the sun, then the shadow describes a parabola; at higher latitudes it describes an ellipse, and at lower a hyperbola.

85 Solar and Lunar Eclipses

To determine the beginning and end of a solar eclipse, together with the maximum fraction of the solar disc that is obscured, if the right ascensions, declinations, and radii of the sun and moon are known for two moments in time sufficiently close to the time of the eclipse.

EXAMPLE. At the famous solar eclipse that occurred at Athens during the Peloponnesian War on August 3, 431 B.C., the magnitudes mentioned had, at 4:30 P.M. and 5:30 P.M. mean Athenian time, the values

$$A_0 = 126° \ 51' \ 52'', \qquad \Delta_0 = 19° \ 23' \ 46'', \qquad R_0 = 15' \ 52'',$$
$$\alpha_0 = 126° \ 40' \ 55'', \qquad \delta_0 = 19° \ 38' \ 58'', \qquad r_0 = 15' \ 38.5''$$

and

$$A_1 = 126° \ 54' \ 21'', \qquad \Delta_1 = 19° \ 23' \ 11'', \qquad R_1 = 15' \ 52'',$$
$$\alpha_1 = 127° \ \ 8' \ 49'', \qquad \delta_1 = 19° \ 24' \ 30'', \qquad r_1 = 15' \ 36.5''.$$

A solar eclipse can only occur at a time when the moon is sufficiently close to the sun on the celestial sphere, i.e., at a time when the differences $a = \alpha - A$ and $d = \delta - \Delta$ between the right ascensions and declinations of the two bodies are sufficiently small.

The spherical cosine theorem gives for the spherical distance z of the midpoints of the two bodies (their central axis) the formula

$$\cos z = \sin \Delta \sin \delta + \cos \Delta \cos \delta \cos a.$$

We replace $\cos z$ and $\cos a$ here by

$$1 - 2 \sin^2 \frac{z}{2} \quad \text{and} \quad 1 - 2 \sin^2 \frac{a}{2}$$

and obtain

$$1 - 2 \sin^2 \frac{z}{2} = \cos d - 2 \cos \Delta \cos \delta \sin^2 \frac{a}{2}.$$

If we now write $1 - 2 \sin^2 (d/2)$ for $\cos d$, we obtain

$$\sin^2 \frac{z}{2} = \cos \Delta \cos \delta \sin^2 \frac{a}{2} + \sin^2 \frac{d}{2}.$$

If we now consider that, according to our assumption, a and d and, therefore, also z are small angles that in no case exceed $1°$, we can substitute the angles themselves for their sine (No. 15) and write

$$z^2 = a^2 \cos \Delta \cos \delta + d^2.$$

If in addition to this we introduce the abbreviations

$$\sqrt{\cos \Delta \cos \delta} = g \quad \text{and} \quad ag = x$$

and substitute y for d, we obtain the simple equation

$$z^2 = x^2 + y^2.$$

The magnitudes a, x, y, and z are most conveniently measured in angular seconds.

If the right ascensions and declinations of the moon and the sun for two moments of time sufficiently close to the time of the eclipse (the first moment being taken as the zero point of time) are known and are, for example, α_0, A_0, δ_0, and Δ_0 for the first moment and α_1, A_1, δ_1, and Δ_1 for the second, then we also know the values a, d, and g, and therefore also $x = ga$ and $y = d$ for these moments in time, and we can calculate from these the hourly increases h and k of x and y. Since the eclipse lasts only a short time, we can assume that the magnitudes x and y change uniformly in the period of time here under consideration and that, consequently, at time t, i.e., at t hours after moment 0,

$$x = x_0 + ht \quad \text{and} \quad y = y_0 + kt.$$

If we introduce these values into the above equation, it assumes the form

$$z^2 = (x_0 + ht)^2 + (y_0 + kt)^2,$$

which permits us to calculate the central axis of the two bodies for any moment t.

The eclipse begins and ends at the moments when the central axis z is equal to the sum s of the two radii R and r. In the period of time under consideration the solar radius does not change ($R = R_0 = R_1$), while the lunar radius exhibits the slight hourly increase $\rho = -2''$, so that

$$r = r_0 + \rho t \quad \text{and} \quad s = R + r = R + r_0 + \rho t = s_0 + \rho t.$$

We therefore obtain for the desired moment t of the beginning (and also the end) of the eclipse the so-called

ECLIPSE EQUATION:

$$(x_0 + ht)^2 + (y_0 + kt)^2 = (s_0 + \rho t)^2.$$

This quadratic equation has two roots for the unknown t; *the smaller value, t', indicates the beginning of the eclipse, and the larger, t'', the end.*

The maximum eclipse occurs at the moment τ in which the central axis z reaches its minimum value ζ. Thus, we have

$$z^2 = z_0^2 + 2mt + n^2t^2,$$

where

$$z_0^2 = x_0^2 + y_0^2, \quad m = x_0h + y_0k, \quad n^2 = h^2 + k^2.$$

If we write

$$z^2 = z_0^2 - \frac{m^2}{n^2} + \left[nt + \frac{m}{n} \right]^2,$$

we see that z attains its minimum value when the bracket disappears. We then have

$$\tau = -\frac{m}{n^2} \quad \text{and} \quad \zeta = \sqrt{z_0^2 - \frac{m^2}{n^2}}.$$

At the moment of the maximum eclipse the moon has advanced over the solar disc by $(R + r - \zeta)/2R$ of the sun's diameter.

The fraction of the solar disc that is covered by the moon at that moment can also be calculated easily from ζ.

Carrying out the computations for the Athenian solar eclipse, we obtain:

$a_0 = -657(-10'\,57'')$,	$a_1 = +868(+14'\,28'')$,
$\log g_0 = 9.97428$,	$\log g_1 = 9.97462$,
$x_0 = -619.2$,	$x_1 = 818.7$,
$y_0 = +912(+15'\,12'')$,	$y_1 = +79(1'\,19'')$,
$h = x_1 - x_0 = 1438$,	$k = y_1 - y_0 = 833$,
$s_0 = 1890.5, \quad s_1 = 1888.5$,	$\rho = s_1 - s_0 = -2$

and the eclipse equation is

$$(-619 + 1438t)^2 + (912 - 833t)^2 = (1890.5 - 2t)^2$$

or

$$2761729t^2 - 3292074t - 2359085 = 0$$

or

$$t^2 - 1.192034t = 0.8542059159.$$

Its roots are

$$t' = -0.50373, \qquad t'' = 1.69576.$$

Converting the decimals into minutes and seconds, we obtain −30 min 13 sec and 1 hr 41 min 45 sec, respectively.

Consequently:

Beginning of eclipse: 3 hr 59 min 47 sec,
End of eclipse: 6 hr 11 min 45 sec.

The length of the eclipse was therefore 2 hr 12 min, the moment of maximum eclipse 5 hr 5 min 46 sec [$2\tau = t' + t''$ gives $\tau = 0.596$]. The central axis of the sun and moon at this moment is obtained from

$$\zeta^2 = (619 - 1438 \cdot 0.596)^2 + (912 - 833 \cdot 0.596)^2;$$

it is

$$\zeta = \sqrt{238^2 + 415.5^2} = 479, \qquad \text{i.e., } 8'.$$

The moon then covers $\frac{1410}{1904}$, i.e., 74% of the central solar diameter and 67% of the solar disc.

Lunar eclipses are treated in a similar way. But here, instead of being concerned with the sun, we are concerned with the so-called *shadow circle*, i.e., the cross section of the conical shadow (the umbra) cast by the sun-illuminated earth at the distance of the moon. The angle radius $\mathfrak{R}$ is equal to $p - \kappa$, where p represents the lunar parallax* and κ represents the half aperture angle of the conical shadow. κ is the excess of the angle radius R over the parallax* P of the sun.

[In the Figure 101, let S be the center of the sun, E the center of the earth, K the apex of the conical shadow, AB the diameter of the shadow circle, *se* a tangent to the periphery of the sun and the earth,

---

* The lunar or solar parallax is the angle radius of the earth on the moon or sun, respectively.

EF the perpendicular to *Ss* from *E*, and thus $\angle EAe = p$, $\angle AEK = \Re$, and $\angle FES = \angle eKE = \kappa$. Since p is an external angle of the triangle *EKA*, we have $p = \Re + \kappa$. It also follows from $\triangle SEF$ that

$$\sin \kappa = \frac{SF}{SE} = \frac{Ss}{SE} - \frac{Ee}{SE}.$$

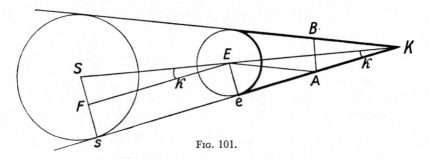

Fig. 101.

Since the minuend of the right side is the sine of the angle radius of the sun and the subtrahend is the sine of the solar parallax, it follows that

$$\sin \kappa = \sin R - \sin P$$

or, because the angle involved is so small (κ is smaller than 16.2′, $R < 16.3′$, and $P < 8.9″$),

$$\kappa = R - P,$$

as was asserted above.]

The right ascension of the center of the shadow circle is the right ascension of the sun increased or diminished by 180° and the declination is the reciprocal value of the solar declination.

In order to take account of the atmospheric refraction, in computing a lunar eclipse the theoretical value for the radius of the shadow circle given above, $\Re = p + P - R$, must be replaced by a value 2% greater.

86   Sidereal and Synodic Revolution Periods

To determine the synodic revolution period of two coplanar rotation rays for which the sidereal revolution periods are known.

A rotation ray is a line segment *AB* of invariable length the end point *B* of which rotates about the starting point *A* in a plane *E* at a constant rate of revolution, while the starting point either remains at rest or describes a curve of plane E. Using a well-known astronomical expression we call the time T in the course of which the rotation ray AB describes one complete revolution of 360° its *sidereal revolution period*.

Let a second rotation ray of the plane E with the starting point a and the end point b have the sidereal revolution period t ($< T$).

We will consider the angle that the two rays form with each other at a given moment of time. The time s at the end of which they once again form the *same* angle we will call the *synodic revolution period* of the two rays or the synodic revolution period of the one ray with respect to the other.

In order to find this we will imagine an auxiliary rotation ray $a'b'$ whose starting point a' always coincides with A and whose direction always agrees with that of ab, and we will now consider the relative rotation of this auxiliary ray with respect to AB. Since the rotation of $a'b'$ (or ab) in the unit time is equal to $360°/t$ and that of AB is $360°/T$, the relative rotation of $a'b'$ with respect to AB in each time unit is

$$(1) \qquad \delta = \left(\frac{1}{t} - \frac{1}{T}\right)360°.$$

If $a'b'$ resumes the same position with respect to AB at the end of s units of time, then $s\delta$ must equal 360° or

$$(2) \qquad \delta = \frac{1}{s}\,360°.$$

From (1) and (2) it follows that

$$\frac{1}{s} = \frac{1}{t} - \frac{1}{T} \quad \text{or} \quad s = \frac{Tt}{T-t},$$

and thus the synodic revolution period s is represented as a function of the two sidereal revolution periods T and t.

This unpretentious problem, the solution to which is also a model of brevity and simplicity, nevertheless possesses noteworthy applications, four of which we will discuss.

PROBLEM 1. The hands of a clock are superimposed one on the other at exactly 12:00; when is the next time they are exactly superimposed one on the other?

Here let AB be the small hand, $ab = Ab$ the big hand, $T = 12$ hr, $t = 1$ hr, thus $s = \frac{12 \cdot 1}{11} = 1\frac{1}{11}$ hr $= 1$ hr 5 min $27\frac{3}{11}$ sec.

The event takes place at 5 min $27\frac{3}{11}$ sec after 1:00.

PROBLEM 2. *From the synodic revolution period ($583\frac{1}{2}$ days) of Venus, determine its sidereal revolution period.*

The sidereal revolution period of a planet is understood to mean the time in which the rotation ray sun–planet makes one complete revolution. The synodic revolution period of the planet is understood to mean the time s at the end of which the three celestial bodies sun, earth, planet are once again in the same position with respect to one another.

Here AB is the rotation ray sun–earth, ab the rotation ray sun–Venus, and $T = 365\frac{1}{4}$ days. The synodic revolution period s of Venus has been determined by observations. Its sidereal revolution period t is obtained from the relation

$$\frac{1}{t} - \frac{1}{T} = \frac{1}{s}$$

as 224.7 days.

PROBLEM 3. *To determine the relation between the solar day and the sidereal day.*

A solar day is the time interval between two successive culminations of the sun, a sidereal day the time interval between two successive culminations of a fixed star or the time interval within which the earth rotates once about its own axis.

Let the midpoint of the sun be S, that of the earth E, a marked point of the earth's equator O. Here AB is the rotation ray SE, ab the rotation ray EO, T is here $365\frac{1}{4}$ days (1 year, the period of time in which $AB = SE$ completes one full revolution of 360°), t the length of a sidereal day, and s the length of a solar day (the period of time at the end of which the ray EO is once again in the same position relative to the sun). From

$$\frac{1}{s} = \frac{1}{t} - \frac{1}{T}$$

we obtain

$$\frac{T}{t} = \frac{T}{s} + 1.$$

T/t represents the number of sidereal days, T/s the number of solar days, that occur in a year. The sought-for relation can accordingly be stated in the following form:

A year contains one more sidereal day than the number of solar days ($365\frac{1}{4}$ solar days, $366\frac{1}{4}$ sidereal days).

PROBLEM 4. *What is the relation between the sidereal and synodic month?*

A sidereal month is the time it takes the rotation ray *EM* (earth–moon) to complete one full revolution. A synodic month is the time interval between two successive new moons (full moons). Here *AB* is the rotation ray *SE*, *ab* the rotation ray *EM*, $T = 365\frac{1}{4}$ days, *t* the length of the sidereal month, *s* the length of the synodic month. The sought-for relation accordingly reads

$$\frac{1}{t} - \frac{1}{s} = \frac{1}{T}.$$

Verbally it can be stated as follows: *The reciprocal of the synodic month subtracted from the reciprocal of the sidereal month is equal to the reciprocal of the sidereal year.*

This can be confirmed for the numerical values:

$t = 27.3217$ days, $s = 29.5306$ days, $T = 365.2564$ days.

87 Progressive and Retrograde Motion of the Planets

When does a planet pass from progressive to retrograde motion (or conversely, from retrograde to progressive motion)?

The planetary orbits, considered as circles on the ecliptic plane, their orbital radii and revolution periods, as well as their positions at a given moment of time serving as the starting point of the time record are assumed to be known.

SOLUTION. The motion of a planet is conventionally called progressive when it travels among the fixed stars of the celestial sphere like the sun, i.e., from west to east, and retrograde when it travels in the opposite direction, i.e., from east to west. The transition from one motion to the other occurs when the planet appears to be stationary for a brief period among the fixed stars, in other words, when the sight-line "earth–planet" retains the same direction for a short period of time.

The earth and the planet have the orbital radii *r* and *R*, respectively, and the revolution periods *u* and *U*, and the orbital radii, which are rotating about the sun, accordingly have the rates of revolution $k = 2\pi/u$ and $K = 2\pi/U$.

The solution to the problem is most conveniently obtained by the vector method. Let *O*, *p*, *P* be the midpoints of the sun, the earth, and the planet, $\mathbf{r} = \overrightarrow{Op}$ and $\Re = \overrightarrow{OP}$ the vectorial distances of the earth and the planet from the sun. The vectors $\mathfrak{r}$ and $\mathfrak{R}$ are "rotational vectors," i.e., vectors with the constant lengths r and R, that rotate in the ecliptic plane E with constant velocities k and K, respectively, about their fixed point of origin O. For the vectors $\dot{\mathfrak{r}}$ and $\dot{\mathfrak{R}}$ of the orbital velocities we again select O as the starting point. The magnitudes of the velocities $\dot{\mathfrak{r}}$ and $\dot{\mathfrak{R}}$ are kr and KR, the directions always perpendicular to the directions of $\mathfrak{r}$ and $\mathfrak{R}$. If we then imagine two vectors $\mathfrak{r}_0$ and $\mathfrak{R}_0$ situated in E, originating at O, and possessing the magnitudes r and R that are always $90°$ in advance of the rotational vectors $\mathfrak{r}$ and $\mathfrak{R}$, then

$$\dot{\mathfrak{r}} = k\mathfrak{r}_0 \quad \text{and} \quad \dot{\mathfrak{R}} = K\mathfrak{R}_0.$$

The vectorial distance of the planet from the earth is $\mathfrak{s} = \overrightarrow{pP} = \overrightarrow{OP} - \overrightarrow{Op} = \mathfrak{R} - \mathfrak{r}$, the relative velocity of the planet with respect to the earth (i.e., the velocity of the planet for an observer on the earth, for whom the earth is at rest) is thus

$$\dot{\mathfrak{s}} = \dot{\mathfrak{R}} - \dot{\mathfrak{r}} = K\mathfrak{R}_0 - k\mathfrak{r}_0.$$

Let the angle by which the vector $\mathfrak{R}$ is in advance of the vector $\mathfrak{r}$ at time 0 be α and at time t let it be ζ. Then

(1) $$\zeta = \alpha + \kappa t,$$

where $\kappa = K - k$ represents the angle by which the vector $\mathfrak{R}$ rotates in advance of the vector $\mathfrak{r}$ in the unit time.

The motion of the planets is then progressive when the vector $\dot{\mathfrak{s}}$ rotates in a counterclockwise direction for an observer at the North Pole and retrograde when it rotates in a clockwise direction for this observer, i.e., in accordance with whether the apex S of the vector $\overrightarrow{OS} = \mathfrak{s} \times \dot{\mathfrak{s}}$ that is perpendicular to E lies above or below the ecliptic plane. Now,

$$\mathfrak{s} \times \dot{\mathfrak{s}} = (\mathfrak{R} - \mathfrak{r}) \times (\dot{\mathfrak{R}} - \dot{\mathfrak{r}}) = (\mathfrak{R} - \mathfrak{r}) \times (K\mathfrak{R}_0 - k\mathfrak{r}_0) = \mathfrak{p} - \mathfrak{q}$$

with

$$\mathfrak{p} = k\mathfrak{R} \times \mathfrak{R}_0 + k\mathfrak{r} \times \mathfrak{r}_0, \qquad \mathfrak{q} = K\mathfrak{r} \times \mathfrak{R}_0 + k\mathfrak{R} \times \mathfrak{r}_0,$$

it being assumed that the vectors $\mathfrak{p}$ and $\mathfrak{q}$ also have their starting point at O. The vector $\mathfrak{p}$ has the magnitude $KR^2 + kr^2$ and lies above E. The vector $\mathfrak{q}$, as may be seen from Figure 102, lies above or below E

accordingly as $\cos \zeta$ is positive or negative, and has the magnitude $(K + k)Rr|\cos \zeta|$. The vector $\mathring{\mathfrak{s}} \times \dot{\mathfrak{s}}$ thus lies above or below E

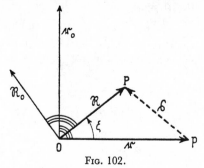

Fig. 102.

accordingly as $KR^2 + kr^2 - (K + k)Rr \cos \zeta$ is positive or negative, i.e., accordingly as

$$\cos \zeta \lessgtr \frac{KR^2 + kr^2}{(K + k)Rr}.$$

Now, according to Kepler's third law,

$$U^2:u^2 = R^3:r^3 \quad \text{or} \quad k^2:K^2 = R^3:r^3,$$

so that the ratio $k:K$ on the right side of the obtained inequality can be replaced by $W^3:w^3$, where $W = \sqrt{R}$, $w = \sqrt{r}$. We thus obtain for this right side the value

$$\frac{w^3 W^4 + W^3 w^4}{(W^3 + w^3)W^2 w^2} = \frac{(W + w)Ww}{W^3 + w^3} = \frac{Ww}{W^2 + w^2 - Ww}$$

$$= \frac{\sqrt{Rr}}{R + r - \sqrt{Rr}},$$

and our conclusion reads:

The motion of a planet is progressive or retrograde accordingly as

$$\cos \zeta \lessgtr \frac{\sqrt{Rr}}{R + r - \sqrt{Rr}}.$$

At the moments when

(2) $$\cos \zeta = \frac{\sqrt{Rr}}{R + r - \sqrt{Rr}},$$

the one type of motion changes into the other.

EXAMPLE. *How many days after upper conjunction does Venus become retrograde?*

Here $r = 149$, $R = 107.5$ million kilometers, k and K, respectively, in degrees are $0.9856°$ and $1.602°$, κ thus equals $0.6164°$ per day, with $\alpha = 180°$ and $\sqrt{Rr}/(R + r - \sqrt{Rr}) = 0.974$. From (1) and (2) we therefore obtain $\cos 0.6164t = -0.974$ and from this $t = 271$ days.

88   Lambert's Comet Problem

To express the time required for a comet to describe an arc of its parabolic orbit by means of the focal radii and the chord connecting the end points of the arc.

Johann Heinrich Lambert (1728–1777) in 1761 published a paper on comet orbits in which may be found the celebrated formula bearing his name; the formula represents the area of a parabolic focal sector as a function of the bounding focal radii and the sector chord.

For the derivation of the Lambert formula we require a formula of the English astronomer Barker, which we will derive first.

We begin with the amplitude equation of a parabola, $y^2 = 4kx$, in which k represents the shortest focal radius, which is commonly known to be one fourth of the parabola parameter.

Let us consider the sector FOP, which is enclosed by the minimum focal radius FO, the focal radius $FP = r$ of an arbitrary point $P(x|y)$, and the parabola arc OP, and in which the angle $OFP = W$ represents the so-called true anomaly of the point P.

Barker's problem is stated thus: *Represent the area of the parabola sector as a function of the anomaly.*

In order to solve the problem we first express the sector area S in terms of x and y. If we drop the perpendicular PQ from P to the axis, S is the difference between the area of the half sector OPQ (cf. No. 56) and the area of the triangle FPQ, so that

$$S = \tfrac{2}{3}xy - \tfrac{1}{2}(x - k)y \quad \text{or} \quad 6S = y(x + 3k).$$

We then express x and y in terms of W. According to the polar coordinate theorem of the parabola, the focal radius is

$$r = \frac{p}{1 + \cos W} = \frac{k}{\cos^2 \dfrac{W}{2}},$$

and consequently,

$$y = r \sin W = 2r \sin \frac{W}{2} \cos \frac{W}{2} = 2k \tan \frac{W}{2}$$

and

$$x = y^2/4k = k \tan^2 \frac{W}{2}.$$

If we introduce Barker's auxiliary magnitude

$$T = \tan \frac{W}{2},$$

we obtain

$$x = kT^2, \qquad y = 2kT$$

(the equation of the parabola in a parametric form), and after substitution of these values into the above area formula, we obtain

$$S = k^2(T + \tfrac{1}{3}T^3).$$

This is *Barker's formula.*

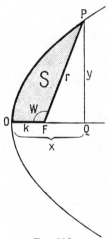

Fig. 103.

W is positive or negative accordingly as P lies above or below the
axis. In the first case, T and S are positive; in the second, negative.

Now for the solution of Lambert's problem!

Let P and P' be two points of the parabola, W and W' their
anomalies, T and T' the corresponding Barker auxiliary magnitudes,
S and S' the areas of the sectors FOP and FOP', with $FP = r$ and
$FP' = r'$ as the focal radii of the two points, $\angle PFP' = 2\zeta$ the angle

between them, $PP' = s$ the connecting chord, and σ the area of the sector PFP' enclosed by the two focal radii.   Let r lie above the axis and r' above or below it; in the first case, let $r' < r$, and thus in both cases $W' < W$.

The area σ is then in both cases the difference $S - S'$.

Now, according to Barker,

$$3S = k^2(3T + T^3), \qquad 3S' = k^2(3T' + T'^3),$$

and consequently,

$$3\sigma = k^2(T - T')[3 + T^2 + T'^2 + TT'].$$

Using the abbreviations J, O, J', O' for

$$\sin \frac{W}{2}, \quad \cos \frac{W}{2}, \quad \sin \frac{W'}{2}, \quad \cos \frac{W'}{2}$$

and i, o for $\sin \zeta$, $\cos \zeta$, we can write the factor in parentheses as

$$(T - T') = \frac{J}{O} - \frac{J'}{O'} = \frac{JO' - OJ'}{OO'} = \frac{i}{OO'},$$

and the factor in square brackets as

$$
\begin{aligned}
[\ ] &= 1 + T^2 + 1 + T'^2 + 1 + TT' \\
&= 1 + \frac{J^2}{O^2} + 1 + \frac{J'^2}{O'^2} + 1 + \frac{JJ'}{OO'} \\
&= \frac{O^2 + J^2}{O^2} + \frac{O'^2 + J'^2}{O'^2} + \frac{OO' + JJ'}{OO'} \\
&= \frac{1}{O^2} + \frac{1}{O'^2} + \frac{o}{OO'}.
\end{aligned}
$$

If we introduce these values and, in accordance with the polar equation, express k/O^2 and k/O'^2 as r and r', respectively, we obtain

$$3\sigma = i(r + r' + o\sqrt{rr'})\sqrt{rr'}.$$

Now,

$$
\begin{aligned}
i^2 &= (JO' - OJ')^2 = J^2O'^2 + O^2J'^2 - 2JOJ'O' \\
&= (1 - O^2)O'^2 + (1 - O'^2)O^2 - 2JOJ'O' \\
&= O^2 + O'^2 - 2OO'(OO' + JJ') = O^2 + O'^2 - 2oOO',
\end{aligned}
$$

and, since $k = rO^2 = r'O'^2$,

$$i = \sqrt{k(r + r' - 2o\sqrt{rr'})}\big/\sqrt{rr'}.$$

If we introduce this value into the equation found for 3σ, we obtain

$$3\sigma = (r + r' + o\sqrt{rr'})\sqrt{k(r + r' - 2o\sqrt{rr'})}.$$

We transform this equation further by introducing the chord s. Its square, according to the cosine theorem, is

$$s^2 = r^2 + r'^2 - 2rr' \cos 2\zeta = r^2 + r'^2 - 2rr'(2o^2 - 1),$$

i.e.,

$$s^2 = (r + r')^2 - 4rr'o^2.$$

From this we obtain

$$4rr'o^2 = (r + r' + s)(r + r' - s).$$

We abbreviate and write

$$v = \sqrt{r + r' + s}, \qquad u = \sqrt{r + r' - s},$$

obtaining

$$r + r' = \frac{v^2 + u^2}{2}, \qquad o\sqrt{rr'} = \pm\frac{uv}{2},$$

where the upper sign applies when the enclosed angle 2ζ is concave and the lower when it is convex.

If we substitute these two values into our last formula for 3σ, it finally yields

$$3\sigma = \sqrt{k}\,\frac{v^2 + u^2 \pm vu}{2}\cdot\frac{v \mp u}{\sqrt{2}} = \sqrt{\frac{k}{8}}\cdot(v^3 \mp u^3)$$

or, in complete form,

$$\sigma = \sqrt{\frac{k}{72}}\left[(r + r' + s)^{1.5} \mp (r + r' - s)^{1.5}\right].$$

This formula represents the parabola sector σ as a function of the two bounding focal radii r *and* r' *and the chord* s *connecting their end points.*

In order to use this formula to determine the time required for a comet to complete its orbital arc, we need only introduce the value found for σ into the Gauss formula of the *Theoria motus,*

$$\frac{2\sigma}{t\sqrt{p}\sqrt{1 + \mu}} = G$$

(cf. No. 96).

Since here $p = 2k$ and the comet mass μ is to be set equal to zero, we have initially

$$Gt\sqrt{k} = \sigma\sqrt{2}$$

and, as a result of substitution,

$$6Gt = (r + r' + s)^{1.5} \mp (r + r' - s)^{1.5}.$$

This remarkable formula contains the solution to the problem posed. It is usually called the *Lambert formula*, although it had already been formulated by Euler.

It states that *the time required by a comet to describe an orbital arc depends only on the arc chord and the sum of the focal radii of the ends of the arc.*

According to Lagrange, Lambert's formula represents the most beautiful and significant discovery in the theory of comet motion. It is, in fact, of fundamental importance for the determination of comet orbits.

This determination is carried out essentially in the following way:

The longitude and latitude of the comet is determined for *three* different moments of time, together with the corresponding longitude and distance of the sun (from the earth). Let r and r' be the respective focal radii of the first and third time of measurement, s the distance between the ends of the focal radii. r' and s are expressed in terms of the known magnitudes and r, and these values are substituted into the Lambert equation, which results in an equation with only one unknown, r. From this equation r is obtained, and then r' and s are found from the previously mentioned expressions. This then gives us the focus and two points of the orbit, so that it is completely determined. When the Gauss formula is applied to one of the points, we obtain the time at which the comet passes the perihelion. After this has been determined, the position of the comet for any moment of time can be obtained from the Gauss formula.

Extremes

At what value of x, *if* x *is a positive variable, will the expression* $\sqrt[x]{x}$ *be at a maximum?*

Jacob Steiner posed this problem in *Crelle's Journal*, vol. XL; it may also be found in his *Works*, vol. 2, p. 423.

SOLUTION. According to the inequality of exponential functions (No. 12),

$$e^{(x-e)/e} \geqq 1 + \frac{x-e}{e},$$

where the equal sign applies only when $x = e$. The inequality is simplified to

$$e^{x/e} \cdot \frac{1}{e} \geqq \frac{x}{e} \quad \text{or to} \quad e^{x/e} \geqq x.$$

Here we extract the xth root and obtain

$$\sqrt[e]{e} \geqq \sqrt[x]{x}.$$

Verbally expressed: *The Euler number* e *is the number yielding the maximum possible value for the expression* $\sqrt[x]{x}$ *for which* x *is a positive variable.*

To inscribe in a given acute-angled triangle the triangle of minimum perimeter.

This celebrated problem stems from I. F. Fagnano, son of the Italian count C. Fagnano (1682–1766), who became famous as a result of his remarkable studies of lemniscate partition.

The following solution of the problem is distinguished by its extreme simplicity. It comes from Fr. Gabriel-Marie, author of the excellent book *Exercices de Géométrie*.

Let the given triangle be *ABC* and let *XYZ* be a triangle inscribed in it, with *X*, *Y*, and *Z* on *BC*, *CA*, and *AB*, respectively. We will initially consider that *Z* is arbitrarily situated on *AB*; we draw its mirror images *H* and *K* on *BC* and *CA*, respectively, and determine the points of intersection *X* and *Y* of the connecting line *HK* with *BC*

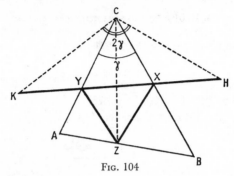

FIG. 104

and *CA*. For a *fixed* point *Z* the triangle *XYZ* thus formed has the smallest perimeter of all the inscribed triangles. In fact: let *X'* and *Y'* be two other points on *BC* and *CA*. Since *ZX'* and *HX'* are mirror images, and also *ZY'* and *KY'*, and naturally also *ZX* and *HX*, as well as *ZY* and *KY*, the perimeters of the two inscribed triangles to be compared can be written as

$$ZXYZ = HX \ + XY \ \ + YK \ = HK,$$
$$ZX'Y'Z = HX' + X'Y' + Y'K = HX'Y'K.$$

However, since the direct path *HK* from *H* to *K* is shorter than the roundabout path *HX'Y'K*, the first triangle possesses a smaller perimeter than the second.

It now merely remains to choose the point *Z* in such manner as to obtain the smallest possible segment *HK* (which represents the perimeter of *XYZ*). Now *CZ* is the mirror image of *CH* and also of *CK*; likewise, $\angle ZCB = \angle HCB$ and $\angle ZCA = \angle KCA$ and thus $\angle HCK = 2\gamma$. Segment *HK* is therefore the base of an isosceles triangle (*HKC*) with a constant apex angle 2γ and the variable leg $s = CZ$; as such it attains a minimum when *CZ* is at a minimum, i.e., when *CZ* is perpendicular to *AB*.

Since we could just as easily have carried out the investigation with *X* or *Y* as with *Z*, *AX* is perpendicular to *BC* and *BY* to *CA*. The points *X*, *Y*, *Z* are thus the base points of the altitudes of the triangle *ABC*.

RESULT: *Of all the triangles that can be inscribed in a given acute-angled triangle, the one with the smallest perimeter is the triangle formed by the base points of the altitudes.*

91   Fermat's Problem for Torricelli

To find the point the sum of whose distances from the vertexes of a given triangle is the smallest possible.

This celebrated problem was put by the French mathematician Fermat (1601–1665) to the Italian physicist Torricelli (1608–1647), the famous student of Galileo, and was solved by the latter in several ways.

The simplest solution is the one obtained by the use of

VIVIANI'S THEOREM: *In an equilateral triangle the sum of the three distances of a point from the sides of a triangle has a value that is independent of the position of the point.*

This value is equal to the altitude of the triangle.

Viviani (1622–1703), an Italian mathematician and physicist, was a student of Galileo and Torricelli.

In Viviani's theorem the distance of a point from a triangle side is reckoned as positive when it is inside the triangle and negative when it is outside.

PROOF.   Let the equilateral triangle have the vertexes P, Q, and R, the side g, the altitude h, and the area J.   If x, y, z are the distances of an arbitrary point O from the sides QR, RP, PQ, then

$$s = x + y + z$$

is the designated sum.

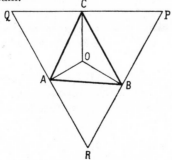

FIG. 105.

Now, the area of the triangle PQR is composed (additively or subtractively) of the three component triangles OQR, ORP, OPQ, so that we obtain the equation

$$\tfrac{1}{2}gx + \tfrac{1}{2}gy + \tfrac{1}{2}gz = J$$

no matter what position the point O may have. From this we obtain directly

$$s = x + y + z = \frac{2J}{g} = h,$$

and thus the auxiliary theorem is proved. Now let ABC be the given triangle. We choose the point O so that the three perpendiculars at A, B, C to AO, BO, CO form an equilateral triangle PQR. Let O' be any other point. Then if $O'A'$, $O'B'$, $O'C'$ are the perpendiculars dropped from O' to QR, RP, PQ, we have

$$A'O' \leqq AO', \qquad B'O' \leqq BO', \qquad C'O' \leqq CO',$$

where, however, the equal sign cannot apply to all three. By addition it follows from this that

(1) $$A'O' + B'O' + C'O' < AO' + BO' + CO'.$$

However, according to the auxiliary theorem as applied to the equilateral triangle PQR,

(2) $$AO + BO + CO \gtreqless A'O' + B'O' + C'O',$$

where the equals sign applies when O' is inside the triangle PQR and the "smaller than" sign when O' is outside. From (2) and (1) we get

$$AO + BO + CO < AO' + BO' + CO',$$

so that $AO + BO + CO$ is the smallest possible sum of the distances.

Since the quadrilaterals $OBPC$, $OCQA$, $OARB$ are circle quadrilaterals, each of the three angles BOC, COA, and AOB is equal to $120°$.

The point we are looking for is accordingly the common point of intersection of the three circle arcs with the chords BC, CA, AB *and the common peripheral angle of* $120°$.

The construction of this point is impossible when one triangle angle, for example, $\sphericalangle ACB = \gamma$ reaches or exceeds $120°$.

In that event C itself is the point O that we are looking for. Specifically, in this case,

$$AC + BC < AU + BU + CU,$$

no matter where the point U may be.

PROOF. We introduce the angles $ACU = \psi$ and $BCU = \varphi$. If U lies in the space enclosed by the angle $ACB = \gamma$, the sum of ψ and φ is equal to γ; if U lies in the space enclosed by the adjacent angle of γ, the difference between these two angles is equal to γ; and, finally, if U lies in the space of the opposite angle from γ, then

$$\psi + \varphi = 360° - \gamma.$$

Let the base points of the perpendiculars dropped from U to AC and BC be F and G. Their distances from C are then

$$x = CU \cos \psi \quad \text{and} \quad y = CU \cos \varphi,$$

with such a distance, e.g., x, being counted as positive when $\cos \psi$ is positive or negative when $\cos \psi$ is negative. In each case then we have

$$AC = AF + x \quad \text{and} \quad BC = BG + y,$$

and accordingly

$$AC + BC = AF + BG + x + y.$$

Now

$$x + y = CU \cos \psi + CU \cos \varphi = CU(\cos \psi + \cos \varphi)$$
$$= 2 \cdot CU \cdot \cos \frac{\psi + \varphi}{2} \cos \frac{\psi - \varphi}{2}.$$

Since, according to the above, one of the two cosines of the right side of this equation has the magnitude $\cos (\gamma/2)$, and this (because $\gamma/2 \geqq 60°$) is smaller than $\frac{1}{2}$, the right side has a maximum magnitude of CU. This yields

$$AC + BC \leqq AF + BG + CU.$$

Since the legs AF and BG of the right triangles AUF and BUG are smaller than the hypotenuses AU and BU, it is certainly true that

$$AC + BC < AU + BU + CU. \qquad \text{Q.E.D.}$$

92 **Tacking Under a Headwind**

How must a sailboat tack with a north wind in order to get north as quickly as possible?

SOLUTION. Let the course of the boat be $O\gamma°N$, and let the sail form the acute angle α with the bearing north and the angle β with the course bearing.

First let us solve the preliminary problem: *Let the maximum speed that a sailboat can make through the wind with the most favorable sail position be C knots; how great a speed can it make when the angle of the sail with the bearing of the wind is α and with the axis of the boat is β?*

Let the pressure exerted upon the sail by the wind when the sail is perpendicular to the wind be P. If the sail forms an angle α differing from 90° with the bearing of the wind, then the wind pressure P' (which works perpendicular to the sail) is smaller. It is reasonable to assume that the wind pressure is now equal to only sin α times P, so that $P' = P \sin \alpha$. This formula, conceived by Lössl, is, however, only approximate.

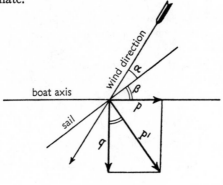

Fig. 106.

We divide P' into two components: one, $p = P' \sin \beta$, in the direction of the boat axis; the other, $q = P' \cos \beta$, perpendicular to it. Of these components p is the only relevant one for the forward motion of the boat. Thus, pressure exercised by the wind on the boat in the course direction has the value

$$p = P \sin \alpha \sin \beta.$$

The velocity c of the boat is proportional to this pressure:

$$c = kp = kP \sin \alpha \sin \beta,$$

where k represents the proportionality constant. For $\alpha = \beta = 90°$ this formula becomes

$$c_{max} = C = kP,$$

so that we can replace kP in the formula by C. The solution to our preliminary problem thus reads

$$c = C \sin \alpha \sin \beta.$$

This formula forms the basis of the solution of the main problem. C is here the velocity that the north wind gives to the boat when it travels due south and the sail is perpendicular to the wind direction. If the boat is to get as far north as possible in a given time, the northerly component c' of the boat's velocity c must be at a maximum. This component is, however,

$$c' = c \sin \gamma = C \cdot \sin \alpha \sin \beta \sin \gamma.$$

Consequently, what is necessary is to choose the three angles α, β, γ, the sum of which is 90°, in such manner as to obtain the maximum *product* for $\sin \alpha \sin \beta \sin \gamma$.

This reduces our task to the following problem:

When is the product of the sines of three angles of a constant concave sum at a maximum?

The solution of this problem is very similar to that of No. 10.

It is based on the theorem: *Of two angle pairs with equal concave sums the pair possessing the higher sine product is the pair with the smaller difference between its angles.*

[It follows from the formulas that $2 \sin X \sin Y = \cos (X - Y) - \cos (X + Y)$, and $2 \sin x \sin y = \cos (x - y) - \cos (x + y)$, where X, Y and x, y represent the two pairs with the common sum

$$X + Y = x + y \quad (\leqq 180°).$$

Since the subtrahends of the right sides are equally great, the larger right side is the one that possesses the greater minuend, i.e., in this case, the one in which the minuend shows the smaller angle difference.]

Let the constant sum of the three variable angles α, β, γ be 3κ ($\leqq 180°$). Now if α, β, γ is such an angle triplet in which none of the angles chances to equal κ, then at least one, let us say α, must necessarily be greater than κ, and another, let us say β, must be smaller than κ. We form a new triplet α', β', γ' such that (1) $\alpha' = \kappa$, (2) the pairs α', β' and α, β possess equal sums, and (3) $\gamma' = \gamma$. According to the above theorem, $\sin \alpha' \sin \beta'$ will then be $> \sin \alpha \sin \beta$, and consequently, $\sin \alpha' \sin \beta' \sin \gamma'$ will also be $> \sin \alpha \sin \beta \sin \gamma$, or

(1) $$\sin \kappa \sin \beta' \sin \gamma' > \sin \alpha \sin \beta \sin \gamma.$$

Since $\beta' + \gamma' = 2\kappa$, the same theorem yields

(2) $$\sin \kappa \sin \kappa \geqq \sin \beta' \sin \gamma'.$$

Combining (1) and (2), we obtain

$$\sin \kappa \sin \kappa \sin \kappa > \sin \alpha \sin \beta \sin \gamma.$$

Consequently:

The product of the sines of three angles of constant concave sum assumes its maximum value when the angles are equal.

The solution to our sailboat problem thus reads $\alpha = \beta = \gamma = 30°$. This means that:

The axis of the boat must form a 60° angle with the bearing north, and the sail must bisect the angle formed by the wind bearing and the boat's axis.

In these optimal positions the northerly motion is equal to exactly $\frac{1}{8}$ the maximum southerly motion.

93   The Honeybee Cell (Problem by Réaumur)

The cell of the honeybee (cf. Figure 107) has the form of a regular hexagonal prism that is sealed at only one end by a regular hexagon *arbpcq*, while at the other end it is sealed by a roof consisting of three congruent rhombuses *PBSC*, *QCSA*, and *RASB* that are inclined toward each other and toward the axis of the prism at equal angles, in such

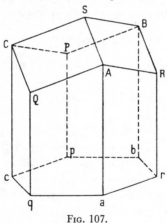

FIG. 107.

manner that the lateral surfaces of the prism are congruent trapezoids (*AarR*, *RrbB*, etc.). The longest side of one such trapezoid is somewhat more than twice as long as the diameter of the inscribed circle of the base surface *arbpcq*. As a result of the regular arrangement of the rhombuses, each of the three rhombus diagonals (*SP*, *SQ*, *SR*) originating at the roof apex S forms the same angle with the axis of the prism as the rhombus plane, and the two planes ABC and PQR are perpendicular to the edges of the prism. Since the obtuse-angled rhombus vertexes abut on each other at S, the diagonals mentioned are the *short* rhombus diagonals.

This singular construction of the honeybee cell suggested to naturalists like Maraldi, Réaumur, and others (at the beginning of the eighteenth century) that the bees had chosen this design in order to save as much as possible in the building material, i.e., in wax. The problem posed by Réaumur in this connection to the Swiss mathematician Koenig can be stated as:

To close a regular hexagonal prism with a roof consisting of three congruent rhombuses in such manner as to obtain a solid of prescribed volume and minimal surface.

SOLUTION. Let the regular hexagonal cross section of the prism have the side $2e$, so that its shorter diagonals $ab = bc = ca = 2d = 2e\sqrt{3}$ and thus also $AB = BC = CA = 2d = 2e\sqrt{3}$. Let the distance of the plane PQR and the apex S of the roof from the plane ABC be x, and let the short rhombus diagonals ($SP = SQ = SR$) be $2y$.

Since the projection from $SR = 2y$ on the axis of the prism is $2x$, and on the plane PQR is $2e$, we obtain the equation

$$(1) \qquad\qquad y^2 = e^2 + x^2.$$

If $\mathfrak{P}$, $\mathfrak{Q}$, $\mathfrak{R}$ are the points at which the prism edges passing through P, Q, R intersect the plane ABC, then $A\mathfrak{R}B\mathfrak{P}C\mathfrak{Q}$ is a regular hexagon with the side $2e$.

First it becomes apparent that the volume of the prism undergoes no change when the rooflike closure that has been described is chosen instead of the plane closure $A\mathfrak{R}B\mathfrak{P}C\mathfrak{Q}$, since as much room is added on the one side of the plane ABC (pyramid $S \cdot ABC$) as is taken away from the other side (the three pyramids $P \cdot BC\mathfrak{P}$, $Q \cdot CA\mathfrak{Q}$, $R \cdot AB\mathfrak{R}$). Only the surface changes with the change in design; the surface decreases by the area $6e^2\sqrt{3}$ of the hexagon $A\mathfrak{R}B\mathfrak{P}C\mathfrak{Q}$, as well as by the area of the six right triangles $P\mathfrak{P}B$, $P\mathfrak{P}C$, $Q\mathfrak{Q}C$, $Q\mathfrak{Q}A$, $R\mathfrak{R}A$, $R\mathfrak{R}B$—together $6ex$—while it increases by the total area of the three rhombuses $PBSC$, $QCSA$, $RASB$, namely $6dy = 6e\sqrt{3}\,y$. The saving in surface area thus obtained is accordingly

$$6e^2\sqrt{3} + 6ex - 6e\sqrt{3}\,y$$

or

$$6e^2\sqrt{3} - 6e[y\sqrt{3} - x],$$

so that it now remains to obtain a minimum value for the expression
in the bracket

$$u = y\sqrt{3} - x$$

by an appropriate choice of x.

Now, if v is understood to be the similarly constructed expression
$x\sqrt{3} - y$, then, as a result of (1),

$$u^2 - v^2 = 2(y^2 - x^2) = 2e^2$$

or

$$u^2 = 2e^2 + v^2.$$

From this it follows that u attains a minimum (specifically $e\sqrt{2}$)
when v is equal to zero, i.e., when

(2) $$y = x\sqrt{3}.$$

From (1) and (2) we obtain

$$x = e\sqrt{\tfrac{1}{2}} \quad \text{and} \quad y = e\sqrt{\tfrac{3}{2}}.$$

The diagonal $SR = 2y = e\sqrt{6}$ is consequently shorter than the
diagonal $AB = 2d = 2e\sqrt{3} = e\sqrt{12}$, so that the three rhombus
angles abutting on one another at S are obtuse.   If we designate the
acute rhombus angle SAR as 2φ, it follows from $\tan \varphi = y/d = 1/\sqrt{2}$
and $\tan 2\varphi = 2 \tan \varphi/(1 - \tan^2 \varphi)$ that $\tan 2\varphi = \sqrt{8}$, $\cos 2\varphi = \tfrac{1}{3}$,
and $2\varphi = 70° 32'$. The obtuse rhombus angle 2Φ is therefore
$109° 28'$.

For the angle μ of the rhombus diagonals SP, SQ, SR with respect
to the axis of the prism we obtain the relation $\tan \mu = 2e/2x = \sqrt{2}$,
and thus $\mu = 90° - \varphi = 54° 44'$.

The angle ν of the rhombuses with respect to the prism cross section
is, finally, $\nu = 90° - \mu = \varphi = 35° 16'$.

Since the tangent of the acute trapezoid angle ($\measuredangle aAR$) has the
value $2e/x = \sqrt{8}$ ($= \tan 2\varphi$), the acute and obtuse angles of the
trapezoid correspond to the acute and obtuse angles, respectively, of
the rhombus.

Particular interest attaches to the angles enclosed between every
two bounding surfaces of the prism. These angles are easily
determined.

To begin with, since the three-sided corners *S*, *P*, *Q*, *R* are congruent and regular (each side is 2Φ), the surface angles belonging to these corners are all equal to each other. Since the four-sided corners *A*, *B*, *C* are also regular and congruent (each side is 2φ), these corners also all have the same surface angle. Now, a *surface angle* of the corner *P* at *p* as $\angle bpc$ equals 120°, and a surface angle of the corner *A* at *a* as $\angle qar$ also equals 120°.

Consequently, *all the surface angles of the prism are* 120° (naturally, with the exception of the right angles forming the base surface).

The angles we have just calculated have in fact been confirmed by actual measurement for the honeybee cell—within the limits of observational error. Of particular interest is the remarkable fact that *every two abutting wax surfaces enclose an angle of* 120°.

94   Regiomontanus' Maximum Problem

At what point of the earth's surface does a perpendicularly suspended rod appear longest? (I.e., at what point is the visual angle at a maximum?)

This problem was posed in 1471 by the mathematician Johannes Müller, called Regiomontanus after his birthplace Königsberg in Franconia, to the Erfurt professor Christian Roder. This problem, which in itself is not difficult, nevertheless deserves special attention as the *first extreme problem* encountered in the history of mathematics since the days of antiquity.

The author of the following simple solution is Ad. Lorsch, who published it in vol. XXIII of the *Zeitschrift für Mathematik und Physik*.

Let *A* be the upper and *B* the lower end point of the rod, *F* the base point of the perpendicular to the earth's surface from *A* (or *B*), so that the segments $FA = a$ and $FB = b$ are known. Since the rod appears to be equally long at *all* the points of a circle on the earth's surface described about *F* as the center, it is sufficient to erect an arbitrary perpendicular g to *FA* at *F* and to seek on this line that runs horizontally on the earth's surface the point *O* at which the visual angle $\omega = \angle AOB$ is a maximum.

First Lorsch shows that the circle of circumscription $\mathfrak{K}$ of the triangle *ABO* is tangent to the line g at *O*. Indeed, if g were not tangent to $\mathfrak{K}$, then $\mathfrak{K}$ would have another point *Q* in common with g besides point *O*, and for each intermediate point *Z* of g between *O* and *Q*, $\angle AZB$ would be greater than the boundary angle of the circle $\mathfrak{K}$ on *AB*, and it would consequently be greater than ω, whereas ω is supposed to be the maximum.

Let us therefore draw the circle $\Re$ that passes through points A and B and is tangent to the line $\mathfrak{g}$; the point of tangency O is the place at which the viewing angle of the rod attains its maximum value ω. Indeed, if P is any point other than O on the line $\mathfrak{g}$, then the angle APB is smaller than the boundary angle of $\Re$ on AB, and consequently smaller than ω. Lorsch also shows the most convenient and quickest method of constructing the circle $\Re$ and/or its midpoint M and radius r. To begin with, the midpoint M lies on the perpendicular bisector of AB, which runs parallel to the line $\mathfrak{g}$ and passes through the midpoint N of AB. Now, in the rectangle $MOFN$ the side FN is equal to the opposite side MO, and is thus equal to r, so that all that is necessary is to mark off from B (or A) the distance FN on the perpendicular bisector in order to obtain, at the resulting point of intersection, the desired midpoint M.

If one wishes to determine the position of O by calculation—using its distance t from F—one need only bear in mind that, according to the tangent theorem, $FO^2 = FA \cdot FB$. This equation immediately gives us $t = \sqrt{ab}$.

An interesting variant of the problem of Regiomontanus is the *Saturn problem*, probably first posed by Hermann Martus, the author of the well-known problem collection:

At what latitude circle of Saturn does the ring appear widest?

Saturn is assumed to be a sphere with a radius of 56,900 km, and the ring is assumed to be a circular ring in the plane of Saturn's equator, having an inner radius of 88,500 km and an outer radius of 138,800 km.

SOLUTION. In Figure 108, let the arc $\mathfrak{M}$ represent a meridian, M the midpoint of Saturn, AB the width of the ring, $MA = a$ being the outer radius, and $MB = b$ the inner radius of the ring, and let $MC = r$ be the equatorial radius of Saturn on MA. Let O be the point situated at the latitude $\varphi = \measuredangle CMO$ at which the ring width appears greatest, so that $\measuredangle AOB = \psi$ is a maximum.

We now apply Lorsch's considerations to our figure and directly obtain the following solution. We draw the circle $\Re$ that passes through the points A and B and is tangent to the meridian $\mathfrak{M}$; the point of tangency O is the place at which the ring width appears to be greatest.

In order to *calculate* the latitude φ of O and the maximum ψ, we examine the right triangles MZF and AZF, in which Z is the center of the circle $\mathfrak{K}$, F the center of AB. From these triangles, with the understanding that ρ is the radius of $\mathfrak{K}$, we obtain

$$\cos \varphi = \frac{MF}{MZ} = \frac{a+b}{2(r+\rho)} \quad \text{and} \quad \sin \psi = \frac{AF}{AZ} = \frac{a-b}{2\rho}.$$

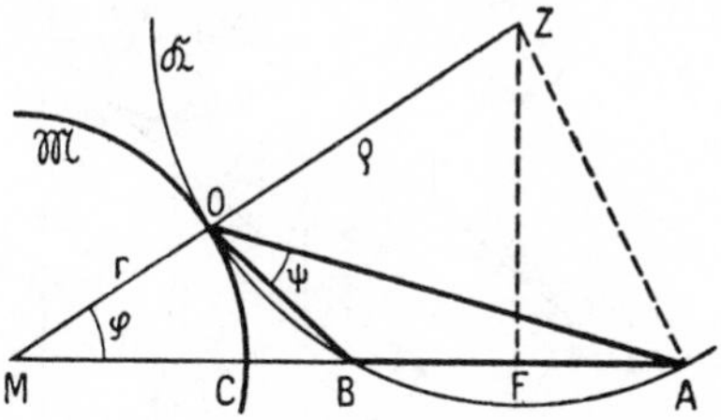

Fig. 108.

The unknown ρ, however, follows from the secant theorem, according to which $MA \cdot MB = MZ^2 - \rho^2$ or $ab = (r+\rho)^2 - \rho^2 = r^2 + 2r\rho$, and consequently $\rho = (ab - r^2)/2r$. If we introduce this into the above, we at length obtain

$$\cos \varphi = \frac{(a+b)r}{ab+r^2} \quad \text{and} \quad \sin \psi = \frac{(a-b)r}{ab-r^2}$$

and from this, $\varphi = 33\frac{1}{2}°$, $\psi = 18\frac{1}{2}°$.

95   The Maximum Brightness of Venus

In what position does the planet Venus appear to have the greatest brilliance?

SOLUTION. Let the midpoints of the sun, earth, and Venus be S, E, V, the radii of the orbits (assumed as circular) of the earth and Venus $SE = a$ and $SV = b$, the variable distance of Venus from the earth $EV = r$, the radius of Venus h. The tangents to Venus from S and E touch Venus along circles I and II, respectively, whose diameters in the plane SEV we will call AB and CD, respectively. Since $AB \perp SV$ and $CD \perp EV$, the angle between the planes of the two circles is equal to the angle $\varphi = SVE$ between their normals VS and VE. The projection of the portion of Venus that is illuminated by the sun and visible from the earth on the plane of circle II consists of the semicircle with the central radius VC and the area $(\pi/2)h^2$ and the projection of the semicircle with the central radius VB, having the area $(\pi/2)h^2 \cos \varphi$. (The area of the projection of a plane surface on a plane is equal to the product of the area of the surface and the cosine of the angle between the two planes.) The radiation from

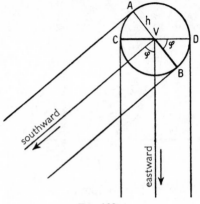

Fig. 109.

Venus to the earth is thus exactly the same as that of a surface at V perpendicular to the rays, with the area

$$J = \tfrac{1}{2}\pi h^2(1 + \cos \varphi).$$

If 1 cm² of this surface at distance 1 develops the illumination intensity c, the entire surface generates the illumination intensity cJ and at the distance $VE = r$ the illumination intensity is

$$\mathfrak{B} = \frac{cJ}{r^2} = \frac{c\pi h^2}{2} \cdot \frac{1 + \cos \varphi}{r^2}.$$

Accordingly, the illumination intensity attains a maximum when the factor

$$f = \frac{1 + \cos \varphi}{r^2}$$

reaches its peak value.

Now, according to the cosine theorem as applied to triangle SEV,

$$\cos \varphi = \frac{r^2 + b^2 - a^2}{2br},$$

and consequently,

$$f = \frac{1}{2br} + \frac{1}{r^2} - \frac{a^2 - b^2}{2br^3}.$$

This expression has the form

$$f = Ax + Bx^2 - Cx^3,$$

where

$$A = \frac{1}{2b}, \qquad B = 1, \qquad C = \frac{a^2 - b^2}{2b}$$

are constants and $x = (1/r)$ is a variable. We must now make the function f of x as great as possible by a suitable choice of x. As the curve of the function shows, f initially grows as $x \; (> 0)$ increases; at a certain point $x = \alpha$ it attains its maximal value, and then declines. For every (positive) $x \neq \alpha$, therefore,

$$Ax + Bx^2 - Cx^3 < A\alpha + B\alpha^2 - C\alpha^3.$$

Accordingly as $x \lessgtr \alpha$, we write this inequality as

$$C(\alpha^3 - x^3) < A(\alpha - x) + B(\alpha^2 - x^2)$$

or

$$C(x^3 - \alpha^3) > A(x - \alpha) + B(x^2 - \alpha^2),$$

and divide both sides by $\alpha - x$ and $x - \alpha$, respectively. From this we find that: The function $C(\alpha^2 + \alpha + x^2)$ lies below the function $A + B(\alpha + x)$ when $x < \alpha$, and above it when $x > \alpha$. Since these two continuous functions increase steadily, they must attain equal values at the point $x = \alpha$, so that

$$C(\alpha^2 + \alpha^2 + \alpha^2) = A + B(\alpha + \alpha).$$

This equation yields

$$\alpha = \frac{B + \sqrt{B^2 + 3CA}}{3C}.$$

If we introduce here the values of A, B, C, we find for the desired distance $r(= 1/\alpha)$ the value

$$r = \sqrt{3a^2 + b^2} - 2b.$$

Now all three sides of the triangle SEV for the optimal position are known $(a:b:r = 1:0.7233:0.4304)$, and the sought-for angular distance $(\angle SEV)$ of Venus from the sun is found to be $39° \, 43.5'$.

96   A Comet Inside the Earth's Orbit

What is the maximum number of days that a comet can remain within the earth's orbit?

We will assume that the earth's orbit is circular and the comet's parabolic, and that the orbital planes coincide.

SOLUTION.   We will select the large half axis of the earth's orbit as the unit length, the mean solar day as the unit time, and we will designate the parabola parameter as $4k$, the base line of the parabola section lying within the earth's orbit as $2y$, the altitude of the section as x, the sector described by the focal radius of the comet within the earth's orbit as S, and finally, the time required to traverse the sector as t.   Then

$$(1) \qquad\qquad y^2 = 4kx$$

according to the amplitude equation of the parabola,

$$(2) \qquad\qquad (x - k)^2 + y^2 = 1$$

according to the circle equation, and

$$(3) \qquad\qquad 3S = y(x + 3k)$$

according to the formula for the area of a parabola section [No. 56. S = the section − triangle = $\frac{4}{3}xy - (x - k)y$].

If $2p$ represents the orbit parameter of a celestial body of mass μ revolving about the sun (the mass of the sun is considered as the unit mass), if t is any time, S the sector described by the body in this time, we can use the *Gauss formula**

$$\frac{2S}{t\sqrt{p}\sqrt{1 + \mu}} = G,$$

where G (the root of the gravitation constant) is the so-called Gauss constant, which has the numerical value of 0.0172021 for the units assumed.

Since the mass of the comet relative to that of the sun is negligible, the Gauss formula is transformed into

$$(4) \qquad\qquad S = Ct\sqrt{k}, \quad \text{with} \quad C = G/\sqrt{2}$$

in our problem.

---

* Gauss, *Theoria motus corporum coelestium in sectionibus conicis solem ambientium* (Hamburg, 1809). (English translation by C. H. Davis reprinted by Dover Publications, 1963.)

From (1) and (2) we find

$$x + k = 1, \qquad y = 2\sqrt{k(1 - k)}$$

and, making use of these values, we obtain from (3)

$$3S = 2\sqrt{k(1 - k)}(1 + 2k).$$

If we introduce here the value for S from (4), it follows that

(5) $\qquad t = c(1 + 2k)\sqrt{1 - k}, \quad$ with $\quad c = \sqrt{8}/3G.$

Since t is to be a maximum, the expression $(1 + 2k)\sqrt{1 - k}$ must be made as great as possible. It therefore remains to select k in such manner that the expression or its square or fourth power, namely,

$$P = (1 + 2k) \cdot (1 + 2k) \cdot (4 - 4k),$$

becomes a maximum. However, since P is a product of factors of constant sum, it attains a maximum (No. 10) when the factors are equally great, thus when

$$1 + 2k = 4 - 4k.$$

This gives us $k = \frac{1}{2}$ and, as a result of (5), $t = 78$.

The sought-for *maximum possible length of stay* is thus 78 days.

97   The Problem of the Shortest Twilight

On what day of the year is the twilight shortest at a place of given latitude?

This problem was posed, but not solved, by the Portuguese Nunes in 1542 in his book *De crepusculis*. Jacob Bernoulli and d'Alembert solved the problem by means of differential calculus, but obtained no simple results. The first elementary solution stems from Stoll (*Zeitschrift für Mathematik und Physik*, vol. XXVIII). The following very simple solution is from Brünnow's *Lehrbuch der sphärischen Astronomie* (Textbook of Spherical Astronomy).

A distinction is made between civil and astronomical twilight. Civil twilight ends when the midpoint of the sun stands $6\frac{1}{2}°$ below the horizon. Approximately at this moment one must turn on one's lights in order to continue working. Astronomical twilight ends when the midpoint of the sun stands $18°$ below the horizon; it is approximately at this time that the astronomer can begin making observations.

It is convenient to choose as the beginning of twilight the moment at which the midpoint of the sun is intersected by the horizon.

Let the latitude of the observation point be φ, the pole distance of the sun p.

The duration of the twilight is measured by the angle d that is formed by the two-hour circle arcs of the nautical triangles determined by the sun for the beginning and end of the twilight. If we superimpose one of these triangles on the other in such manner that the two pole distances coincide, the angle between the two latitude complements b (now having in common only the world pole P) represents

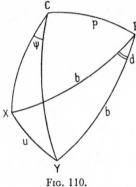

Fig. 110.

the duration d of the twilight. In this position let the triangles be PCX and PCY, with $PC = p$, $PX = PY = b = 90° - \varphi$, $CX = 90°$, $CY = 90° + h$ (h is to be understood as representing the depth of the sun below the horizon at the end of the twilight), and $\angle XPY = d$. Moreover, let $XY = u$ and $\angle XCY = \psi$.

From the isosceles triangle PXY it follows, according to the cosine theorem, that

$$(1) \qquad\qquad \cos d = \frac{\cos u - \sin^2 \varphi}{\cos^2 \varphi}.$$

Consequently, d becomes a minimum or $\cos d$ a maximum when $\cos u$ is at a maximum.

From the triangle CXY it follows, however, that

$$\cos u = \cos CX \cos CY + \sin CX \sin CY \cos \psi$$

or, since $\cos CX = 0$, $\sin CX = 1$, $\sin CY = \cos h$, that

$$\cos u = \cos h \cos \psi.$$

Thus, $\cos u$ attains its greatest possible value when $\cos \psi$ is a maximum, i.e., when

$$\psi = 0.$$

On the day of the shortest twilight, point X accordingly falls on the side CY, and the base $XY = u$ of the isosceles triangle PXY is h. At the same time we find from (1) for the minimum duration $\mathfrak{d}$ of the twilight

$$\cos \mathfrak{d} = \frac{\cos h - \sin^2 \varphi}{\cos^2 \varphi}$$

or, in accordance with the two formulas

$$\cos \mathfrak{d} = 1 - 2 \sin^2 \frac{\mathfrak{d}}{2}, \qquad \cos h = 1 - 2 \sin^2 \frac{2}{h},$$

(I) $$\sin \frac{\mathfrak{d}}{2} = \frac{\sin \dfrac{h}{2}}{\cos \varphi}.$$

To find the *corresponding* declination of the sun δ, we express the cosine of the angle $\omega = \angle PCX = \angle PCY$ twice in accordance with the cosine theorem and set the resulting values equal to each other.

It follows from $\triangle PCX$ (since $\cos CX = 0$, $\sin CX = 1$) that

$$\cos \omega = \frac{\sin \varphi}{\sin P},$$

from $\triangle PCY$ (since $\cos CY = -\sin h$, $\sin CY = \cos h$) that

$$\cos \omega = \frac{\sin \varphi + \cos p \sin h}{\sin p \cos h}.$$

Equalizing, we obtain

$$\sin \varphi \cos h = \sin \varphi + \cos p \sin h$$

or

$$-\cos p \sin h = \sin \varphi (1 - \cos h)$$

or

$$-\cos p \cdot 2 \sin \frac{h}{2} \cos \frac{h}{2} = \sin \varphi \cdot 2 \sin^2 \frac{h}{2}$$

or, finally,

$$\cos p = -\sin \varphi \tan \frac{h}{2}.$$

Because of the minus sign, the pole distance p is an obtuse angle for northern latitudes, the sun's declination δ is thus *southerly* and

$$\text{(II)} \qquad\qquad \sin \delta = \sin \varphi \tan \frac{h}{2}.$$

The shortest twilight duration is determined by (I) and the southerly declination of the sun for the day on which that twilight occurs is given by (II).

From the declination the sought-for day can be found by means of the nautical almanac.

This datum is also found with sufficient accuracy if the familiar formula

$$\text{(2)} \qquad\qquad \sin \delta = \sin \varepsilon \sin l$$

is used; here δ represents the sun's declination, l the angular distance of the sun from the autumnal or vernal equinox, and ε the inclination of the ecliptic ($23° 27'$). Since the above-mentioned angular distance changes at an average daily rate of $m = 59.1'$, the sought-for information varies by $n = l/m$ days from the 23rd of September or from the 21st of March.

For Leipzig, for example, ($\varphi = 51° 20.1'$) we find, from (II), $\delta = 7° 6.2'$, then from (2), $l = 18° 6.3'$, and then $n = 18.4$. The shortest twilight in Leipzig thus falls on October 11 and March 3.

98   Steiner's Ellipse Problem

Of all the ellipses that can be circumscribed about (inscribed in) a given triangle, which one has the smallest (largest) area?

"Dans le plan, la question des polygones d'aire maximum ou minimum inscrits ou circonscrits à une ellipse ne présente aucune difficulté. Il suffit de projeter l'ellipse de telle manière qu'elle devienne un cercle, et l'on est ramené à une question bien connue de géométrie élémentaire"* (Darboux, *Principes de Géométrie analytique*, p. 287).

---

* Translation: "In a plane the question of polygons of maximum or minimum area inscribed in or circumscribed about an ellipse offers no difficulty. All that is necessary is to project the ellipse in such manner that it is transformed into a circle, and the problem is reduced to a well-known question of elementary geometry".

The solution of the problem is based on the two auxiliary theorems:

I. *Of all the triangles inscribed in a circle the one possessing the maximum area is the equilateral.*

II. *Of all the triangles that can be circumscribed about a circle the one possessing the minimum area is the equilateral.*

PROOF OF I. We call the circle diameter d, the sides and angles of an inscribed triangle p, q, r and α, β, γ, respectively, the area of the triangle J. Then

$$J = \tfrac{1}{2} pq \sin \gamma$$

and

$$p = d \sin \alpha, \qquad q = d \sin \beta,$$

and consequently,

$$J = \tfrac{1}{2} d^2 \cdot \sin \alpha \sin \beta \sin \gamma.$$

According to No. 92, the product of the sines $\sin \alpha \sin \beta \sin \gamma$ of the three angles α, β, γ of constant sum (180°) is at a maximum when

$$\alpha = \beta = \gamma (= 60°),$$

i.e., when the triangle is equilateral. The area of this maximal triangle is $\tfrac{3}{16}\sqrt{3}d^2$, thus $\sqrt{27}/4\pi$ of the area of the circle.

PROOF OF II. If we designate the sides of an arbitrary circumscribed triangle PQR as p, q, r, then the tangents to the circle from the vertexes P, Q, R are $x = s - p$, $y = s - q$, $z = s - r$, where s represents half the perimeter of the triangle

$$\left(s = \frac{p + q + r}{2} = x + y + z \right).$$

The area J of the triangle and the radius ρ of the inscribed circle are given by the well-known formulas

$$J = \rho s \quad \text{and} \quad J = \sqrt{xyzs} \quad \text{(Hero of Alexandria)}.$$

These give us

$$s\rho^2 = xyz.$$

Making use of the formula $J = \rho s$, we write this equation in the following two ways:

$$\text{(1)} \qquad \frac{1}{yz} + \frac{1}{zx} + \frac{1}{xy} = \frac{1}{\rho^2},$$

$$\text{(2)} \qquad \frac{1}{yz} \cdot \frac{1}{zx} \cdot \frac{1}{xy} = \frac{1}{J^2 \rho^2}.$$

We now introduce the new unknowns

$$u = \frac{1}{yz}, \qquad v = \frac{1}{zx}, \qquad w = \frac{1}{xy}$$

and obtain

$$u + v + w = \frac{1}{\rho^2}, \qquad uvw = \frac{1}{J^2 \rho^2}.$$

Since J is supposed to be a minimum and ρ is constant, uvw must attain a maximum.

A product uvw of numbers u, v, w of constant sum ($u + v + w = $ const.) reaches a maximum, however (No. 10), when the numbers are equal to each other: $u = v = w$. The circumscribed triangle therefore becomes smallest when $yz = zx = xy$, i.e., when $x = y = z$, i.e., when $p = q = r$, which proves II.

We find that the area of the smallest circumscribed triangle is four times that of the maximum inscribed triangle, i.e., $\sqrt{27}\,\rho^2$, and for the ratio of this area to the area of the circle we obtain the improper fraction $\sqrt{27}/\pi$.

Now for the *solution of the ellipse problem*! Let $\mathfrak{E}$ be any ellipse circumscribed about (inscribed in) the given triangle abc, f its surface area, δ the area of the triangle abc. We consider $\mathfrak{E}$ as the normal projection of a circle $\mathfrak{K}$, whose surface area we will call F. In the projection the inscribed (circumscribed) triangle ABC of the circle, possessing an area we will call Δ, corresponds to the inscribed (circumscribed) triangle abc of the ellipse. If μ represents the cosine of the angle between the plane of the circle and the plane of the ellipse, then the normal projection of every surface lying in the plane of the circle is the μ-multiple of the surface. This gives us the formulas

$$f = \mu F, \qquad \delta = \mu \Delta.$$

Since δ is constant, f attains a minimum (maximum) when the quotient f/δ or the equal quotient F/Δ reaches a minimum (maximum). The latter quotient, however, according to auxiliary theorem I. (II.) reaches its minimal (maximal) value $4\pi/\sqrt{27}$ ($\pi/\sqrt{27}$) when the triangle ABC is equilateral.

To establish more exactly the ellipse determined by this condition, we make use of the properties of a normal projection: 1. *Parallelism is not annulled by projection.* 2. *The ratio between parallel segments is maintained in projection:* in particular, the ratio of two segments of the same line is not altered.

Now, the center M of the circle is the point of intersection of the medians of the equilateral triangle ABC and the diameter through C bisects the chords of the circle parallel to AB. Consequently, the point of intersection of the medians of the triangle abc is the center point m of the sought-for ellipse, and the ellipse diameter through c bisects the ellipse chords parallel to the side ab, so that ab and mc are *conjugate* directions of the ellipse. Now, since the circle radius MK parallel to the circle chord (tangent) AB is equal to $1/\sqrt{3}(\sqrt{3}/6)$ of AB, the ellipse half diameter mk parallel to the ellipse chord (tangent) ab is also equal to $1/\sqrt{3}(\sqrt{3}/6)$ of ab.

RESULT. Of all the ellipses that can be circumscribed about (inscribed in) a given triangle abc, the one with the smallest (greatest) area is the ellipse whose midpoint m is the point of intersection of the medians of the triangle abc and from which the ellipse half diameter to c (to the center of ab) and the ellipse half diameter parallel to ab, $mk = ab/\sqrt{3}(ab/2\sqrt{3})$, are conjugate half diameters. The area of the ellipse thus characterized—the so-called Steiner ellipse—is

$$\frac{4\pi}{\sqrt{27}} \left(\frac{\pi}{\sqrt{27}} \right) \quad \text{of the area of the triangle.}$$

This ellipse can be constructed easily in accordance with No. 42.

99   Steiner's Circle Problem

Of all isoperimetric plane surfaces (i.e., those having equal perimeters) the circle has the greatest area.

And conversely:

Of all plane surfaces with equal area the circle has the smallest perimeter.

This fundamental double theorem was first proved by J. Steiner (*Crelle's Journal*, vol. XVIII; also in Steiner's *Gesammelte Werke*, vol. II). Steiner even provided several proofs. Here we will consider only the one that is based upon the Steiner symmetrization principle.

First we will prove the second half of the theorem.

It is obviously sufficient to limit our considerations to convex surfaces, i.e., those surfaces in which the line segment connecting two arbitrary points of the surface belongs completely to the surface.

We will first prove the auxiliary theorem:

Of all trapezoids with common base lines and altitudes the isosceles trapezoid is the one the sum of whose legs is smallest.

Let $ABCD$ be an arbitrary trapezoid with the base lines BC and AD, the legs AB and CD. Let the mirror image of B on the perpendicular bisector of AD be B', let the center of CB' be C_0. On the extension of

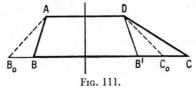

Fig. 111.

CB we set $BB_0 = CC_0$ and obtain the isosceles trapezoid AB_0C_0D, which has base lines and altitude in common with the given trapezoid, and consequently also the same area.

If we extend DC_0 by its own length to H, we obtain the parallelogram $DCHB'$, in which the diagonal DH is shorter than the sum of the sides DC and CH:

$$DH < DC + CH.$$

However, since $DH = 2 \cdot DC_0 = DC_0 + AB_0$ and $CH = DB' = AB$, we obtain

$$AB_0 + DC_0 < AB + DC.$$

Thus, the isosceles trapezoid has the smallest leg sum.

Now let $\mathfrak{F}$ be the surface having the smallest perimeter for the given area J; let the perimeter be u.

We draw an arbitrary line $\mathfrak{g}$ and divide $\mathfrak{F}$ by perpendiculars to $\mathfrak{g}$ into trapezoids $ABCD$ that we select so narrow that the arc-shaped legs AB and CD can be considered as rectilinear. From the points of intersection of the dividing lines $\ldots AD, BC, \ldots$ with $\mathfrak{g}$ we mark off on the dividing lines on both sides of $\mathfrak{g}$ the half chords $\ldots AD, BC, \ldots$, as a result of which we obtain the points $\ldots A', D', B', C', \ldots$ and the trapezoids $\ldots, A'B'C'D', \ldots$. The new trapezoid $A'B'C'D'$ is isosceles and possesses equal base lines and altitude with $ABCD$, so that the area is also the same. This gives us

(1) $A'B' + C'D' \leqq AB + CD,$

in which the equals sign applies only when $ABCD$ is also isosceles.

Our method enables us to obtain from $\mathfrak{F}$ a new surface $\mathfrak{F}'$ with the symmetry axis $\mathfrak{g}$, having the same area as $\mathfrak{F}$ and a perimeter, therefore, that cannot be smaller than u. Thus, the equals sign in (1) must always apply. All trapezoids *ABCD* are therefore isosceles, and the perpendicular bisector of *BC* is an axis of symmetry of $\mathfrak{F}$.

The surface $\mathfrak{F}$ of minimal perimeter therefore possesses an axis of symmetry in every direction.

But such a surface must be a circle!

PROOF. Let I and II be two mutually perpendicular symmetry axes of $\mathfrak{F}$, M their point of intersection. Let the mirror image of an *arbitrary* point P of $\mathfrak{F}$ on I be P_1, and let the mirror image of P_1 on II be $P' \equiv P_{12}$. Then PMP' is a straight line and

$$MP' = MP,$$

i.e., the point M is a midpoint of the surface.

Now $\mathfrak{F}$ can only have one midpoint. Indeed, if N were a second midpoint, then extending PM by its own length, we would first arrive at P'; next, extending $P'N$ by its own length, we would arrive at a new point P'' of $\mathfrak{F}$; then extending $P''M$ by its own length, we would arrive at a point P''' of $\mathfrak{F}$; extending $P'''N$ by itself, we would then come to still another point of $\mathfrak{F}$, etc. If these operations are represented graphically it will be observed that in this manner we would end up at some arbitrary distance beyond the drawing paper (on which $\mathfrak{F}$ lies), which is naturally absurd. Thus, $\mathfrak{F}$ has only the one midpoint M.

It follows from this, further, that: This M must belong to *each* axis of symmetry of $\mathfrak{F}$.

Indeed, if M does not lie on the axis of symmetry $\mathfrak{a}$ of $\mathfrak{F}$, then we can draw the mirror images m and p of M and of an arbitrary surface point P on $\mathfrak{a}$, extend pM by its own length to the surface point p', and draw the mirror image p'' of p' on $\mathfrak{a}$. Now, since p'' is a point of $\mathfrak{F}$, Pmp'' is a straight line, and $mp'' = mP$, this would mean that $\mathfrak{F}$ had a *second* midpoint, m, and this is impossible.

Thus, all the axes of symmetry intersect at M.

Now let F be a fixed *boundary* point of $\mathfrak{F}$ and P an arbitrary boundary point of $\mathfrak{F}$. Since the perpendicular bisector of FP is an axis of symmetry of $\mathfrak{F}$, it passes through M. Therefore,

$$MP = MF;$$

i.e., *all* the boundary points of $\mathfrak{F}$ are equidistant from M, and the surface $\mathfrak{F}$ is a circle.

Consequently, of all surfaces of equal area the circle has the smallest perimeter.

We now state conversely:

Of all isoperimetric surfaces the circle has the greatest area.

PROOF. Let the perimeter f of an arbitrary surface $\mathfrak{F}$, which is not a circle, be equal to the perimeter k of the circular surface $\mathfrak{K}$. Let the area of $\mathfrak{F}$ be F and that of $\mathfrak{K}$ be K.

Now, if $F \geqq K$, we will consider the circular surface $\mathfrak{K}'$, concentric to $\mathfrak{K}$, of area $K' = F$, and we will let its perimeter be k'. Since $\mathfrak{K}'$ covers $\mathfrak{K}$,

$$(2) \hspace{4cm} k' \geqq k.$$

However—since the surfaces $\mathfrak{K}'$ and $\mathfrak{F}$ have the same area—according to the theorem proved above, $k' < f$ or

$$(3) \hspace{4cm} k' < k.$$

The inequalities (2) and (3) contradict each other, however, and thus the assumption that $F \geqq K$ must be false. Consequently, $F < K$. Q.E.D.

The foregoing Steiner proof of the major isoperimetric theorem for the circle has certain weaknesses. The same is true of the proof of the major isoperimetric theorem for the sphere, presented in the following section.

The reader may learn how these weaknesses can be eliminated and the Steiner proof formulated in a completely rigorous fashion by consulting the excellent book *Kreis und Kugel* (Circle and Sphere) by W. Blaschke. Unfortunately, we cannot go into these interesting investigations because of lack of space.

100   Steiner's Sphere Problem

Of all solids of equal surface the sphere possesses the maximum volume.

Of all solids of equal volume the sphere possesses the smallest surface. (Steiner, *Crelle's Journal*, vol. XVIII; Steiner, *Gesammelte Werke*, vol. II.)

As in No. 99, we will prove the second part of the theorem first.

Naturally, we will consider only convex solids, i.e., those solids in which the line segment connecting two arbitrary points on the solid belongs completely to the solid.

Steiner's proof is based on the principle of symmetrization and the theorem:

Of all triangular prisms whose parallel edges AA', BB', CC' *have the prescribed lengths* h, k, l *and lie on three given lines, the prism with the plane of symmetry normal to the edges possesses the smallest base surface sum* ABC + A'B'C'.

PROOF. We will designate the distances of the edges from one another as a, b, c, so that

$$\mathfrak{A} = \tfrac{1}{2}a(k + l), \qquad \mathfrak{B} = \tfrac{1}{2}b(l + h), \qquad \mathfrak{C} = \tfrac{1}{2}c(h + k)$$

are the areas of the three trapezoidal prism faces. These areas are given magnitudes. We extend CB and $C'B$ to the point of intersection P, and CA and $C'A'$ to the point of intersection Q, and obtain

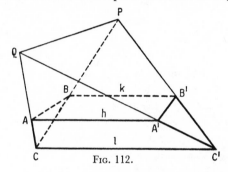

FIG. 112.

the tetrahedron $CC'PQ$ in which for brevity we will call the surfaces $CC'P$ and $CC'Q$ "lateral surfaces" and the surfaces CPQ and $C'PQ$ "top surfaces."

We determine the relations between the areas J, J', $\mathfrak{P}$, $\mathfrak{Q}$ of the tetrahedron bounding surfaces CPQ, $C'PQ$, $CC'P$, $CC'Q$, on the one hand, and the areas Δ, Δ', $\mathfrak{A}$, $\mathfrak{B}$, $\mathfrak{C}$ of the prism bounding surfaces ABC, $A'B'C'$, $BB'C'C$, $CC'A'A$, $AA'B'B$, on the other.

From the ray theorem it follows that

$$(1) \qquad \frac{CP}{CB} = \frac{C'P}{C'B'} = \frac{l}{\lambda} \quad \text{and} \quad \frac{CQ}{CA} = \frac{C'Q}{C'A'} = \frac{l}{\mu},$$

where λ is the difference between l and k, and μ is the difference between l and h. Now, since the areas of similar triangles are in the same proportion to each other as the squares of homologous sides, we obtain the relations

$$\frac{\mathfrak{P}}{\mathfrak{P} - \mathfrak{A}} = \frac{l^2}{k^2} \quad \text{and} \quad \frac{\mathfrak{Q}}{\mathfrak{Q} - \mathfrak{B}} = \frac{l^2}{h^2}.$$

From these we obtain

(2) $$\mathfrak{P} = \alpha\mathfrak{A}, \qquad \mathfrak{Q} = \beta\mathfrak{B},$$

with

$$\alpha = \frac{l^2}{l^2 - k^2} \quad \text{and} \quad \beta = \frac{l^2}{l^2 - h^2}.$$

Moreover, since the areas of two triangles with a common angle are to each other as the products of the adjacent sides of this angle, we obtain

$$\frac{J}{\Delta} = \frac{CP \cdot CQ}{CA \cdot CB} \quad \text{and} \quad \frac{J'}{\Delta'} = \frac{C'P \cdot C'Q}{C'A' \cdot C'B'},$$

and consequently as a result of (1),

(3) $$J = \kappa\Delta \quad \text{and} \quad J' = \kappa\Delta',$$

where κ is the *constant* $l^2/\lambda\mu$.

From (2) it follows that the areas $\mathfrak{P}$ and $\mathfrak{Q}$ of the lateral surfaces of the tetrahedron are constant no matter where the prism edges AA', BB', CC' happen to lie, and from (3), that the sum S of the areas J and J' of the top surfaces of the tetrahedron is κ times the sum Σ of the areas Δ and Δ' of the base surfaces of the prism:

(4) $$S = \kappa\Sigma.$$

We will now prove the auxiliary theorem: *Of all tetrahedrons with two fixed corners* C, C' *and two movable corners* P *and* Q *that lie on the fixed lines* I *and* II *parallel to* CC', *the tetrahedron in which* P *and* Q *lie on the perpendicular bisector plane of* CC' *is the one possessing the smallest area sum* S *of its top surfaces* CPQ *and* C'PQ.

To begin with, it is clear that the tetrahedrons concerned all have the same volume V. (The base surface $CC'P$ has the constant area $\mathfrak{P}$ and the corresponding apex Q lies on a fixed parallel to the plane $CC'P$.)

We draw through the center M of CC' the plane E normal to CC' and designate its points of intersection with the lines I and II as p and q. Let P and Q be two (other) points anywhere on I and II.

We now express the tetrahedron volume V, first using the tetrahedron $CC'pq$ and then the tetrahedron $CC'PQ$.

For this purpose we construct at C and C' on the top surfaces Cpq and $C'pq$ perpendiculars running toward the inside* of these surfaces and designate their point of intersection on E as O.

We will select the common length of the two perpendiculars as our unit length. The perpendiculars from O to the top surfaces CPQ and $C'PQ$ and to the planes $\text{I} \cdot CC'$ and $\text{II} \cdot CC'$ we will designate as x, x', m, n, the common area of the lateral surfaces $CC'p$ and $CC'P$ as $\mathfrak{P}$, that of the lateral surfaces $CC'q$ and $CC'Q$ as $\mathfrak{Q}$, and, finally, the areas of the top surfaces Cpq, $C'pq$, CPQ, $C'PQ$ as i, i', J, J'. We then obtain for the volume V of the tetrahedrons $CC'pq$ and $CC'PQ$ the formulas

$$3V = i + i' + m\mathfrak{P} + n\mathfrak{Q} \quad \text{and} \quad 3V = xJ + x'J' + m\mathfrak{P} + n\mathfrak{Q},$$

respectively [where x, x', m, and n, respectively, are positive or negative accordingly as O lies on the inside or outside of the bounding surfaces CPQ, $C'PQ$, $\text{I} \cdot CC'$, and $\text{II} \cdot CC'$, respectively]. It follows from this that

$$xJ + x'J' = i + i'.$$

If we consider that the *perpendicular* x (x') from O to the plane CPQ $(C'PQ)$ is shorter than the *oblique line* OC (OC'), we see that x and x' are proper fractions. The left side of the last equation is therefore smaller than $J + J'$ and consequently also

$$i + i' < J + J',$$

which proves the auxiliary theorem.

We now go back to (4). Since, according to the auxiliary theorem, S becomes a minimum when P and Q lie on E, and, as a result of (4), Σ and S attain a minimum at the same time, then Σ attains a minimum when the prism bounding surfaces ABC and $A'B'C'$ are symmetrical with respect to E. Q.E.D.

NOTE. The preceding proof assumes that one prism edge (I) differs from the other two. This limitation is of no importance, since it is immediately apparent that the theorem is true in the case $h = k = I$.

The continuation of the proof for the major isoperimetric theorem is similar to that in No. 99.

Let $\mathfrak{K}$ be the solid that for a given volume V has the smallest surface; let the latter be O.

---

* The inside of a bounding surface of a tetrahedron is the side on which the tetrahedron is situated.

We choose an arbitrary plane E and divide $\mathfrak{K}$ by perpendiculars to E into triangular prisms $ABCA'B'C'$, which we assume to be so narrow that the bounding triangles ABC and $A'B'C'$ belonging to the surface of $\mathfrak{K}$ can be considered as plane triangles. From the points of intersection of the perpendiculars $\ldots AA', BB', CC', \ldots$ with E we mark off on the perpendiculars on both sides of E the halves of the segments $\ldots AA', BB', CC', \ldots$, as a result of which we obtain the points $\ldots, a, a', b, b', c, c', \ldots$. The new prism $abca'b'c'$ possesses the symmetry plane E normal to the edges and, according to the above prism theorem, possesses a smaller base surface sum than $ABCA'B'C'$:

(5) $$abc + a'b'c' \leqq ABC + A'B'C',$$

in which the equals sign applies only if the prism $ABCA'B'C'$ also possesses a symmetry plane normal to the edges.

By means of our procedure we obtain from $\mathfrak{K}$ a new solid $\mathfrak{K}'$ with the symmetry plane E, possessing the same volume V as $\mathfrak{K}$ and a surface that consequently cannot be smaller than O. Therefore, the equals sign in (5) must always apply. All prisms $ABCA'B'C'$ therefore possess one plane of symmetry normal to the edges, the perpendicular bisector plane of AA'.

The solid $\mathfrak{K}$ having the smallest surface thus possesses a parallel symmetry plane for every plane.

Such a solid must, however, be a sphere!

PROOF. Let I, II, III be three symmetry planes of $\mathfrak{K}$ that are normal to each other, M their point of intersection. Let the mirror image of an *arbitrary* point P of $\mathfrak{K}$ on I be P_1, let the mirror image of P_1 on II be P_{12}, let that of P_{12} on III be $P_{123} \equiv P'$. Then PMP' is a straight line and

$$MP' = MP,$$

i.e., the point M is a midpoint of $\mathfrak{K}$.

Now, $\mathfrak{K}$ can have only *one* midpoint. (Proof as in No. 99.)

It then follows from this that M must lie on every symmetry plane of $\mathfrak{K}$.

Indeed, if M does not belong to the symmetry plane Δ of $\mathfrak{K}$, then we can draw the mirror images m and p of M and of an arbitrary point P of the solid on Δ, extend pM by its own length to the point p' of the solid, and draw the mirror image p'' of p' on Δ. Now, since p'' is a point of $\mathfrak{K}$, Pmp'' is a straight line, and $mp'' = mP$, this would result in a *second* midpoint, m, for $\mathfrak{K}$, which is impossible.

All the symmetry planes, therefore, intersect at M.

Now let F be a fixed point and P an arbitrary point of the *surface* of $\mathfrak{K}$. Since the perpendicular bisector plane of FP is the symmetry plane of $\mathfrak{K}$, it passes through M. Therefore,

$$MP = MF;$$

i.e., *all* the surface points of $\mathfrak{K}$ are equidistant from M, and the solid $\mathfrak{K}$ is a sphere.

Of all solids of equal volume the sphere thus has the smallest surface.

We now state conversely:

Of all solids of equal surface the sphere has the greatest volume.

PROOF. Let the surface O of an arbitrary solid $\mathfrak{K}$, which is not a sphere, be equal to the surface o of the sphere $\mathfrak{k}$. Let the volume of $\mathfrak{K}$ be V and that of $\mathfrak{k}$ be v.

Let us assume $V \geqq v$; then let us consider the sphere $\mathfrak{k}'$ concentric to $\mathfrak{k}$, having the area $v' = V$ and the surface o'. Since $\mathfrak{k}$ lies on $\mathfrak{k}'$,

$$(6) \qquad\qquad o' \geqq o.$$

However—since the solids $\mathfrak{k}'$ and $\mathfrak{K}$ have the same volume—according to the previously proved theorem, $o' < O$, or

$$(7) \qquad\qquad o' < o.$$

The inequalities (6) and (7) contradict each other. The assumption $V \geqq v$ must therefore be false, and $v > V$, as we asserted.

Index of Names

A CATALOG OF SELECTED
DOVER BOOKS
IN ALL FIELDS OF INTEREST

A CATALOG OF SELECTED DOVER
BOOKS IN ALL FIELDS OF INTEREST

LASERS AND HOLOGRAPHY, Winston E. Kock. Sound introduction to burgeoning field, expanded (1981) for second edition. 84 illustrations. 160pp. 5⅜ × 8¼. (EUK)   24041-X Pa. $3.50

FLORAL STAINED GLASS PATTERN BOOK, Ed Sibbett, Jr. 96 exquisite floral patterns—irises, poppie, lilies, tulips, geometrics, abstracts, etc.—adaptable to innumerable stained glass projects. 64pp. 8¼ × 11.   24259-5 Pa. $3.50

THE HISTORY OF THE LEWIS AND CLARK EXPEDITION, Meriwether Lewis and William Clark. Edited by Eliott Coues. Great classic edition of Lewis and Clark's day-by-day journals. Complete 1893 edition, edited by Eliott Coues from Biddle's authorized 1814 history. 1508pp. 5⅜ × 8½.
21268-8, 21269-6, 21270-X Pa. Three-vol. set $22.50

ORLEY FARM, Anthony Trollope. Three-dimensional tale of great criminal case. Original Millais illustrations illuminate marvelous panorama of Victorian society. Plot was author's favorite. 736pp. 5⅜ × 8½.   24181-5 Pa. $10.95

THE CLAVERINGS, Anthony Trollope. Major novel, chronicling aspects of British Victorian society, personalities. 16 plates by M. Edwards; first reprint of full text. 412pp. 5⅜ × 8½.   23464-9 Pa. $6.00

EINSTEIN'S THEORY OF RELATIVITY, Max Born. Finest semi-technical account; much explanation of ideas and math not readily available elsewhere on this level. 376pp. 5⅜ × 8½.   60769-0 Pa. $5.00

COMPUTABILITY AND UNSOLVABILITY, Martin Davis. Classic graduate-level introduction th theory of computability, usually referred to as theory of recurrent functions. New preface and appendix. 288pp. 5⅜ × 8½. 61471-9 Pa. $6.50

THE GODS OF THE EGYPTIANS, E.A. Wallis Budge. Never excelled for richness, fullness: all gods, goddesses, demons, mythical figures of Ancient Egypt; their legends, rites, incarnations, etc. Over 225 illustrations, plus 6 color plates. 988pp. 6⅛ × 9¼. (EBE)   22055-9, 22056-7 Pa., Two-vol. set $20.00

THE I CHING (THE BOOK OF CHANGES), translated by James Legge. Most penetrating divination manual ever prepared. Indispensable to study of early Oriental civilizations, to modern inquiring reader. 448pp. 5⅜ × 8½.
21062-6 Pa. $6.50

THE CRAFTSMAN'S HANDBOOK, Cennino Cennini. 15th-century handbook, school of Giotto, explains applying gold, silver leaf; gesso; fresco painting, grinding pigments, etc. 142pp. 6⅛ × 9¼.   20054-X Pa. $3.50

AN ATLAS OF ANATOMY FOR ARTISTS, Fritz Schider. Finest text, working book. Full text, plus anatomical illustrations; plates by great artists showing anatomy. 593 illustrations. 192pp. 7⅞ × 10¾.   20241-0 Pa. $6.50

EASY-TO-MAKE STAINED GLASS LIGHTCATCHERS, Ed Sibbett, Jr. 67 designs for most enjoyable ornaments: fruits, birds, teddy bears, trumpet, etc. Full size templates. 64pp. 8¼ × 11.   24081-9 Pa. $3.95

TRIAD OPTICAL ILLUSIONS AND HOW TO DESIGN THEM, Harry Turner. Triad explained in 32 pages of text, with 32 pages of Escher-like patterns on coloring stock. 92 figures. 32 plates. 64pp. 8¼ × 11.   23549-1 Pa. $2.95

SMOCKING: TECHNIQUE, PROJECTS, AND DESIGNS, Dianne Durand. Foremost smocking designer provides complete instructions on how to smock. Over 10 projects, over 100 illustrations. 56pp. 8¼ × 11. 23788-5 Pa. $2.00

AUDUBON'S BIRDS IN COLOR FOR DECOUPAGE, edited by Eleanor H. Rawlings. 24 sheets, 37 most decorative birds, full color, on one side of paper. Instructions, including work under glass. 56pp. 8¼ × 11. 23492-4 Pa. $3.95

THE COMPLETE BOOK OF SILK SCREEN PRINTING PRODUCTION, J.I. Biegeleisen. For commercial user, teacher in advanced classes, serious hobbyist. Most modern techniques, materials, equipment for optimal results. 124 illustrations. 253pp. 5⅜ × 8½. 21100-2 Pa. $4.50

A TREASURY OF ART NOUVEAU DESIGN AND ORNAMENT, edited by Carol Belanger Grafton. 577 designs for the practicing artist. Full-page, spots, borders, bookplates by Klimt, Bradley, others. 144pp. 8⅜ × 11¼. 24001-0 Pa. $5.95

ART NOUVEAU TYPOGRAPHIC ORNAMENTS, Dan X. Solo. Over 800 Art Nouveau florals, swirls, women, animals, borders, scrolls, wreaths, spots and dingbats, copyright-free. 100pp. 8⅛ × 11. 24366-4 Pa. $4.00

HAND SHADOWS TO BE THROWN UPON THE WALL, Henry Bursill. Wonderful Victorian novelty tells how to make flying birds, dog, goose, deer, and 14 others, each explained by a full-page illustration. 32pp. 6½ × 9¼. 21779-5 Pa. $1.50

AUDUBON'S BIRDS OF AMERICA COLORING BOOK, John James Audubon. Rendered for coloring by Paul Kennedy. 46 of Audubon's noted illustrations: red-winged black-bird, cardinal, etc. Original plates reproduced in full-color on the covers. Captions. 48pp. 8¼ × 11. 23049-X Pa. $2.25

SILK SCREEN TECHNIQUES, J.I. Biegeleisen, M.A. Cohn. Clear, practical, modern, economical. Minimal equipment (self-built), materials, easy methods. For amateur, hobbyist, 1st book. 141 illustrations. 185pp. 6⅛ × 9¼. 20433-2 Pa. $3.95

101 PATCHWORK PATTERNS, Ruby S. McKim. 101 beautiful, immediately useable patterns, full-size, modern and traditional. Also general information, estimating, quilt lore. 140 illustrations. 124pp. 7⅞ × 10¾. 20773-0 Pa. $3.50

READY-TO-USE FLORAL DESIGNS, Ed Sibbett, Jr. Over 100 floral designs (most in three sizes) of popular individual blossoms as well as bouquets, sprays, garlands. 64pp. 8¼ × 11. 23976-4 Pa. $2.95

AMERICAN WILD FLOWERS COLORING BOOK, Paul Kennedy. Planned coverage of 46 most important wildflowers, from Rickett's collection; instructive as well as entertaining. Color versions on covers. Captions. 48pp. 8¼ × 11. 20095-7 Pa. $2.50

CARVING DUCK DECOYS, Harry V. Shourds and Anthony Hillman. Detailed instructions and full-size templates for constructing 16 beautiful, marvelously practical decoys according to time-honored South Jersey method. 70pp. 9¼ × 12¼. 24083-5 Pa. $4.95

TRADITIONAL PATCHWORK PATTERNS, Carol Belanger Grafton. Cardboard cut-out pieces for use as templates to make 12 quilts: Buttercup, Ribbon Border, Tree of Paradise, nine more. Full instructions. 57pp. 8¼ × 11. 23015-5 Pa. $3.50

25 KITES THAT FLY, Leslie Hunt. Full, easy-to-follow instructions for kites made from inexpensive materials. Many novelties. 70 illustrations. 110pp. 5⅜ × 8½.
22550-X Pa. $2.25

PIANO TUNING, J. Cree Fischer. Clearest, best book for beginner, amateur. Simple repairs, raising dropped notes, tuning by easy method of flattened fifths. No previous skills needed. 4 illustrations. 201pp. 5⅜ × 8½. 23267-0 Pa. $3.50

EARLY AMERICAN IRON-ON TRANSFER PATTERNS, edited by Rita Weiss. 75 designs, borders, alphabets, from traditional American sources. 48pp. 8¼ × 11.
23162-3 Pa. $1.95

CROCHETING EDGINGS, edited by Rita Weiss. Over 100 of the best designs for these lovely trims for a host of household items. Complete instructions, illustrations. 48pp. 8¼ × 11. 24031-2 Pa. $2.25

FINGER PLAYS FOR NURSERY AND KINDERGARTEN, Emilie Poulsson. 18 finger plays with music (voice and piano); entertaining, instructive. Counting, nature lore, etc. Victorian classic. 53 illustrations. 80pp. 6½ × 9¼. 22588-7 Pa. $1.95

BOSTON THEN AND NOW, Peter Vanderwarker. Here in 59 side-by-side views are photographic documentations of the city's past and present. 119 photographs. Full captions. 122pp. 8¼ × 11. 24312-5 Pa. $6.95

CROCHETING BEDSPREADS, edited by Rita Weiss. 22 patterns, originally published in three instruction books 1939-41. 39 photos, 8 charts. Instructions. 48pp. 8¼ × 11. 23610-2 Pa. $2.00

HAWTHORNE ON PAINTING, Charles W. Hawthorne. Collected from notes taken by students at famous Cape Cod School; hundreds of direct, personal *apercus*, ideas, suggestions. 91pp. 5⅜ × 8½. 20653-X Pa. $2.50

THERMODYNAMICS, Enrico Fermi. A classic of modern science. Clear, organized treatment of systems, first and second laws, entropy, thermodynamic potentials, etc. Calculus required. 160pp. 5⅜ × 8½. 60361-X Pa. $4.00

TEN BOOKS ON ARCHITECTURE, Vitruvius. The most important book ever written on architecture. Early Roman aesthetics, technology, classical orders, site selection, all other aspects. Morgan translation. 331pp. 5⅜ × 8½. 20645-9 Pa. $5.50

THE CORNELL BREAD BOOK, Clive M. McCay and Jeanette B. McCay. Famed high-protein recipe incorporated into breads, rolls, buns, coffee cakes, pizza, pie crusts, more. Nearly 50 illustrations. 48pp. 8¼ × 11. 23995-0 Pa. $2.00

THE CRAFTSMAN'S HANDBOOK, Cennino Cennini. 15th-century handbook, school of Giotto, explains applying gold, silver leaf; gesso; fresco painting, grinding pigments, etc. 142pp. 6⅛ × 9¼. 20054-X Pa. $3.50

FRANK LLOYD WRIGHT'S FALLINGWATER, Donald Hoffmann. Full story of Wright's masterwork at Bear Run, Pa. 100 photographs of site, construction, and details of completed structure. 112pp. 9¼ × 10. 23671-4 Pa. $6.95

OVAL STAINED GLASS PATTERN BOOK, C. Eaton. 60 new designs framed in shape of an oval. Greater complexity, challenge with sinuous cats, birds, mandalas framed in antique shape. 64pp. 8¼ × 11. 24519-5 Pa. $3.50

CHILDREN'S BOOKPLATES AND LABELS, Ed Sibbett, Jr. 6 each of 12 types based on *Wizard of Oz, Alice,* nursery rhymes, fairy tales. Perforated; full color. 24pp. 8¼ × 11. 23538-6 Pa. $3.50

READY-TO-USE VICTORIAN COLOR STICKERS: 96 Pressure-Sensitive Seals, Carol Belanger Grafton. Drawn from authentic period sources. Motifs include heads of men, women, children, plus florals, animals, birds, more. Will adhere to any clean surface. 8pp. 8½ × 11. 24551-9 Pa. $2.95

CUT AND FOLD PAPER SPACESHIPS THAT FLY, Michael Grater. 16 colorful, easy-to-build spaceships that really fly. Star Shuttle, Lunar Freighter, Star Probe, 13 others. 32pp. 8¼ × 11. 23978-0 Pa. $2.50

CUT AND ASSEMBLE PAPER AIRPLANES THAT FLY, Arthur Baker. 8 aerodynamically sound, ready-to-build paper airplanes, designed with latest techniques. Fly *Pegasus, Daedalus, Songbird,* 5 other aircraft. Instructions. 32pp. 9¼ × 11¼. 24302-8 Pa. $3.95

SIDELIGHTS ON RELATIVITY, Albert Einstein. Two lectures delivered in 1920-21: *Ether and Relativity* and *Geometry and Experience.* Elegant ideas in non-mathematical form. 56pp. 5⅜ × 8½. 24511-X Pa. $2.25

FADS AND FALLACIES IN THE NAME OF SCIENCE, Martin Gardner. Fair, witty appraisal of cranks and quacks of science: Velikovsky, orgone energy, Bridey Murphy, medical fads, etc. 373pp. 5⅜ × 8½. 20394-8 Pa. $5.95

VACATION HOMES AND CABINS, U.S. Dept. of Agriculture. Complete plans for 16 cabins, vacation homes and other shelters. 105pp. 9 × 12. 23631-5 Pa. $4.95

HOW TO BUILD A WOOD-FRAME HOUSE, L.O. Anderson. Placement, foundations, framing, sheathing, roof, insulation, plaster, finishing—almost everything else. 179 illustrations. 223pp. 7⅞ × 10¾. 22954-8 Pa. $5.50

THE MYSTERY OF A HANSOM CAB, Fergus W. Hume. Bizarre murder in a hansom cab leads to engrossing investigation. Memorable characters, rich atmosphere. 19th-century bestseller, still enjoyable, exciting. 256pp. 5⅜ × 8. 21956-9 Pa. $4.00

MANUAL OF TRADITIONAL WOOD CARVING, edited by Paul N. Hasluck. Possibly the best book in English on the craft of wood carving. Practical instructions, along with 1,146 working drawings and photographic illustrations. 576pp. 6½ × 9¼. 23489-4 Pa. $8.95

WHITTLING AND WOODCARVING, E.J Tangerman. Best book on market; clear, full. If you can cut a potato, you can carve toys, puzzles, chains, etc. Over 464 illustrations. 293pp. 5⅜ × 8½. 20965-2 Pa. $4.95

AMERICAN TRADEMARK DESIGNS, Barbara Baer Capitman. 732 marks, logos and corporate-identity symbols. Categories include entertainment, heavy industry, food and beverage. All black-and-white in standard forms. 160pp. 8⅛ × 11. 23259-X Pa. $6.95

DECORATIVE FRAMES AND BORDERS, edited by Edmund V. Gillon, Jr. Largest collection of borders and frames ever compiled for use of artists and designers. Renaissance, neo-Greek, Art Nouveau, Art Deco, to mention only a few styles. 396 illustrations. 192pp. 8⅜ × 11¼. 22928-9 Pa. $6.00

THE MURDER BOOK OF J.G. REEDER, Edgar Wallace. Eight suspenseful stories by bestselling mystery writer of 20s and 30s. Features the donnish Mr. J.G. Reeder of Public Prosecutor's Office. 128pp. 5⅜ × 8½. (Available in U.S. only)
24374-5 Pa. $3.50

ANNE ORR'S CHARTED DESIGNS, Anne Orr. Best designs by premier needlework designer, all on charts: flowers, borders, birds, children, alphabets, etc. Over 100 charts, 10 in color. Total of 40pp. 8¼ × 11.
23704-4 Pa. $2.50

BASIC CONSTRUCTION TECHNIQUES FOR HOUSES AND SMALL BUILDINGS SIMPLY EXPLAINED, U.S. Bureau of Naval Personnel. Grading, masonry, woodworking, floor and wall framing, roof framing, plastering, tile setting, much more. Over 675 illustrations. 568pp. 6½ × 9¼.
20242-9 Pa. $8.95

MATISSE LINE DRAWINGS AND PRINTS, Henri Matisse. Representative collection of female nudes, faces, still lifes, experimental works, etc., from 1898 to 1948. 50 illustrations. 48pp. 8⅜ × 11¼.
23877-6 Pa. $2.50

HOW TO PLAY THE CHESS OPENINGS, Eugene Znosko-Borovsky. Clear, profound examinations of just what each opening is intended to do and how opponent can counter. Many sample games. 147pp. 5⅜ × 8½.
22795-2 Pa. $2.95

DUPLICATE BRIDGE, Alfred Sheinwold. Clear, thorough, easily followed account: rules, etiquette, scoring, strategy, bidding; Goren's point-count system, Blackwood and Gerber conventions, etc. 158pp. 5⅜ × 8½.
22741-3 Pa. $3.00

SARGENT PORTRAIT DRAWINGS, J.S. Sargent. Collection of 42 portraits reveals technical skill and intuitive eye of noted American portrait painter, John Singer Sargent. 48pp. 8¼ × 11⅛.
24524-1 Pa. $2.95

ENTERTAINING SCIENCE EXPERIMENTS WITH EVERYDAY OBJECTS, Martin Gardner. Over 100 experiments for youngsters. Will amuse, astonish, teach, and entertain. Over 100 illustrations. 127pp. 5⅜ × 8½.
24201-3 Pa. $2.50

TEDDY BEAR PAPER DOLLS IN FULL COLOR: A Family of Four Bears and Their Costumes, Crystal Collins. A family of four Teddy Bear paper dolls and nearly 60 cut-out costumes. Full color, printed one side only. 32pp. 9¼ × 12¼.
24550-0 Pa. $3.50

NEW CALLIGRAPHIC ORNAMENTS AND FLOURISHES, Arthur Baker. Unusual, multi-useable material: arrows, pointing hands, brackets and frames, ovals, swirls, birds, etc. Nearly 700 illustrations. 80pp. 8⅜ × 11¼.
24095-9 Pa. $3.75

DINOSAUR DIORAMAS TO CUT & ASSEMBLE, M. Kalmenoff. Two complete three-dimensional scenes in full color, with 31 cut-out animals and plants. Excellent educational toy for youngsters. Instructions; 2 assembly diagrams. 32pp. 9¼ × 12¼.
24541-1 Pa. $4.50

SILHOUETTES: A PICTORIAL ARCHIVE OF VARIED ILLUSTRATIONS, edited by Carol Belanger Grafton. Over 600 silhouettes from the 18th to 20th centuries. Profiles and full figures of men, women, children, birds, animals, groups and scenes, nature, ships, an alphabet. 144pp. 8⅜ × 11¼.
23781-8 Pa. $4.95

SURREAL STICKERS AND UNREAL STAMPS, William Rowe. 224 haunting, hilarious stamps on gummed, perforated stock, with images of elephants, geisha girls, George Washington, etc. 16pp. one side. 8¼ × 11.   24371-0 Pa. $3.50

GOURMET KITCHEN LABELS, Ed Sibbett, Jr. 112 full-color labels (4 copies each of 28 designs). Fruit, bread, other culinary motifs. Gummed and perforated. 16pp. 8¼ × 11.   24087-8 Pa. $2.95

PATTERNS AND INSTRUCTIONS FOR CARVING AUTHENTIC BIRDS, H.D. Green. Detailed instructions, 27 diagrams, 85 photographs for carving 15 species of birds so life-like, they'll seem ready to fly! 8¼ × 11.   24222-6 Pa. $2.75

FLATLAND, E.A. Abbott. Science-fiction classic explores life of 2-D being in 3-D world. 16 illustrations. 103pp. 5⅜ × 8.   20001-9 Pa. $2.00

DRIED FLOWERS, Sarah Whitlock and Martha Rankin. Concise, clear, practical guide to dehydration, glycerinizing, pressing plant material, and more. Covers use of silica gel. 12 drawings. 32pp. 5⅜ × 8½.   21802-3 Pa. $1.00

EASY-TO-MAKE CANDLES, Gary V. Guy. Learn how easy it is to make all kinds of decorative candles. Step-by-step instructions. 82 illustrations. 48pp. 8¼ × 11.
23881-4 Pa. $2.50

SUPER STICKERS FOR KIDS, Carolyn Bracken. 128 gummed and perforated full-color stickers: GIRL WANTED, KEEP OUT, BORED OF EDUCATION, X-RATED, COMBAT ZONE, many others. 16pp. 8¼ × 11.   24092-4 Pa. $2.50

CUT AND COLOR PAPER MASKS, Michael Grater. Clowns, animals, funny faces...simply color them in, cut them out, and put them together, and you have 9 paper masks to play with and enjoy. 32pp. 8¼ × 11.   23171-2 Pa. $2.25

A CHRISTMAS CAROL: THE ORIGINAL MANUSCRIPT, Charles Dickens. Clear facsimile of Dickens manuscript, on facing pages with final printed text. 8 illustrations by John Leech, 4 in color on covers. 144pp. 8⅜ × 11¼.
20980-6 Pa. $5.95

CARVING SHOREBIRDS, Harry V. Shourds & Anthony Hillman. 16 full-size patterns (all double-page spreads) for 19 North American shorebirds with step-by-step instructions. 72pp. 9¼ × 12¼.   24287-0 Pa. $4.95

THE GENTLE ART OF MATHEMATICS, Dan Pedoe. Mathematical games, probability, the question of infinity, topology, how the laws of algebra work, problems of irrational numbers, and more. 42 figures. 143pp. 5⅜ × 8½. (EBE)
22949-1 Pa. $3.50

READY-TO-USE DOLLHOUSE WALLPAPER, Katzenbach & Warren, Inc. Stripe, 2 floral stripes, 2 allover florals, polka dot; all in full color. 4 sheets (350 sq. in.) of each, enough for average room. 48pp. 8¼ × 11.   23495-9 Pa. $2.95

MINIATURE IRON-ON TRANSFER PATTERNS FOR DOLLHOUSES, DOLLS, AND SMALL PROJECTS, Rita Weiss and Frank Fontana. Over 100 miniature patterns: rugs, bedspreads, quilts, chair seats, etc. In standard dollhouse size. 48pp. 8¼ × 11.   23741-9 Pa. $1.95

THE DINOSAUR COLORING BOOK, Anthony Rao. 45 renderings of dinosaurs, fossil birds, turtles, other creatures of Mesozoic Era. Scientifically accurate. Captions. 48pp. 8¼ × 11.   24022-3 Pa. $2.50

THE BOOK OF WOOD CARVING, Charles Marshall Sayers. Still finest book for beginning student. Fundamentals, technique; gives 34 designs, over 34 projects for panels, bookends, mirrors, etc. 33 photos. 118pp. 7¾ × 10⅝.   23654-4 Pa. $3.95

CARVING COUNTRY CHARACTERS, Bill Higginbotham. Expert advice for beginning, advanced carvers on materials, techniques for creating 18 projects— mirthful panorama of American characters. 105 illustrations. 80pp. 8⅜ × 11.
24135-1 Pa. $2.50

300 ART NOUVEAU DESIGNS AND MOTIFS IN FULL COLOR, C.B. Grafton. 44 full-page plates display swirling lines and muted colors typical of Art Nouveau. Borders, frames, panels, cartouches, dingbats, etc. 48pp. 9⅜ × 12¼.
24354-0 Pa. $6.95

SELF-WORKING CARD TRICKS, Karl Fulves. Editor of *Pallbearer* offers 72 tricks that work automatically through nature of card deck. No sleight of hand needed. Often spectacular. 42 illustrations. 113pp. 5⅜ × 8½.   23334-0 Pa. $3.50

CUT AND ASSEMBLE A WESTERN FRONTIER TOWN, Edmund V. Gillon, Jr. Ten authentic full-color buildings on heavy cardboard stock in H-O scale. Sheriff's Office and Jail, Saloon, Wells Fargo, Opera House, others. 48pp. 9¼ × 12¼.
23736-2 Pa. $3.95

CUT AND ASSEMBLE AN EARLY NEW ENGLAND VILLAGE, Edmund V. Gillon, Jr. Printed in full color on heavy cardboard stock. 12 authentic buildings in H-O scale: Adams home in Quincy, Mass., Oliver Wight house in Sturbridge, smithy, store, church, others. 48pp. 9¼ × 12¼.   23536-X Pa. $4.95

THE TALE OF TWO BAD MICE, Beatrix Potter. Tom Thumb and Hunca Munca squeeze out of their hole and go exploring. 27 full-color Potter illustrations. 59pp. 4¼ × 5½. (Available in U.S. only)   23065-1 Pa. $1.75

CARVING FIGURE CARICATURES IN THE OZARK STYLE, Harold L. Enlow. Instructions and illustrations for ten delightful projects, plus general carving instructions. 22 drawings and 47 photographs altogether. 39pp. 8⅜ × 11.
23151-8 Pa. $2.50

A TREASURY OF FLOWER DESIGNS FOR ARTISTS, EMBROIDERERS AND CRAFTSMEN, Susan Gaber. 100 garden favorites lushly rendered by artist for artists, craftsmen, needleworkers. Many form frames, borders. 80pp. 8¼ × 11.
24096-7 Pa. $3.50

CUT & ASSEMBLE A TOY THEATER/THE NUTCRACKER BALLET, Tom Tierney. Model of a complete, full-color production of Tchaikovsky's classic. 6 backdrops, dozens of characters, familiar dance sequences. 32pp. 9⅜ × 12¼.
24194-7 Pa. $4.50

ANIMALS: 1,419 COPYRIGHT-FREE ILLUSTRATIONS OF MAMMALS, BIRDS, FISH, INSECTS, ETC., edited by Jim Harter. Clear wood engravings present, in extremely lifelike poses, over 1,000 species of animals. 284pp. 9 × 12.
23766-4 Pa. $9.95

MORE HAND SHADOWS, Henry Bursill. For those at their 'finger ends,'' 16 more effects—Shakespeare, a hare, a squirrel, Mr. Punch, and twelve more—each explained by a full-page illustration. Considerable period charm. 30pp. 6½ × 9¼.
21384-6 Pa. $1.95

READY-TO-USE BORDERS, Ted Menten. Both traditional and unusual inter-changeable borders in a tremendous array of sizes, shapes, and styles. 32 plates. 64pp. 8¼ × 11. 23782-6 Pa. $3.50

THE WHOLE CRAFT OF SPINNING, Carol Kroll. Preparing fiber, drop spindle, treadle wheel, other fibers, more. Highly creative, yet simple. 43 illustrations. 48pp. 8¼ × 11. 23968-3 Pa. $2.50

HIDDEN PICTURE PUZZLE COLORING BOOK, Anna Pomaska. 31 delightful pictures to color with dozens of objects, people and animals hidden away to find. Captions. Solutions. 48pp. 8¼ × 11. 23909-8 Pa. $2.25

QUILTING WITH STRIPS AND STRINGS, H.W. Rose. Quickest, easiest way to turn left-over fabric into handsome quilt. 46 patchwork quilts; 31 full-size templates. 48pp. 8¼ × 11. 24357-5 Pa. $3.25

NATURAL DYES AND HOME DYEING, Rita J. Adrosko. Over 135 specific recipes from historical sources for cotton, wool, other fabrics. Genuine premodern handicrafts. 12 illustrations. 160pp. 5⅜ × 8½. 22688-3 Pa. $2.95

CARVING REALISTIC BIRDS, H.D. Green. Full-sized patterns, step-by-step instructions for robins, jays, cardinals, finches, etc. 97 illustrations. 80pp. 8¼ × 11. 23484-3 Pa. $3.00

GEOMETRY, RELATIVITY AND THE FOURTH DIMENSION, Rudolf Rucker. Exposition of fourth dimension, concepts of relativity as Flatland characters continue adventures. Popular, easily followed yet accurate, profound. 141 illustrations. 133pp. 5⅜ × 8½. 23400-2 Pa. $3.00

READY-TO-USE SMALL FRAMES AND BORDERS, Carol B. Grafton. Graphic message? Frame it graphically with 373 new frames and borders in many styles: Art Nouveau, Art Deco, Op Art. 64pp. 8¼ × 11. 24375-3 Pa. $3.50

CELTIC ART: THE METHODS OF CONSTRUCTION, George Bain. Simple geometric techniques for making Celtic interlacements, spirals, Kellstype initials, animals, humans, etc. Over 500 illustrations. 160pp. 9 × 12. (Available in U.S. only) 22923-8 Pa. $6.00

THE TALE OF TOM KITTEN, Beatrix Potter. Exciting text and all 27 vivid, full-color illustrations to charming tale of naughty little Tom getting into mischief again. 58pp. 4¼ × 5½. (USO) 24502-0 Pa. $1.75

WOODEN PUZZLE TOYS, Ed Sibbett, Jr. Transfer patterns and instructions for 24 easy-to-do projects: fish, butterflies, cats, acrobats, Humpty Dumpty, 19 others. 48pp. 8¼ × 11. 23713-3 Pa. $2.50

MY FAMILY TREE WORKBOOK, Rosemary A. Chorzempa. Enjoyable, easy-to-use introduction to genealogy designed specially for children. Data pages plus text. Instructive, educational, valuable. 64pp. 8¼ × 11. 24229-3 Pa. $2.50

Prices subject to change without notice.

Available at your book dealer or write for free catalog to Dept. GI, Dover Publications, Inc., 31 East 2nd St. Mineola, N.Y. 11501. Dover publishes more than 175 books each year on science, elementary and advanced mathematics, biology, music, art, literary history, social sciences and other areas.